TWO-CENTER EFFECTS IN ION-ATOM COLLISIONS

A Symposium in Honor of M. Eugene Rudd

TWO-CENTER EFFECTS IN ION-ATOM COLLISIONS

A Symposium in Honor of M. Eugene Rudd

Lincoln, NE May 1994

EDITORS
Timothy J. Gay
Anthony F. Starace
University of Nebraska

American Institute of Physics

AIP CONFERENCE PROCEEDINGS 362

Woodbury, New York

Authorization to photocopy items for internal or personal use, beyond the free copying permitted under the 1978 U.S. Copyright Law (see statement below), is granted by the American Institute of Physics for users registered with the Copyright Clearance Center (CCC) Transactional Reporting Service, provided that the base fee of $6.00 per copy is paid directly to CCC, 222 Rosewood Drive, Danvers, MA 01923. For those organizations that have been granted a photocopy license by CCC, a separate system of payment has been arranged. The fee code for users of the Transactional Reporting Service is: 1-56396-342-6/ 95 /$6.00.

© 1996 American Institute of Physics

Individual readers of this volume and nonprofit libraries, acting for them, are permitted to make fair use of the material in it, such as copying an article for use in teaching or research. Permission is granted to quote from this volume in scientific work with the customary acknowledgment of the source. To reprint a figure, table, or other excerpt requires the consent of one of the original authors and notification to AIP. Republication or systematic or multiple reproduction of any material in this volume is permitted only under license from AIP. Address inquiries to Office of Rights and Permissions, 500 Sunnyside Boulevard, Woodbury, NY 11797-2999; phone 516-576-2268; fax: 516-576-2499; e-mail: rights@aip.org.

L.C. Catalog Card No. 96-83379
ISBN 1-56396-342-6
DOE CONF- 9405351

Printed in the United States of America

CONTENTS

Photographs vii
Preface xi
Biographical Sketch of M. Eugene Rudd xiii

INVITED PAPERS

I. SADDLE-POINT ELECTRONS

Ionization in Low Energy Ion-Atom Collisions (Theory) 5
    S. Yu. Ovchinnikov, J. H. Macek, and S. V. Passovets
Current Status of the Saddle-Point Model 19
    T. J. Gay
Saddle Electron Emission from Ion-Atom Collisions at Intermediate
Projectile Energies 29
    S. G. Suárez
Saddle Point Electrons in Slow Ion-Atom Collisions 41
    M. Pieksma, S. Yu. Ovchinnikov, J. van Eck, W. B. Westerveld, and A. Niehaus

II. TARGET AND PROJECTILE ELECTRON INTERACTIONS

Target and Projectile Ionization in Ion-Atom Collisions:
Theoretical Aspects 59
    S. T. Manson
Two-Center Effects in Electron Spectra from Ion-Atom Collisions 69
    P. Richard
Two-Center Effects in the Ejected-Electron Spectra in Ion-Atom Collisions ... 84
    D. R. Schultz, C. O. Reinhold, and R. E. Olson
Passive and Active Electrons in the Electron Loss Process 103
    E. C. Montenegro and W. E. Meyerhof

III. HIGHLY-CHARGED PROJECTILES AND TWO-CENTER DESCRIPTIONS

Classical-Quantum Correspondence for Ionization in Fast
Ion-Atom Collisions 115
    J. Burgdörfer and C. O. Reinhold
Close-Coupling Methods: A Critical Evaluation 135
    C. D. Lin
The CDW-EIS Method 147
    R. D. Rivarola, P. D. Fainstein, and V. H. Ponce

One- and Two-Center Electron Emission in Energetic Ion-Atom
Collisions 163
    N. Stolterfoht

IV. M. E. RUDD'S CONTRIBUTIONS TO ATOMIC PHYSICS

Proton-Atom Collisions: Contributions of M. E. Rudd 181
    L. H. Toburen
Electron Transfer to Continuum States. 193
    J. H. Macek
M. E. Rudd's Contributions to Autoionizing States in Ion-Atom Collisions 205
    A. K. Edwards
Modeling Ionization Cross Sections: Two Decades of Dreams Come True 214
    Y-K. Kim

CONTRIBUTED PAPERS

Two-Center Effects in Electron Excitation 229
    C. A. Ramírez and R. D. Rivarola
Evidence of Double Scattering Processes in Ion-Atom Ionization............. 233
    S. Suárez, W. Cravero, R. Barrachina, W. Meckbach, R. Maier,
    M. Tobisch, and K. O. Groeneveld
Electron Plate-Impact Distortions in Electron Spectroscopy 237
    V. D. Irby
Charge State Distributions in Copper Following Multiple Ionization
and One-Electron Capture in Collisions with H^+ Ions..................... 241
    M. B. Shah, C. J. Patton, J. Geddes, and H. B. Gilbody
Radial Dose Distributions in the Delta-Ray Theory of Track Structure 245
    F. A. Cucinotta, R. Katz, J. W. Wilson, and R. R. Dubey
Elastic Scattering Model of the Binary Encounter Electrons
in Ion-Atom Collisions... 266
    C. P. Bhalla and S. R. Grabbe
Auger Electron Spectroscopy of Free Argon Clusters. 274
    A. Knop, D. N. McIlroy, P. A. Dowben, and E. Rühl
Charge Transfer and Ionization in Ion Collisions with Circular
Rydberg Atoms 281
    D. M. Homan, M. J. Cavagnero, and D. A. Harmin

Author Index.. 287

One of Professor Rudd's early laboratories (circa 1941).

CONFRERENCE PARTICIPANTS

1. Dimitri Krebtukov
2. Gordon Gallup
3. Sam Cipolla
4. Duane Jaecks
5. Lisa Wiese
6. Gordon Berry
7. Phil Altick
8. Mark Gealy
9. Lew Cocke
10. Brian Moudry
11. Ben Birdsey
12. Yang-Soo Chung
13. Marc Pieksma
14. Ted Jorgensen
15. Mitio Inokuti
16. Cheng Pan
17. Chander Bhalla
18. Pat Richard
19. Eugen Merzbacher
20. Serge Ovchinnikov
21. Horst Schmidt-Bocking
22. Yong-Ki Kim
23. Ron McKnight
24. Henry Valk
25. George Kerby
26. Ken Trantham
27. Don Madison
28. Jiyun Kuang
29. Sergio Suárez
30. Tibor Vajnai
31. Bob DuBois
32. Larry Toburen
33. Victor Irby
34. John Risley
35. Nico Stolterfoht
36. Mansukh Shah
37. Dave Schultz
38. Eduardo Montenegro
39. Carl Bailey
40. Roberto Rivarola
41. Steve Manson
42. Chii-Dong Lin
43. Joe Macek
44. Siegbert Hagmann
45. Tony Starace
46. Gene Rudd
47. Alan Edwards
48. Tim Gay
49. Walter Meyerhof
50. Joachim Burgdörfer
51. Orhan Yenen
52. Ilya Fabrikant

Participants not in the photograph:

Minqi Bao
Dean Homan
Marty Johnston
Robert Katz
Roger Kirby
David McIlroy

1
2
3
4
5
6
7
8
9
10
11
12
13
14
15
16
17
18
19
20
21
22
23
24
25
26
27
28
29
30
31
32
33
34
35
36
37
38
39
40
41
42
43
44
45
46
47
48
49
50
51
52

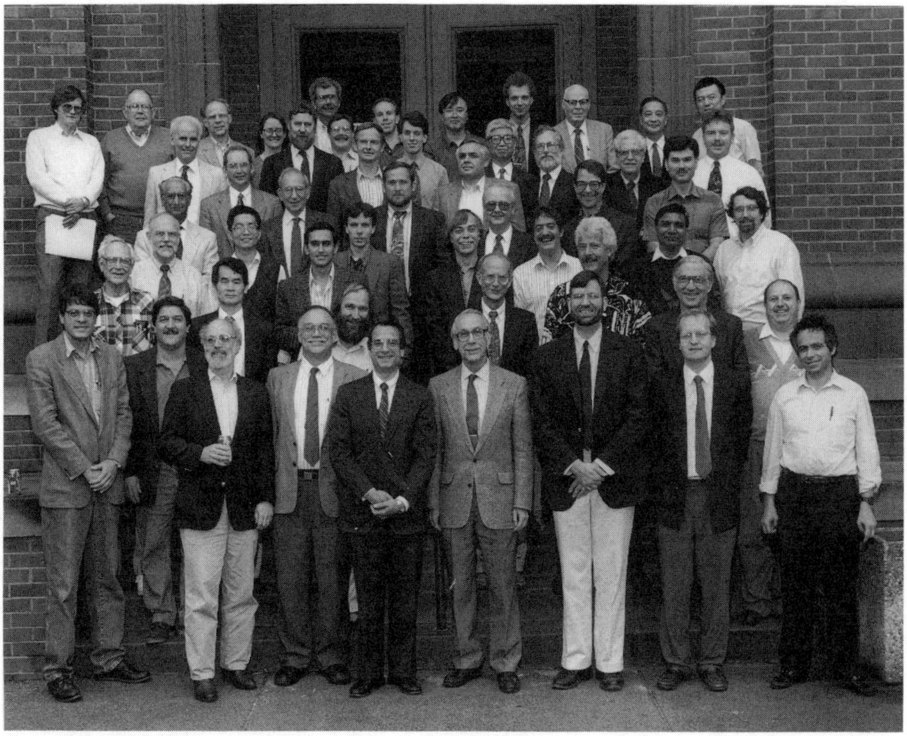

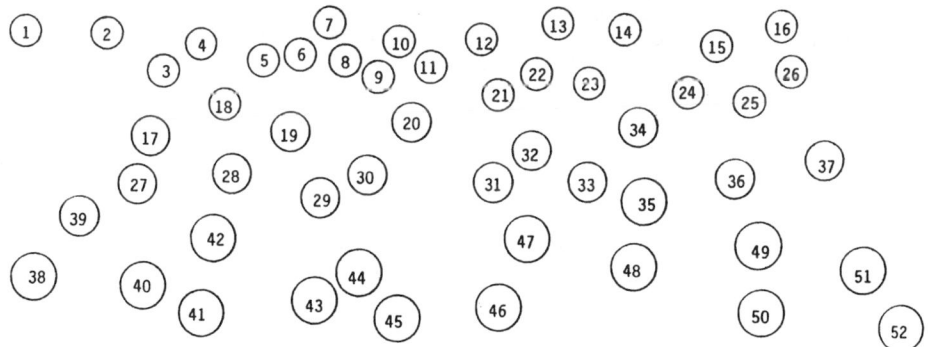

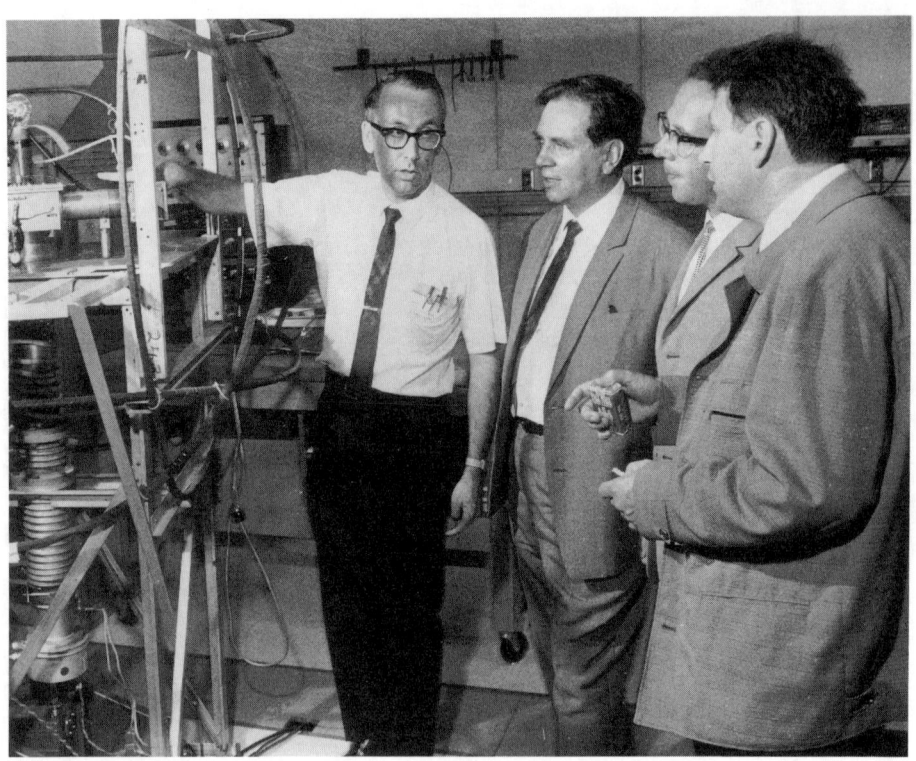

Visiting Soviet scientists viewing Nebraska apparatus in 1969.
Left-to-right: Rudd, Yu.N.Demkov (Moscow),V.V.Afrosimov
(Leningrad),S.V.Bobashev(Leningrad).

PREFACE

Following a distinguished career in atomic physics, sketched in the accompanying biography, M. Eugene Rudd retired from the faculty of the University of Nebraska in the spring of 1993. In order to celebrate his many research accomplishments, a two day scientific conference, held on 13–14 May 1994, was organized in Lincoln. It was felt that the special spirit that Rudd brought to the lab would be highlighted by a conference designed to analyze in detail a subject of current research in which he was a pioneer: two center-effects in ion-atom ionizing collisions.

Prior to 1980, primarily as a result of the experiments of Rudd and his co-workers, a standard view of ionizing collisions had developed. This view held that almost all the ionized electrons could be associated with either the ionized target or the receding projectile. The former, having mostly small momenta, were produced in "soft" collisions with the projectile, and emerged almost isotropically from the collision region. The remaining electrons had velocities similar to that of the projectile, and formed a "cusp" distribution about its velocity vector in the forward direction. Beginning in the mid-1980's however, it became increasingly clear that electronic trajectories that were manifestly determined by a combination of the action of the two Coulomb centers, both target and projectile acting together, might be observed. The topic of this meeting was to consider such effects in detail.

The speakers were chosen because of their expertise in this area, and included a number of Professor Rudd's students and immediate colleagues. Eugen Merzbacher delivered the banquet address on "The Crisis at the *Physical Review*." In addition to the four scientific sessions for invited speakers, a poster session was held with 17 contributions.

We intend that this *Festschrift* will serve as a useful summary of the current status of this field, and as a good starting point for researchers interested in studying this topic. We hope that it will also give the scientific community at large an indication of the influence of M. E. Rudd's career, and of the high esteem in which his colleagues hold both it and him.

T. J. Gay
A. F. Starace
Lincoln, Nebraska

Biographical Sketch of M. Eugene Rudd

Robert Katz and Duane H. Jaecks

Department of Physics and Astronomy
University of Nebraska-Lincoln, Lincoln, NE 68588-0111

M. Eugene Rudd was born in 1927, in Fargo, North Dakota, of Norwegian ancestry, and has retained a lifelong interest in Norwegian arts and culture. While in high school he worked in a photographic studio, developing what turned out to be a lifelong interest in its technical aspects. After graduating from high school he served for 18 months in the U. S. Army Signal Corps where he was able to put his knowledge of photography to use. He obtained his undergraduate education at Concordia College, Moorhead, Minnesota, graduating in 1950. (In 1992, Concordia College presented him with an honorary D.Sc. in recognition of his distinguished career in science and education.) He was married in 1953 to Eileen L. Hovland, with whom he had three children, Eric, Nancy and Leif. His graduate studies began at the University of Buffalo, where he was a teaching assistant from 1950-1954. He completed his course work and examinations for the Ph.D., but opportunities for research were limited so in 1954, upon receiving an invitation from Concordia College, he took an M.A. and accepted a job first as Assistant, then Associate Professor and Professor of Physics at Concordia College, in the years 1954-1965.

In 1958, while he was a faculty member at Concordia, the College received a surplus Cockroft-Walton accelerator from Iowa State University through two Concordia alumni. A separate building was constructed for it, and Gene assisted Prof. Carl Bailey in installing it. In 1960, wishing to finish his Ph.D. and wanting to gain experience on an accelerator similar to the Concordia machine, he chose to work with Prof. Ted Jorgensen at the University of Nebraska, who had returned from wartime research in nuclear physics at Los Alamos to initiate a program of research in atomic collisions. With his Ph.D. studies at Nebraska completed in 1962 while on leave from Concordia, Rudd was made Professor of Physics at Concordia College. At Concordia he built an apparatus to make additional measurements of the type he had made at Nebraska, of the angular and energy distribution of electrons produced by proton impact on gases, and secured a contract from the Atomic Energy Commission in 1963 to support the work. During the course of these measurements he discovered fine structure in the electron spectrum from helium due to autoionization from doubly excited states. This led to a National Science Foundation research grant in 1965. After improving the resolution of the instrument to what was near the state of the art in electron spectroscopy at that time, he was able to identify previously unobserved Rydberg series and studied Auger electron spectra as well. As a result of this pioneering work, the study of autoionizing emissions from ion-atom collisions soon became a separate, well-developed field of study.

When Professor Jorgensen began to think about retirement, Henry Valk, who was then department chairman, began a national search to find someone to head up the atomic collisions program at Nebraska. To his pleasant surprise the name that was continually advanced was Gene Rudd, whose work even then at Concordia was being supported by NSF (and has since been continuously supported by NSF, representing one of the longest-funded atomic physics programs in the history of this agency). Gene returned to Nebraska as Associate Professor of Physics in 1965, and was made Professor in 1968, a position which he has held since with distinction. In 1970-72 he was Acting Chairman of the Physics Deprtment on the departure of Henry Valk to become Dean at Georgia Tech.

Gene's research in atomic collision physics has been characterized by originality and thoroughness. He has written review articles and chapters in the Encyclopedia of Physics and elsewhere, and is a co-author on a book on atomic collisions. He has published more than 75 refereed research papers, many of which were of seminal importance. He was the first to study systematically the dependence of the probability for electron emission in ion-atom collisions on the incident ion's energy, and on the ejected electron's energy and angle of emission. In 1964 he made the first experimental observation of doubly excited atomic states produced by heavy particle impact on rare gas target ions with sufficient resolution to identify the states. In 1968 with Ted Jorgensen he made the first experimental observation of "Doppler shifts"of ejected electron spectra. In 1970 came his most famous "first", the observation that a significant component of the ejected electrons travel with a velocity equal to that of the incident ion (a mechanism of ionization called "electron capture to the continuum"). In a collaboration with Joe Macek, the electron promotion model was used to describe the ejection of high energy electrons in low energy collisions.

Rudd's pioneering work continues. He has begun the first experimental measurements of proton impact on atomic hydrogen. This process is fundamental because it involves only 3 particles, and hence has been well studied theoretically. It is fundamental in astrophysics since protons, electrons and hydrogen atoms are among the primary constituents of stars. His work is fundamental in other areas as well. Electrons ejected in atomic collisions in matter produce additional ionizations, and break molecular bonds. These processes produce radiation damage. Rudd's data has led to further investigations of biological effects of heavy ions in human tissue, of the effects of incident cosmic rays on computers in satellites, and to a host of other phenomena. It has been said by Eugen Merzbacher, former president of the American Physical Society, that Rudd's name among the authors of a research paper inspires general confidence in the results and vouches for their reliabilty. His painstaking data compilations have been of immense value to atomic and radiation effects physicists.

Rudd's work has been well recognized. He is a Fellow of the American Physical Society, and was chairman of the Division of Electron and Atomic Physics of the American Physical Society in 1980. He was also a member of the Committee on Atomic and Molecular Science of the National Academy of Sciences in 1980, the organizer and chairman of a symposium on the History of Spectroscopy, and Chairman of the Report Committee on Secondary Electron spectra of the International Commission on Radiation Units and Measurements 1989-present. At the University of Nebraska he received the Outstanding Scientist Award from Sigma Xi in 1973, and the Burlington-Northern Distinguished Teacher-Scholar Award in 1991.

In his teaching and service to the University, Rudd has brought to bear the same care, attention to detail, and enthusiasm for continuous improvement that he brings to his research. He served for three years (1978-81) on the College of Arts and Sciences Task Force on the Liberal Arts, which instituted the current liberal education requirements for the College. He then developed both two-semester and one-semester "Liberal Arts Physics" courses which he has taught for 12 semesters since 1980. Since no suitable texts for these courses could be purchased, he developed his lecture notes into a book. In the past five years Rudd has developed two other new courses: a physics course on scientific revolutions for the UNL Honors Program; and a course on "Issues in Science and Religion," which stemmed from his service on the Area Studies Committee for Religious Studies in the College of Arts and Sciences.

Rudd's interest in teaching not only the laws of physics but its history has led to other novel teaching contributions. In the Fall of 1978, he and Duane Jaecks held an exhibition of historically significant scientific instruments in the Sheldon Art Gallery. In 1987 he organized an exhibition in Love Library entitled "Light and Color: An Exhibition of Notable Books from the History of Optics," whose annotated catalog drew praise from 3 faculty members in Harvard's History of Science Department. More recently he wrote a history of physics and astronomy at UNL entitled "Science on the Great Plains" which has been published as a "Nebraska Studies" book by the Univesity of Nebraska Press. His collection of historically significant scientific books and instruments is extensive.

There was also an overarching teaching component to Professor Rudd's research. His research students developed a wide range of experimenal as well as critical thinking skills. Because of this outstanding background his research students have gone on to contribute to a wide range of fields. Some of these areas include university teaching and research, governmental and industrial research and development, and medical physics. But most telling of Rudd's success as a teacher are the comments of students who got to know him well. One undergraduate wrote:"I worked for Dr. Rudd for approximately a year and a half. I consider that time as the best learning experience of my college career." A former doctoral student wrote that Rudd's strengths as a teacher are "interest, dedication, and caring." Undergraduates write that "his enthusiasm is contagious" and that "he encourages students to find connections between physics and life."

A most apt description of Gene Rudd has been given by one of his associates in the area of religious studies, Professor John D. Turner, Cotner Professor of Religious Studies, Classics and History at UNL. "Eugene Rudd is truly a scholar of the Renaissance variety, vastly well-rounded, on the cutting edge of his field, yet with a solid grasp of the history of his own discipline and of the many intellectual giants on whose accomplishments it now stands. A humble, unassuming and thoroughly decent human being. Eugene is the paragon of a university professor in the highest sense of the term, an inspiration to all who have known him."

INVITED PAPERS

I. SADDLE-POINT ELECTRONS

Ionization in Low Energy Ion-Atom Collisions (Theory)

S. Yu. Ovchinnikov*, J. H. Macek, and S. V. Passovets*

Department of Physics and Astronomy, University of Tennessee, Knoxville, TN 37996-1501
and
Oak Ridge National Laboratory, Post Office Box 2008
Oak Ridge, TN 37831-6373

In low energy atomic collision processes, the main contribution to the cross section often comes from "hidden crossings" of potential energy curves in the complex internuclear coordinate plane. Ionization of hydrogen by bare proton has been calculated at intermediate energies (E/A=5-25 keV). In addition to total cross section the electron energy spectra are discussed.

I. GENERAL PROPERTIES OF HIDDEN CROSSINGS

In the adiabatic approximation the quasimolecule formed by the electron moving in the field of the two nuclei is described by a set of adiabatic potential energy curves (1). The state of the quasimolecule evolving adiabatically at internuclear separations R can be read off from a diagram of these potential curves. Electronic transitions leading to capture, excitation and ionization all depend on the breakdown of the adiabatic approximation (1). At low velocities these breakdowns usually occur only locally and can be traced to the local analytical properties of the potential energy curves. Well-known examples are the situations studied by Landau and Zener (2), and Demkov (3). In the Landau-Zener case the adiabatic potential energy curves come close to each other without actually becoming degenerate. Instead they show a typical avoided crossing behaviour, simplest to describe by involving a diabatic representation with approximately linear potential curves that actually do cross (2). For the Demkov case, the adiabatic potential curves show an abrupt repulsion which can easily be described by a set of diabatic potential curves that run with constant distance from each other (3). Atomic units are used throughout.

The basic concepts of complex path integration were articulated by Demkov (4) and developed rigorously by Solov'ev (5). To set this picture in its most general context, consider a Schrödinger equation of the form

$$\left[-\frac{1}{2M}\frac{d^2}{dR^2} + \mathcal{H}(R;x)\right]\Psi = E\Psi,\qquad(1)$$

where R is a reaction coordinate, x represents a set of dimensionless coordinates orthogonal to R (4), $\mathcal{H}$ is the Hamiltonian for the the system minus the kinetic energy operator corresponding to the reaction coordinate R, and M is a parameter with dimensions of mass. Differential equations such as Eq.(1) may be integrated along any path in the complex R-plane to connect solutions at some initial $R = R_i$ to solutions at a final $R = R_f$. This flexibility in the choice of R is used to select some path in the complex plane along which good approximations to the exact Ψ are evident. The structure of Eq.(1) suggests that an appropriate path can be found by computing the eigenvalues of $\mathcal{H}(R; x)$. These eigenvalues $\epsilon_n(R)$ are interpreted in terms of a single function $\epsilon(R)$ which is single-valued on a multisheeted Riemann surface.

Mathematical studies of these generic situations have shown that in an asymptotic approximation one can associate a transition probability

$$P(v_0) = \exp\{-\frac{2}{v_0}\Delta\}, \qquad (2)$$

with the single passage of the quasimolecule through a region of the internuclear distance where such a localized breakdown of the adiabatic approximation occurs. Here v_0 is the impact velocity and Δ is the generalized Massey parameter, which can be expressed in terms of the local analytical structure of the adiabatic potential curves.

As was pointed out already by Landau and Lifshitz (6), the Landau-Zener and Demkov models given above are just special cases of a more general situation where the probability of a transition from one potential curve to another can be written in the general form (2) with a Massey parameter Δ which only depends on local analytical properties of the potential curves. The two mechanisms for excitation and ionization were actually discovered in this way by Solov'ev (7) and Ovchinnikov and Solov'ev (8). They showed that the adiabatic potential curves of the one-electron diatomic molecular ion have two different series of "hidden crossings" in the complex plane of the internuclear distance R, denoted by S and T by Ovchinnikov and Solov'ev (8).

There are two kinds of series of branch points in the two-Coloumb-center problem which are associated with hidden crossings: the S-series and the T-series. The branch points of the S_{lm}-series connect the pairs of potential energy curves ϵ_{nlm} and $\epsilon_{n+1\,l\,m}$ having the same angular quantum numbers, and radial quantum numbers differing by unity for all $n \geq l+1$. The set of branch points forms an infinite series localized in a small region Ω of the complex R-plane and has a limit point. In the neighborhood of the region Ω the energy surface $\epsilon(R)$ has the form of an infinite "spiral staircase" or a corkscrew, part of which is shown in Fig.1 (left). When we go around the region Ω we transit from a given energy eigenvalue to the neighboring higher or lower energy eigenvalue (depending on direction of rotation). The boundary of the continuum is reached after an infinite number of loops clockwise around the Ω or after one counter-clockwise turn from the lowest energy eigenvalue. Figure 1 (right) shows this part of a Riemann surface.

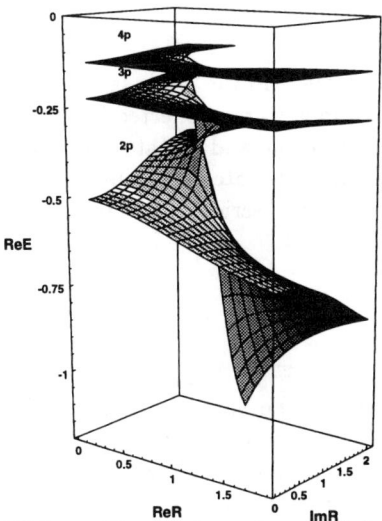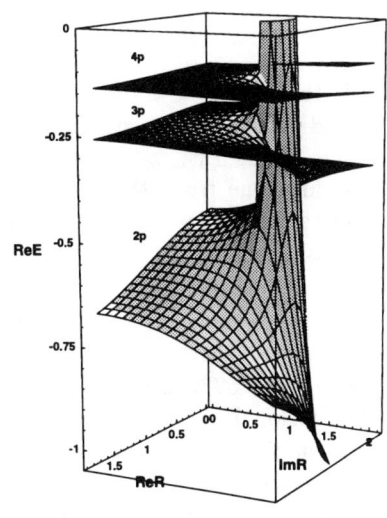

FIG. 1. Plot of the real part of the energy function $\epsilon(R)$ vs. complex R. The plot represents a Riemann surface connected by branch points of an S-series. The values along the real axis are the energy eigenvalues $\epsilon_n(R)$ of H_2^+. The right figure shows a "back" view of an extended Riemann surface.

An approximate expression for the limit point of the S-series (8) is

$$R_S = \frac{1}{Z}\left\{\left(l+\frac{1}{2}\right)^2 - \frac{(m+1)^2}{2} \pm i(m+1)\sqrt{2\left(l+\frac{1}{2}\right)^2 - \frac{(m+1)^2}{4}}\right\}, \quad (3)$$

where l and m are the quantum numbers of the united atom and $Z = Z_1 + Z_2$ is the sum of the charges. The S-series of branch points is related to the appearance of the centrifugal barrier in the quasiradial equation. The centrifugal barrier is a feature of the united atom which appears when the nuclei are close. The effective potential for $R = 0$ is

$$V_{eff}(r, R=0) = -\frac{Z}{r} + \frac{l(l+1)}{2r^2}. \quad (4)$$

So for $l = 0$ there is no centrifugal barrier and as, a consequence, there is no $S_{s\sigma}$-series. The coordinates of the branch points of the S_{lm}-series yield the limits of validity of the united atom approximation.

The branch points of the T-series are located at a larger internuclear distance R than the S-series. They are associated with the touching of the level by the top of the barrier in the quasiangular equation and prescribe the limits of the validity of the separated atom approximation. Therefore, to classify the $T_{n_1 n_2 m}$-series we use the quantum numbers of the separated atoms. The branch points of the $T_{n_1 n_2 m}$-series are uniformly on a steep straight line and connect the given energy level ϵ_{nlm} with energy levels $\epsilon_{n+k\,l+k\,m}$ having the same radial quantum number for all k. The greater the values of k, the farther from the real R-axis lies the corresponding branch point.

In the case $Z_1 = Z_2 = 1$ the system is symmetric and the levels split up into g- and u-states. Only states with the same symmetry are connected by branch points of the $T_{n_1 n_2 m}$-series. The branch points of u- and g-states alternate. The first one from the real R axis belongs to a g-state. An approximate expression for the position of the branch point of a T series closest to the real R axis is (9)

$$R_T \approx \frac{\pi^2}{2} n_g^2 + i\frac{\pi}{4} n_g \{\ln(16 n_g) + 2\ln[\ln(8\pi n_g) - 2 + \gamma]\} \tag{5}$$

where $\gamma \approx 0.5772$ is Euler's constant and $n_g = n_2 + \frac{m}{2} + \frac{7}{8} + (n_1 + \frac{m+1}{2})/(\sqrt{2}\pi)$. Conventional energy eigenvalues $\epsilon_n(R)$ represent the values of the function $\epsilon(R)$ on the real axis for different sheets n. The energy eigenvalues for H_2^+ are computed using the program of Solov'ev (7). Figure 2 shows a Riemann surface for the σ_g eigenfunctions of H_2^+ constructed by plotting $\Re\epsilon(R)$ vs. $S = \sqrt{R}$.

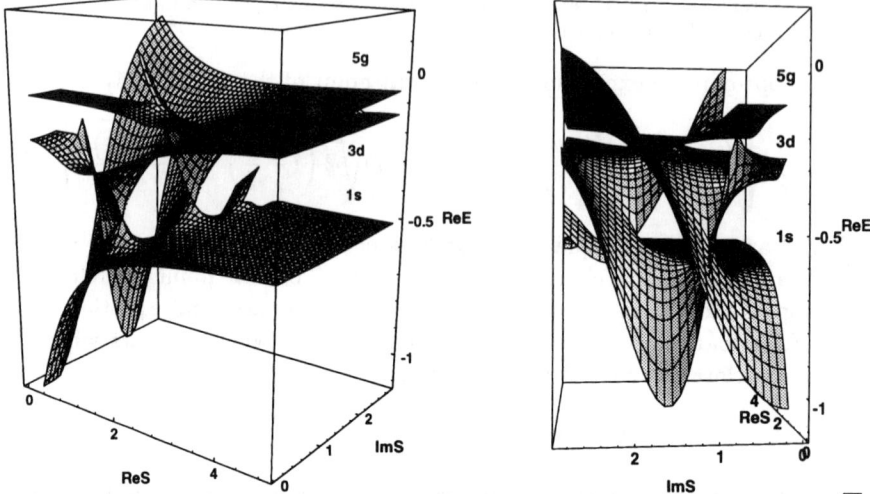

FIG. 2. Plot of the real part of the energy function $\epsilon(R)$ vs. complex $S = \sqrt{R}$. The plot represents a Riemann surface connected by branch points of an S-series. The values along the real axis are the energy eigenvalues $\epsilon_n(R)$ of H_2^+. The right figure shows a "back" view of the Riemann surface.

Different sheets join at branch points. The branch points of interest here are called T-series branch points in Ref. (8) and are exhibited more clearly by plotting the real part of the effective principal quantum number $n(R) = 1/\sqrt{-\epsilon(R)}$ vs. $S = \sqrt{R}$ as in Fig. 3.

Figure 3 (right) shows the surface from the "back" side with the $\Im R > 0$ region foremost. Notice that the complex structure on the real axis merges at the branch points with a remarkably flat region that extends to infinite distance. In this region the different sheets are well represented by the expression

$$\epsilon_{n_1 n_2 |m|}(R) = -(4Z-1)/R - i(n_1' + 1/2 + |m|/2)4\sqrt{2Z}R^{-3/2}$$
$$+ (n_2' + 1/2 + |m|/2)4\sqrt{Z}R^{-3/2} \quad (6)$$

with a magnetic quantum number $|m| = 0$ representing σ states. The charge Z of the nuclei equals unity for H_2^+. The quantum numbers n_1' and n_2' differ from the usual labels n_1, n_2 of the H_2^+ electronic wave functions on the real axis. Both n_2 and n_2' refer to motion perpendicular to the internuclear axis, but n_2' refers to motion in an attractive harmonic oscillator potential. Similarly, both n_1 and n_1' refer to motion parallel to the internuclear axis, but n_1' represents motion in an inverted harmonic oscillator potential; hence the factor of i in Eq.(6).

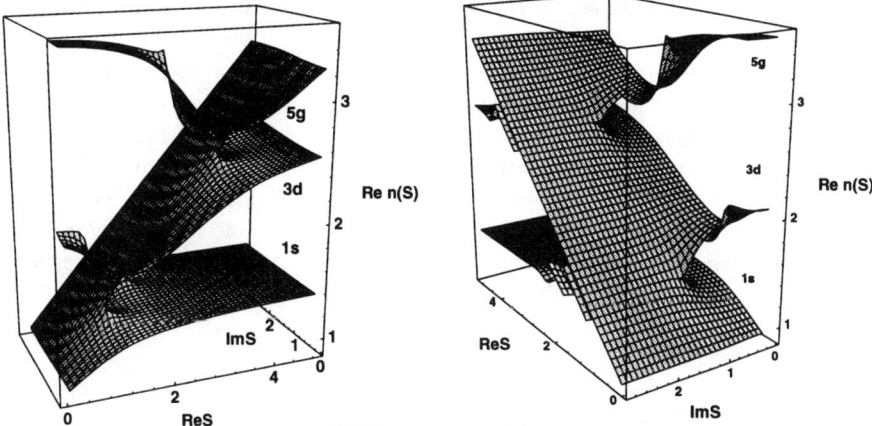

FIG. 3. Plot of $n(R) = 1/\sqrt{\epsilon(R)}$ vs. $S = \sqrt{R}$. The left figure shows a "front" view of the surface with the real axis foremost. The right figure shows a "back" view with a sloping flat region to the left of an infinite series of branch points.

The existence of a hidden crossing point R_c between potential energy curves $\epsilon_1(R)$ and $\epsilon_2(R)$ leads, in the simplest case, to the possibility of writing the Massey parameter Δ as

$$\Delta = Im \int_{ReR_c}^{R_c} (\epsilon_1(R) - \epsilon_2(R)) v_0/v(R) dR \quad (7)$$

where the integral is performed from the real R-axis, e.g., from the point ReR_c, and where $V(R)$ is the impact-parameter-dependent radial velocity.

II. MECHANISMS OF IONIZATION IN $H^+ - H$ COLLISIONS

In the classical picture the mechanism of ionization can be pictured generally as follows: As the potential barrier increases in the collision, the electrons near the top of the barrier are slowed down and are then collected and promoted to the continuum as the top of the barrier further rises. There are two types of potential barriers in atomic collisions as seen by the electron. One is the centrifugal barrier around both nuclei which increases when the nuclei approach each other, and the other is the potential barrier between the two nuclei which increases as they recede from each other.

We will consider the ionization mechanism in the following collision process (10)

$$H^+ + H \longrightarrow H^+ + H^+ + e \ . \tag{8}$$

The entrance channel of this reaction correlates with the quasimolecular states $1s\sigma$ and $2p\sigma$. At an infinite internuclear distance these states are degenerate, so initially 50% of the atoms are in the $1s\sigma$ state and the other 50% are in the $2p\sigma$ state. The quasimolecule H_2^+ has no avoided crossings but only hidden ones (8).

There are therefore two different ionization mechanisms. One is via an S-promotion when the nuclei approach each other. The S-series are associated with unstable trajectories of electrons on top of the centrifugal barrier. These trajectories lie on the internuclear axis. First, let us consider ionization via S-promotion which occurs when the nuclei approach each other. There are two possible channels of ionization via the S-series. One of them goes via $S_{2p\sigma}$ (for which the coordinate is $R_{2p\sigma} = (0.74, 1.01)$), the other one goes via $S_{3d\sigma}$ ($R_{3d\sigma} = (2.74, 1.74)$). The scheme of ionization is via the $S_{p\sigma}$-series

$$2p\sigma \longrightarrow 3p\sigma \longrightarrow ... \longrightarrow np\sigma \longrightarrow ... \tag{9}$$

and via the $S_{d\sigma}$-series

$$3d\sigma \longrightarrow 4d\sigma \longrightarrow ... \longrightarrow nd\sigma \longrightarrow ... \ . \tag{10}$$

The probability of ionization via an S-promotion is given by

$$P_{lm}^{(S)} = \prod_{i=0}^{\infty} P_{ilm} \ , \tag{11}$$

where P_{ilm} is the transition probability between the potential energy curves E_{ilm} and E_{i+1lm}. The probability is determined by $\Delta_{lm}^S = \sum_{i=0}^{\infty} \Delta_{ilm}$ which is the sum of the Massey parameters and equals

$$P_{lm}^{(S)} = e^{-\frac{2}{v_0}\Delta_{lm}^S} \ . \tag{12}$$

Initially the $3d\sigma$ state is populated due to the $1s\sigma$-$3d\sigma$ transition determined by the first branch point of the T_{000}-series at a large internuclear distance ($\sim$ 5 a.u.). The probability of ionization via S-promotion is

$$P^{(S)} = \frac{1}{2}P_{p\sigma}^{(S)}(1 - P_{p\sigma}^{(S)}) + \frac{1}{2}P_{d\sigma}^{(S)}(1 - P_{d\sigma}^{(S)})P_{000} \ . \tag{13}$$

The terms $(1-P)$ in Eq.(13) describe the possibility of recapture of the electrons promoted via an S-promotion, and the factor $\frac{1}{2}$ is associated with the initial population of the $1s\sigma$ and $2p\sigma$ states.

The other mechanism is via a so-called super T^s-promotion when the nuclei are receding. With a super $T^s_{n_1 m}$-series we mean a series constructed from the first branch points of the $T_{n_1 i m}$-series having the same n_1 and m and $i = 0, 1, ..., n_2, ...$. The T-series are related to unstable trajectories of the electrons on the top of the barrier between the nuclei. The electrons are moving perpendicular to the internuclear axis between the nuclei. The electrons promoted to the continuum via a super T^s-promotion are called 'saddle-point' electrons. This reflects the fact that the electrons are picked up in the saddle region of the potential energy. The electrons will be located at the saddle point of the collision system between the nuclei. In the case of equal charges, their velocities will be one-half of the velocity of the incoming particles. Due to the relative velocity of the "$v/2$-electrons", ionization can occur at a finite internuclear distance.

At an internuclear distance of about 5 a.u. these states have common T_{000}-series branch points. The first branch point connects the states $1s\sigma$ and $3d\sigma$. In the second one the states $2p\sigma$ and $4f\sigma$ are connected.

One can show that there are two main sequences for the super $T^s_{n_2 m}$-promotion. One of these sequences is associated with the first branch points of the $T_{0n_2 0}$-series. Therefore the transition scheme is

$$1s\sigma_g \xrightarrow{T_{000}} 3d\sigma_g \xrightarrow{T_{010}} 5g\sigma_g \xrightarrow{T_{020}} ... \xrightarrow{T_{0n_2 0}} ... \qquad (14)$$

The other branch points belonging to the same $T_{0n_2 0}$-series connect higher states directly (e.g. $1s\sigma_g \xrightarrow{T_{000}} 5g\sigma_g$), but are not taken into account because of the fact that they lie farther away from the real R-axis, which results in a very low transition probability. For the same reason, we do not take into account the chain of hidden crossings starting from the $2p\sigma_u$ state (for which the transition scheme would be $2p\sigma_u \xrightarrow{T_{000}} 4f\sigma_u \xrightarrow{T_{010}} ...$).

The second sequence originates from rotational coupling between the $2p\sigma$ and the $2p\pi$ states at a small internuclear distance. Here ionization occurs via the super T^s_{01}-promotion,

$$2p\pi_u \xrightarrow{T_{001}} 4f\pi_u \xrightarrow{T_{011}} 6h\pi_u \xrightarrow{T_{021}} ... \xrightarrow{T_{0n_2 1}} .... \qquad (15)$$

In this separation, the S-promotion is related to the sum of direct ionization and capture to the continuum, as can be represented by a two-center close-coupling expansion calculation. The T-promotion is related to the saddle point mechanism for ionization, as represented by the third-center contribution in a three-center close-coupling expansion calculation.

The results of our total ionization cross sections are shown in Fig. 4, together with the calculated contributions from the S- and T- promotions which are comparable in the energy region shown. The latter is further divided into

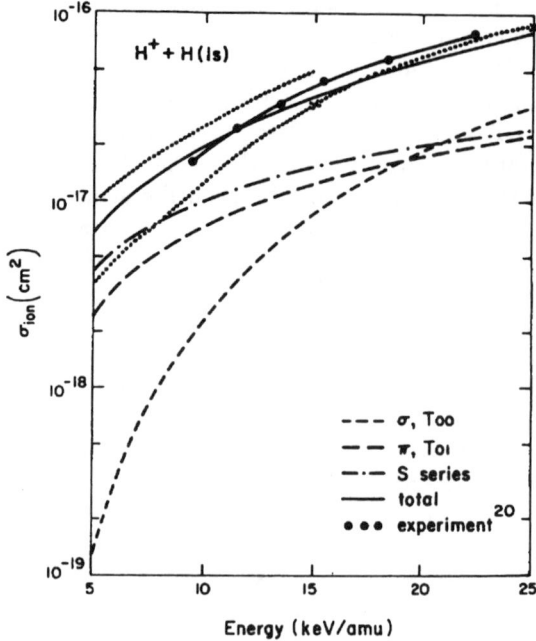

FIG. 4. Comparison of theoretical and experimental total ionization cross sections in 5-25 keV H(1s) + H$^+$ collisions. Contributions to the ionization cross sections from the S- and T-promotions are also indicated: short and long dashed lines for the super T_{00}^s and T_{01}^s-promotions, respectively; dashed-dotted lines, the S-promotion; the solid line is the total ionization cross section. Experiment, from Ref.(14). Other theoretical results: dotted line, Ref.(11); dotted line, Ref.(12); crosses, Ref.(13).

contributions from the σ and π components. We note that the π component dominates the T-promotion at energies below 10 keV. The calculated total ionization cross sections are also compared with the experimental data of Shah et al. (14) and other calculations based on the close-coupling methods (11–13). The present results are in good agreement with the experimental data.

III. RADIAL DECOUPLING

It is widely believed that adiabatic transition probabilities involving one active electron in ion-atom collisions vanish exponentially with decreasing ion velocity v as $\exp(-\Delta/v)$ where Δ is some constant not equal to zero (5,15). An alternative power law dependence for cross sections in the adiabatic regime was first proposed by Born and Fock (16) but was later criticized by Landau (6) who demonstrated that the exponential law followed from the available theory of adiabatic transitions. Such theory ignores the transfer of momentum from the motion of the incident ion to electrons during the transition. Later developments incorporated momentum transfer (17), but it is still considered that momentum transfer does not alter the low velocity limit (18,19), even though it is recognized that such transfer is difficult to formulate in a way that is consistent with adiabatic basis states. A rigorous formulation of electron translation compatible with the adiabatic representation has been given by Solov'ev (5,20), and more rigorous demonstrations of the exponential law are now available. Furthermore, as quoted in Ref. (15), the exponential law generally is in accordance with a large amount of experimental data. Despite the considerable body of evidence supporting the exponential law, we show that the adiabatic representation of Ref. (20) implies a power law $\sigma \propto v^2$ for the cross section σ at low velocities. The usual assumption that v is less than the mean electron velocity $<v_e>$ in the initial state but greater than $<v_e>/\sqrt{M}$, where M is the reduced mass of the ion-atom system, is understood.

The exponential law is based upon Landau's (6) formulation of semi-classical transition probabilities. Transition amplitudes are computed by integrating the difference in the energy eigenvalues in the initial and final channels around the closest-lying point t_1 in the complex time t plane where the energy difference $\Delta E_{fi}(t)$ vanishes. The corresponding nonadiabatic coupling matrix element has a pole at $t = t_1$, where t_1 is usually a complex number. Since t_1 always has an imaginary part for states which are nondegenerate at zero internuclear separation R, and since this difference appears in the exponential factor $\exp[\int_{t_0}^{t_1} \Delta \epsilon_{fi}(t)dt]$, the exponential law follows immediately. In contrast, Ref. (16) assumed a power law dependence for the nonadiabatic coupling matrix element and obtained a power law dependence for the cross section. Here we show, on very general grounds, that the nonadiabatic coupling matrix elements between states with identical orbital angular momentum in the united atom must have a $1/R$ singularity unrelated to the more familiar singularities at a complex internuclear distance R. These latter singularities are well understood in terms of the degeneracy of adiabatic energy eigenvalues $\epsilon_n(R)$ at specific values of complex R (6,5).

In Solov'ev's representation, the adiabatic basis states $\phi_i(R;q)$ are eigenstates of the electronic Hamiltonian at fixed R written in terms of electron coordinates scaled by R, i.e. in terms of $q = r/R$. The nonadiabatic coupling matrix elements are

$$M_{fi}(R) = <\phi_f|\frac{\partial}{\partial R}|\phi_i>, \qquad (16)$$

where the inner product implies integration over q and the partial derivatives are taken holding scaled variables constant (20,5,21). Near the point $R = 0$ the basis states are functions only of the variable $r = Rq$ aside from an R-dependent normalization constant (22)

$$\phi(R \to 0; q) = R^{-3/2}\phi(qR). \qquad (17)$$

It follows immediately that the nonadiabatic coupling matrix element is given by

$$\lim_{R \to 0} M_{fi}(R) = C_{fi}/R \tag{18}$$

where

$$C_{fi} = <\phi_f | \mathbf{r} \cdot \nabla \phi_i >_{\mathbf{r}}, \tag{19}$$

and the subscript **r** indicates that the integrations implied in Eq.(19) are over the usual electron coordinates **r**. The constant C_{fi} is nonzero for initial and final states with the same united atom orbital angular momentum quantum numbers l, m. It follows from Ref. (16) that the cross section for transitions between such states has a power law dependence v^n for some value of n, rather than the standard exponential law. We now show that $n = 2$.

In the time-dependent representation with $R = R(t) = \sqrt{\rho^2 + v^2 t^2}$, and R is sufficiently small, the equations for the coefficients $a_{fi}(t, \rho, v)$ of time-dependent theory are

$$i \frac{da_{fi}(t, \rho, v)}{dt} = \epsilon_f + i \frac{v^2 t}{R^2} \sum_j C_{fj} a_{ji}(t, \rho, v) \tag{20}$$

where ϵ_j represents the value of j'th electronic eigenenergy at $R = 0$ and where the summation goes over the complete set of united atom eigenstates. When these equations are written in terms of the scaled impact parameter $\rho' = \rho/v$, the velocity no longer appears explicitly. It follows that the coefficients are functions of ρ' and t only, so that the transition probability $P(\rho, v) = |a(t \to \infty, \rho/v)|^2$ is a function of ρ/v only. Then cross sections $\sigma_{fi}(v)$ for transitions $i \to f$ are given by

$$\sigma_{fi}(v) = \int_0^\infty P(\rho, v) \rho d\rho = v^2 \int_0^\infty P(\rho') \rho' d\rho' \tag{21}$$

which shows that the low energy cross sections are proportional to v^2. Thus for transitions where the united atom orbital angular momenta ℓ, m do not change, the low velocity limit is the power law v^2 rather than $\exp(-\Delta/v)$, as expected on the basis of Landau's theory.

It should be noted that the v^2 limit is a strictly quantum effect, since, in the classical theory of electron motion, orbits at fixed R can be written in terms of scaled variables q and $d\tau = dt\sqrt{R}$. As R decreases to zero without limit, a given classical orbit simply contracts in accordance with the length scale parameter R. In contrast, quantal electronic eigenstates do not contract without limit as $R \to 0$. Rather, at some distance, the adiabatic eigenstate ceases to contract and becomes a function of the electron coordinate $\mathbf{r} = \mathbf{q}R$, but is otherwise independent of R. The adiabatic wavefunction therefore retains its spatial extent even though the scale parameter R shrinks to zero.

In effect, electron motion decouples from, and becomes nearly independent of, the coordinate R. It is this nonclassical "decoupling" behaviour that leads to the v^2 limit for excitation and ionization cross sections.

There remains the question of whether this behaviour is an artifact of Solov'ev's representation. Experimental evidence on this matter is highly desirable. Measurement or computer simulation of this quantity provides a direct test of the decoupling theory. Ionization of H by H^+ impact is one of interest since the standard theory for this process predicts an exponential dependence (10). Preliminary data, together with very approximate computations of the decoupling contribution (26), suggest that the latter contribution is indeed present. Finally, it should be noted that theories which employ adiabatic basis states modified by *ad hoc* momentum transfer factors (27,28) may also predict nonexponential velocity dependences, including a possible power law at zero velocity.

IV. STURMIAN REPRESENTATION

The low velocity ($E < 2$ keV) ionization data (26) and studies of exact models point to the need for a more general framework than the adiabatic approximation. A brief review of this framework is given below.

The time-dependent Schrödinger equation in Solov'ev's representation has the form

$$i\frac{\partial \psi}{\partial t} - [H_0 + tV]\psi = 0 \;. \qquad (22)$$

We define the Sturmian basis set

$$[H_0 + t_n(E)V]\phi_n(E, x) = E\phi_n(E, x) \qquad (23)$$

with $E > 0$ and with outgoing wave boundary conditions. The symbol x represents a set of coordinates. Then we write

$$\psi = \sum_n \int_{L_i} dE \exp(-iEt)\phi_n(E, x) A_n(E) \qquad (24)$$

where the integral is taken along a path L_i determined by initial conditions. By the usual Laplace transform manipulations, one obtains a set of coupled equations for the expansion coefficients $A_n(E)$

$$-i\frac{\partial A_n(E)}{\partial E} - t_n(E)A_n(E) - i\sum_{n'} <\phi_n(E)|V|\frac{\partial \phi_{n'}(E)}{\partial E}> A_{n'}(E) = 0 \;. \qquad (25)$$

We are working with a class of problems for which the coupling terms are small. To first approximation, we neglect them to obtain

$$A_n^{(0)}(E) = \exp\left[i\int^E t_n(E')dE'\right]. \tag{26}$$

The single Sturmian includes enough dynamics to determine *all* transition amplitudes (approximately). Indeed, there are some exactly solvable problems where one Sturmian gives the exact solution(e.g. if V is separable). Thus we have the remarkable result that while the set of Sturmians are not complete in that they may have only one member for a fixed E, they are nonetheless complete enough to solve the original time-dependent Schrödinger equation.

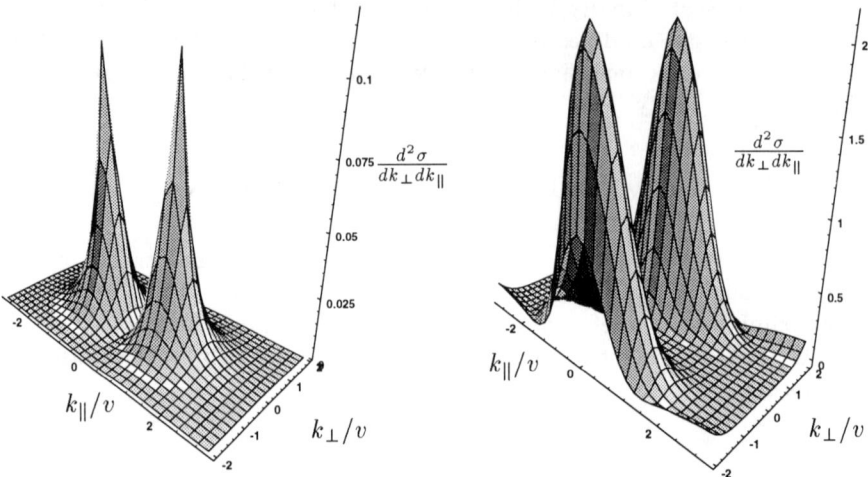

FIG. 5. The double differential cross section (a) for S-promotion, (b) for T-promotion.

V. ENERGY DISTRIBUTIONS

It is not possible to separate the contribution from the two mechanisms by measuring only the total ionization cross section. But it is possible to use the fact that the two mechanisms have different spectra for ejected electrons:

(1) The spectrum associated with the S-mechanism is peaked on the center-of-mass and has two peaks at $k_\perp = 0$ and $k_\parallel = \pm v$. Figure 5 (left) shows a spectrum related to the $S_{p\sigma}$-promotion for $v = 0.4$ a.u. The energy distribution of the fast electrons is exponential.

(2) The T-mechanism spectrum at the energy threshold (which occurs at an infinite separation) is a δ-function at the origin of the frame of reference of the saddle point of the potential. Due to the translational motion of the saddle point, ionization can occur at a finite internuclear distance. This will lead to a spectrum which, at medium energies, will be a rather broad peak

centered on the top of the barrier. Figure 5 (right) shows a spectrum related to the T_{01}-promotion for $v = 0.4$ a.u. The two peaks are associated with the π-symmetry of the T_{01}-promotion.

ACKNOWLEDGMENTS

Support for this research by the National Science Foundation under Grant No PHYS-9222489 is gratefully acknowledged. Support for collaboration with the Ioffe Physical Technical Institute, St. Petersburg, Russia is provided by the National Science Foundation under Grant No. PHY-9213953. This research was sponsored by the Division of Chemical Sciences, U.S. Department of Energy, under Contract No. DE-AC05-84OR21400 managed by Martin Marietta Energy Systems, Inc.

REFERENCES

* Permanent Address: Ioffe Physical Technical Institute, St. Petersburg, Russia.
1. E.E.Nikitin and S.Ya.Umanskii, *Theory of Slow Atomic Collisions* (Springer-Verlag, Berlin-Heidelberg, 1984).
2. L.D.Landau, Phys. Z. Sowjetunion **1**, 88 (1932); C.Zener, Proc. R. Soc. London A **137**, 696 (1932).
3. Yu.N.Demkov, Sov. Phys.–JETP **18**, 138 (1964).
4. Yu. N. Demkov, Proceedings of invited talks of the V ICPEAC, Leningrad, USSR, July (1967)(Published by Joint Institute for Laboratory Astrophysics, Boulder, Colorado, 1968), p 186.
5. E. A. Solov'ev, Sov. Phys.-Usp. **32**, 228 (1989) and E. A. Solov'ev and S. I. Vinitski, J. Phys. B **18**, L557 (1985).
6. L.D.Landau and E.M.Lifshitz, *Quantum Mechanics:Non-Relativistic Theory*, 3rd ed. (Pergamon Press, Oxford, England, 1965).
7. E. A. Solov'ev, Sov. Phys. -JETP.. **54**, 893 (1981).
8. S.Y. Ovchinnikov and E.A. Solov'ev, Sov. Phys.–JETP, **63**, 538 (1986).
9. S. Yu. Ovchinnikov, Phys. Rev. A **42**, 3865 (1990).
10. M. Pieksma and S. Y. Ovchinnikov, J. Phys. B **24**, 2699 (1991).
11. W. Fritsch and C.D. Lin, Phys. Rev. A **27**, 3361 (1983).
12. T. Winter and C.D. Lin, Phys. Rev. A **29**, 3071 (1982).
13. R. Shakeshaft, Phys. Rev. A **18**, 1930 (1978).
14. M.B. Shah, D.S. Elliott and H.B. Gilbody, J. Phys. B **20**, 2481 (1987).
15. N. F. Mott and H. S. W. Massey, *The Theory of Atomic Collisions* (Clarendon, Oxford, 1949), 2nd ed., p. 291.
16. M. Born and V. Fock, Zeit. für Physik **51**, 12 (1928).
17. D.R. Bates and R. McCarroll, Proc. Roy. Soc. **A245**, 175 (1958).
18. D. R. Bates, in *Atomic and Molecular Processes*, ed. by D. R. Bates (Academic Press, New York 1962) p. 550.
19. J. B. Delos, Rev. Mod. Phys. **53**, 287 (1981).
20. E. A. Solov'ev and S. I. Vinitski, J. Phys. B **18**, L557 (1985).
21. U. Fano, J. Macek, K. Jergian, and M. Cavegnero, Phys. Rev. A **35**, 3940 (1993).

22. In accordance with common practice, a small term in the electronic Hamiltonian proportional to v^2 is omitted. The v^2 power law is also obtained when this term is included. A derivation including the v^2 term in the Hamiltonian employs frame transformations between adiabatic and united atom states and will be given elsewhere.
23. T. P. Grozdanov and E. A. Solov'ev, Phys. Rev. A, **42**, 2703 (1990).
24. A. Ovchinnikova and J. H. Macek, Bulletin of the American Physical Society, **38**, 1159(1993).
25. J. H. Macek and S. Y. Ovchinnikov, Bulletin of the American Physical Society, **38**, 1171 (1993).
26. M. Pieksma, J. van Eck, W. B. Westerfeld, and A. Niehaus, International Conference on the Physics of Electronic and Atomic Collisions (ICPEAC), Abstracts of contributed papers, ed. by T. Andersen, B. Fastrup, F. Folkmann and H. Knudsen, (Aarhus University, Denmark) p. 462.
27. S. B. Schneiderman and A. Russek, Phys. Rev. **181**, 311 (1969).
28. J. Vaaben and K. Taulbjerg, J. Phys. B **14**, 1815 (1981).

Current Status Of The Saddle-Point Model

T. J. Gay

*Behlen Laboratory of Physics, University of Nebraska
Lincoln, Nebraska 68588-0111*

Abstract. The current status of evidence for saddle-point electrons is discussed critically. Applications of the saddle-point model to the Barkas effect, ionizing collisions involving highly-charged projectiles, and proton-H collisions are considered.

INTRODUCTION

The saddle-point model of ionizing ion-atom collisions is based on the idea that when a charged projectile ionizes target electrons, some of the ejected electrons find themselves on the transient, moving saddle-point of Coulomb potential with a velocity that matches that of the saddle point. Feeling no force at this position, they "ride" the saddle out of the collision volume and are thus ionized. In the case of the prototypical $H^+ + H$ system, or in any proton - neutral target collision, the saddle point moves at half the projectile velocity, so these electrons are often referred to as "v/2" electrons.

This model has been the topic of numerous papers [1-21], and has proven to be useful for the insights it provides into the general problem of collisional ionization. It has also been applied successfully in the analysis of scaling laws for prediction of total ionization cross sections [6,19] and specific phenomena such as the Barkas effect [22-25]. But the model, its implications and interpretation, and some of the experimental evidence cited to support it have been controversial, and no consensus or comprehensive picture yet exists about the nature of saddle-point effects.

Prior to 1980, primarily as a result of the work of Rudd and co-workers [26], a standard view of ionizing collisions had developed. This view held that almost all the ionized electrons (excluding those produced in violent binary encounters) could be associated either with the ionized target or the receding projectile. The former, mostly having small momenta, were produced in "soft" collisions with the projectile and emerged almost isotropically from the collision region. (Electrons resulting from autoionization of the target were also included in this group.) The remaining ionized electrons had velocities very similar to that of the projectile, and formed a "cusp" distribution about its velocity vector in the forward direction. Based on these considerations, close-coupling calculations of ionization cross sections, for example, employed basis states that were centered either on the target, the projectile, or both. Born-approximation calculations, of course, are inherently one-center approaches to the ionization problem.

© 1996 American Institute of Physics

In 1984, however, Winter and Lin [27] showed that their two-center close-coupling calculations of H^++H total ionization cross sections, which were much smaller than those measured experimentally below 25 keV, could be improved substantially by including basis states centered at a point mid-way between the two charge centers. This improvement implied that a large fraction of the ionized electrons were more appropriately associated not with the two charge centers but with the Coulomb saddle region of the collision. This result was consistent with an observation made earlier by Olson [28] that in classical trajectory Monte-Carlo (CTMC) calculations of ionizing collisions, an enhancement of electrons at v/2 was apparent. By 1985, various laboratories had begun to search for physical manifestations of "two-center effects," i.e. those in which ionized electrons felt roughly equal forces from each of the collision's charge centers. Conceptually, the simplest example of such an effect is the existence of saddle-point electrons.

Evidence for Saddle-Point Electrons

A brief critical analysis and overview of the various evidence for saddle-point electrons follows. The list is comprehensive, but not chronological. Not discussed is the work of Stolterfoht et al.[29], van der Straten and Morgenstern [30], Arcuni [31], and Swensen et al.[32]. These experiments dealt with departures from a Born-approximation ionization picture[29] and post-collision interactions between electrons produced in target autoionization and the receding projectile[30-32]. They thus dealt with two-center effects, but not with saddle-point ionization.

Forward-ejected electrons in slow H^++H ionizing collisions

Winter and Lin found that the disparity between their two-center expansion calculation and the measured cross sections increased with decreasing collision energy, implying that the "saddle-point cross section" will be largest for low energies. Indeed, Wannier threshold theory says that at the ionization threshold, all electrons will emerge on the saddle point [9]. The Winter and Lin work implies that the saddle-point mechanism will dominate other processes below 5 keV.

In an elegant experiment done at Utrecht and reported at this conference, Pieksma et al. [14,20] have studied H^++H collisions between 1 and 6 keV, and found that v/2 electrons dominate the ejected electron spectra above 2 keV. An example of this is shown in Fig.1. While a classical interpretation of these results points directly to saddle-point emission, the quantum-mechanical view is more complicated, involving electron promotion through hidden crossings in the complex space of internuclear distance. The theoretical interpretation of the experimental results is complicated by contributions from "S," or "direct" ionization processes, and by the fact that the electron spectrometer used in these experiments detects electrons emitted into the full forward hemisphere. Nonetheless, the spectra display clear maxima at v/2 for incident proton energies above 2 keV and would appear to be the cleanest signature of a saddle-point mechanism yet demonstrated.

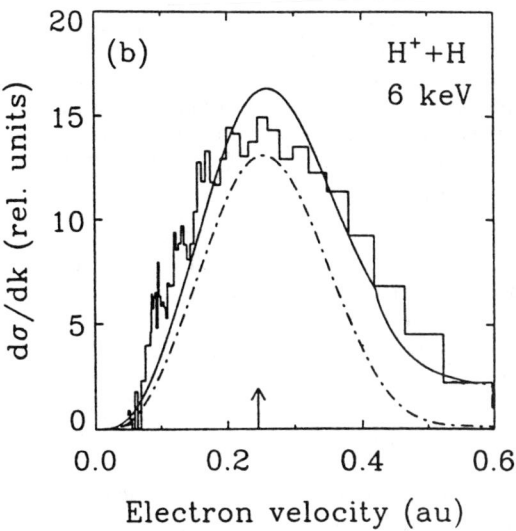

Figure 1. Relative electron yield vs. electron velocity for 6 keV H⁺+H collisions. Projectile velocity divided by two is indicated by the arrow. The dash-dot curve indicates the theoretical prediction for the saddle-point electron contribution; the solid curve is the total yield prediction, fitted with one height parameter to the experimental results. Data and theory of ref. [20].

Saddle-point electrons at intermediate collision energies

In 1986, Olson et al. [2] reported experimental data and CTMC calculations for doubly-differential electron ejection cross sections at 17° in H⁺+He collisions. (It should be pointed out that these data were substantially similar to earlier measurements made by Rudd et al.) The incident energy range they studied was 60 to 200 keV. Plotting the data in velocity space, they showed that most of the electrons ejected in the forward direction have speeds roughly midway between zero and that of the projectile. Thus these electrons experience, over most of their ionizing trajectories, comparable forces from the He⁺ ion and the receding proton. Olson et al. [2,4] chose to call these mid-speed electrons "saddle-point" electrons because of their proximity, both in space and velocity space, to the actual Coulomb saddle point. By dividing the post-collision volume into three regions (see Table I), they were able to show that almost 60% of the ionized electrons are emitted in the middle saddle-point region for 60 keV proton energy.

TABLE I. Populations estimated using CTMC calculations (ref. 4) of the three spatial regions associated with the ionized target, Coulomb saddle point, and the receding projectile. Binning assignments were made when the projectile was at 50 a.u. $\hat{z}$, with the He$^+$ at the origin. The boundaries of the three regions are the planes at $z = 50 \times \frac{1}{3}$ a.u. and $50 \times \frac{2}{3}$ a.u. Antiproton projectile calculation are in parentheses.

E(keV)	Flux fraction %		
	Target region	Midpoint region	Projectile region
60	19.1±0.9	57.8±1.5	23.1±1.0
100	24.5±0.8	59.4±1.3	16.1±0.7
200	46.8±1.4	47.6±1.4	5.6±0.5
(250)	(67.8±2.8)	(31.8±1.9)	(0.5±0.2)
250	50.7±2.1	45.9±2.1	3.4±0.5

The major implication of this work was that the single-center view of ionizing collisions is incomplete and, in the energy range below about 150 keV, qualitatively misleading. An ancillary lesson was that plotting doubly-differential electron spectra in velocity space elucidates several key aspects of the ionization physics. This paper proved to be controversial for several reasons (see, e.g., references 10 and 15). First, the data exhibited no narrow spectral feature at v/2, and the spectral maxima shifted from 0.85v to 0.3v as the proton energy increased over the energy range investigated. (A similar energy-dependant peak shift occurs in the data of Pieksma et al.[20]). Thus the objection was made that no clean signature of a saddle-point mechanism was apparent. This objection is valid, but can be resolved by reemphasizing that the papers of Pieksma et al.[20] and Olson et al.[2] are really claiming different things. At low energy a specific saddle-point mechanism has been identified and the spectra bear out the predictions of a calculation based on this mechanism. At the higher energies investigated by Olson et al., the picture is less precise, but is still based on the idea that most of the forward-ejected electrons are influenced about equally by both the target ion and the receding projectile, i.e. that they live in a region close to the saddle, and that this is apparent from the data once it is plotted in velocity space. A subset of these electrons (whose fraction grows smaller with increasing projectile energy) are those actually stranded on the saddle point and which have a speed precisely half that of the projectile.

A second objection was that by plotting the data in velocity space, mid-velocity maxima were artificially being introduced in the spectra [10,12,13,15]. For example, if one plots the differential cross sections for ejection at 90° in velocity space, a maximum occurs (typically below 0.4v;[15]), but this maximum can hardly be attributed to saddle-point ionization. This concern has validity to the extent that one insists on identifying spectral maxima with a specific saddle-point mechanism. But such an identification is dubious at best in the intermediate energy regime. Nonetheless, if we ask the question "Where are most of the forward-ejected electrons (e.g., with $\theta \leq 15°$) at, say 100 keV?", the answer is: roughly midway between the target and the projectile. This important point is obscured when the data are plotted differentially in energy or in velocity-vector space[12].

While ejected-electron spectra exhibit maxima at non-zero velocity values in velocity space at all ejection angles [18], maxima are only observed in the equivalent energy-differential spectra only for emission angles ≤15° [4]. Gay et al. [4,12] used this fact to answer objections of the second kind discussed above, claiming that since no kinematic effect was present in the energy-differential spectra, the existence of such maxima at the forward angles represented a "clear signature of saddle-point phenomena." Meckback et al. [15], however, have disagreed with this characterization, claiming instead that the energy maxima between 10° and 15° are simply the "backwash" or "remnants" of the charge-transfer-to-the-continuum (CTC) cusp. Herein lies a semantic difficulty. Any backwash from CTC would generally satisfy the rather general definition of saddle-point electrons as advanced by Olson et al. [2], but would not be characteristic of "true" v/2 electrons that originate on or near the saddle-point.

But another question arises in this context. Visible on some 0° spectra, both calculated and experimental, are secondary mid-velocity maxima below the CTC cusp. Such maxima cannot be the backwash of CTC, but could in turn be the saddle-point parents of the maxima at 10°. Unambiguous mid-velocity maxima in $d\sigma/(dEd\Omega)$ at 0° would constitute strong evidence for the importance of saddle-point ionization, but not all spectra exhibit this effect. It is most obvious in the data of Gibson and Reid [33; see Fig.2 below], and in some CTMC calculations [figs. 2 and 3 of ref. 7]. It is also evident in some spectra from Meckback's group [see, e.g., Fig.5 of ref. 8; Fig.6 of ref. 18], but is absent from others. Experiments done at Nebraska by Gealy have failed to find it [34]. Experiments done at 0° are difficult and beam contamination is a pernicious problem. Thus careful experiments need to be done to determine if this second maximum is real.

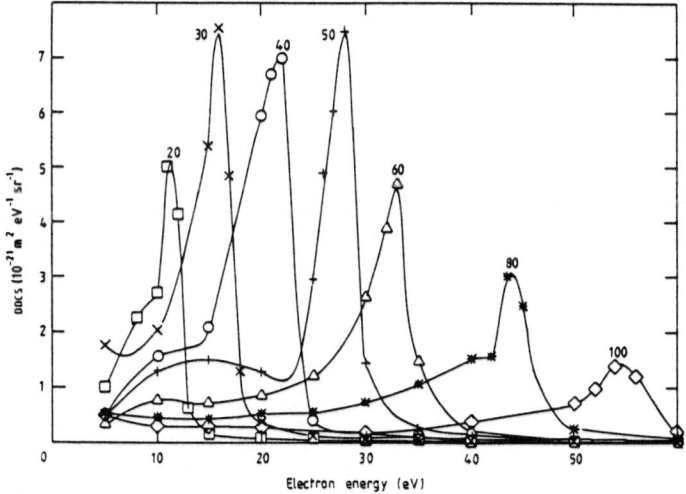

Figure 2. Doubly-differential cross sections in energy for 0° emission of electrons in H⁺+He collisions at proton energies between 20 keV and 100 keV [33]. An apparent mid-velocity secondary maximum, evident between 40 and 60 keV, disappears at higher energies, and, if present, is not resolved at lower ones. See also refs. 7,8, and 18.

Projectile charge-dependence of mid-velocity maxima.

For a given projectile velocity v, the velocity of the saddle point depends on the final projectile (Z_p) and target (Z_t) charges as

$$v_{SP} = \frac{v}{1 + (Z_p / Z_t)^{1/2}} \quad . \tag{1}$$

If saddle-point ionization is an important mechanism, one would expect the electron distribution maximum to shift for isotachic projectiles of different charge. Irby et al. [3] tested this idea using $^3He^{++}$ and H^+ projectiles at specific energies from 60 to 120 keV/amu. They obtained the somewhat counterintuitive result (albeit one predicted by eq.1) that the He^{++} peak cross sections at 17° occurred at lower velocities than did those for isotachic H^+. The experimental situation is unclear on this point, however. Both the Argentinean group [10,13] and DuBois [17] have carefully remeasured these spectra and find no shift. On the other hand, Gay et al. [12] have also redone the experiment and confirm qualitatively that a shift is present. Moreover, Irby et al.[16], using C^{1+}, C^{2+}, and C^{3+} on He have seen charge-dependent shifts, in the direction predicted by a saddle-point hypothesis, that have been qualitatively confirmed by DuBois[21].

Several points must be made here. First, neither CTMC [3] nor CDW-EIS [10] calculations predict a projectile-charge-dependent shift. If such an effect was truly present, one would certainly expect at least a fully classical calculation such as the CTMC to see it. It should also be noted, though, that Burgdörfer et al. [5,11] have seen such shifts in their short-range model calculations in both one and three dimensions. Experimentally, Gay et al. [12] have shown how beam contamination could, in principle, produce an artificial shift, but measured the effect to be negligible for their data. DuBois [17] tried to mimic an artificial shift by intentionally contaminating his beam, but could not reproduce the results of Irby et al.[3] or Gay et al. [12]. Irby has pointed out that spurious scattering of electrons from analyzer back plates can mask shifts that are present [ref. 16 and these proceedings], but Bernardi and Meckback have shown that this effect is negligible in their case [35]. So the situation is confused.

The carbon-projectile experiments [16,21] raise an interesting question about collisions in which at least one of the post-collision charge centers has at least one bound electron. If one makes the (only partially justified) assumption that no electrons are lost from the carbon, the shifts with projectile charge can be viewed in one of two ways. A saddle-point model simply explains them in terms of eq.1. But one can also argue that the changes in spectra are not peak shifts *per se*, but are caused by a reduction in the low-energy electron production cross sections as projectile charge is increased [see Fig.3 of ref.21]. This effect is caused by increased screening of the bare carbon nucleus as the net charge is decreased; with lower projectile charge the more sharply impulsive ionizing events will become more important relative to the "soft" large-impact-parameter collisions that dominate for higher charge states.

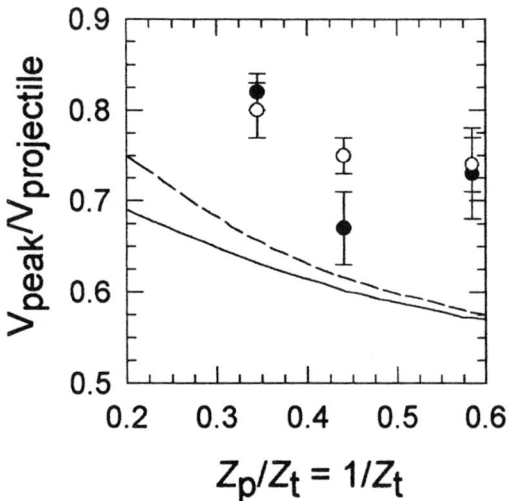

Figure 3. Position of peak in velocity space for 10° emission in 50 keV (open circles) and 100 keV (solid circles) proton impact ionization. Left two data points are for Ne targets, the middle two are for Ar, and the righthand two are for He. $Z_p=1$; Z_t is given by Slater's rules [36]. Solid curve corresponds to the geometric saddle-point velocity; the dashed curve is taken from the one-dimensional short-range potential calculation of ref. [5]. Data of ref. 12 (see text).

Let us consider, however, the experiments of Gay *et al.* [12] in which protons ionized He, Ne, and Ar. Here the projectile charge remains constant, but an effective charge for the target becomes +1 only asymptotically. During the first stages of ionization, the different targets have different effective charges, which can be very crudely estimated using Slater's rules [36]. (While effective charge of this type is really dependent on the distance between the escaping electron and the target, and on the momentum transfer to the ionized electron, any attempt to refine the effective charge concept is probably not warranted, given the crude nature of the model.) Thus Fig. 3 shows that target-dependent variations of the spectral maxima can be viewed as a saddle-point shift. The key point here, however, is that the screening model, used by DuBois [21] to explain the shifts in the carbon data, predicts shifts opposite those observed when applied to the (transient) effective charge of the target. In summary, the saddle-point shift data at this point are inconclusive. An ideal experiment to test these ideas would involve ionization of atomic hydrogen by a series of bare projectiles at low impact velocities.

Ridge electrons.

In addition to the "longitudinal" evidence for saddle-point ionization provided by electrons emitted at small forward angles, there can also be "transverse" effects.

Prior to the work of Olson *et al.*[2], Meckback *et al.*[1] presented the first detailed discussion of two-center effects with regard to mid-velocity electrons. Considering velocities in the region around v/2 for forward-scattered electrons, they found cusps in the electron distribution at 0° as the electron ejection angle was varied. They termed these electrons "Wannier-ridge" electrons, and attributed the ridge- or cusp-like distribution to the transverse compression effect of the saddle region. The existence of a cusp, as such, at 0° (as opposed to a smooth maximum) has since been called into question [37], and there is still controversy regarding this point. Moreover, the data presented by Meckback *et al.* was for 170 keV H$^+$+He collisions, where it is unlikely that significant saddle-point ionization occurs. Nonetheless, analogous ridge-like features at 0° (see, e.g., Fig. 8 of ref.[18]) are a general feature of ionized electron spectra, and can be attributed reasonably to transverse saddle-point focusing effects.

Applications of the Saddle-Point Model

We now consider three applications of the saddle-point model. The first has already been mentioned: the use of a third, mid-region expansion center in close coupling calculations of total ionization cross sections [27]. The fact that this technique is efficacious primarily at low energies, in conjunction with the Wannier threshold picture of saddle-point ionization, implies that "true" (mid-point stranding) saddle-point ionization can be a dominant mechanism only below 25keV, at least for singly-charged projectiles [14]. The more general notion of mid-velocity ionization [2], however, can be used to understand total ionization cross section scaling laws and the Barkas effect.

Ionization cross section scaling laws

When highly-charged ions such as C^{q+} and O^{q+} ionize H or He, it has been shown experimentally that the maximum values of the total ionization cross sections, σ_{max}, and the incident projectile energies at which these cross sections occur, E_{max}, are given by the scaling laws:

$$E_{max} = aq^{0.65} \times 10^4 \text{ eV/amu}$$
$$\text{and} \quad \sigma_{max} = bq^{1.3} \times 10^{-16} \text{cm}^2, \quad (2)$$

where q is the projectile charge and a and b are phenomenologically determined constants [38]. These *ad hoc* equations have been used extensively in models of fusion plasmas. Using a simple, classical saddle point picture, Irby[6] was able to derive them from first principles. Essentially, he argued that at the energy where σ_{max} occurs, the saddle-point velocity ought to match the average target electron velocity. Moreover, the corresponding value of σ_{max} should be proportional to πR^2, where R is the distance between the saddle point and the target nucleus. This distance is determined in turn by picking the collisional impact parameter such that target electrons have enough energy to traverse the saddle region of the Coulomb potential. More recently, Janev *et al.*[19] have used the ideas of hidden crossings and superpromotion in saddle-point ionization to predict the specific energy-dependence of the ionization cross sections below their peak values.

The Barkas effect

Originally, the term "Barkas effect" was used to describe the fact that when K mesons decay in matter, the equally energetic π^+ and π^- particles produced in the decay travelled different distances before stopping [39]. More recently, the term has come to refer to the differences in electromagnetic interaction between a particle of matter and its environment as compared with its antimatter equivalent [25]. It is generally true that negatively-charged particles have lower stopping powers in matter than their positively-charged equivalents. At high energy, this has been attributed to a "polarization" effect, in which swift positive projectiles draw target electrons toward them, making ionization events more likely [39]. In this manner, the positively-charged particle loses energy more quickly and thus experiences a higher stopping power. This effect is typically less than 1%. At lower energies, though, the ionization cross sections for, e.g., protons and antiprotons can be as large as 40% [22,23,40]. Olson and co-workers [23-25] have shown that the difference in ionization cross sections for proton and antiproton projectiles is due almost entirely to production of electrons on the middle saddle region (see Table 1). Thus for initial energies of the order of 1 MeV [41], the difference in proton and antiproton ranges is due primarily to saddlepoint effects. It is not clear for higher energies whether polarization or saddle-point effects dominate matter-antimatter range differences.

ACKNOWLEDGEMENTS

The author would like to thank H. G. Berry, J. Burgdörfer, R. D. DuBois, C. R. Garibotti, M. W. Gealy, V. D. Irby, R. E. Olson, C.O. Reinhold, and M. E. Rudd for numerous useful discussions of this topic. This work was supported by funding from the University of Missouri, the University of Nebraska, and the Department of Energy, Office of Fusion Energy (Grant # DE-FG02-84ER53188).

REFERENCES

1. W. Meckbach, P. R. Focke, A. R. Goñi, S. Suárez, J. Macek, and M. G. Menendez, Phys. Rev. Lett. **57**, 1587 (1986).
2. R. E. Olson, T. J. Gay, H. G. Berry, E. B. Hale, and V. D. Irby, Phys. Rev. Lett. **59**, 36 (1987).
3. V. D. Irby, T. J. Gay, J. Wm. Edwards, E. B. Hale, M. L. McKenzie, and R. E. Olson, Phys. Rev. A(RC) **37**, 3612 (1988).
4. T. J. Gay, H. G. Berry, E. B. Hale, V. D. Irby, and R. E. Olson, Nucl. Instrum. and Meth. **B31**, 336 (1988).
5. J. Burgdörfer, J. Wang, and A. Bárány, Phys. Rev. A(RC) **38**, 4919 (1988).
6. V. D. Irby, Phys. Rev. A **39**, 54 (1989).
7. C. O. Reinhold and R. E. Olson, Phys. Rev. A **39**, 3861 (1989).

8. G. C. Bernardi, S. Suárez, P. D. Fainstein, C. R. Garibotti, W. Meckbach, and P. Focke, Phys. Rev. A **40**, 6863 (1989).
9. G. Bandarage and R. Parson, Phys. Rev. A **41**, 5878 (1990).
10. G. Bernardi, P. Fainstein, C. R. Garibotti, and S. Suárez, J. Phys. B **23**, L139 (1990).
11. J. Wang, J. Burgdörfer, and A. Bárány, Phys. Rev. A **43**, 4036, (1991).
12. T. J. Gay, M. W. Gealy, and M. E. Rudd, J. Phys. B **23**, L823 (1990).
13. G. Bernardi, S. Suárez, P. Fainstein, C. Garibotti, W. Meckbach, and P. Focke, J. Phys. B **23**, L829 (1990).
14. M. Pieksma and S. Y. Ovchinnikov, J. Phys. B **24**, 2699 (1991).
15. W. Meckbach, S. Suárez, P. Focke, and G. Bernardi, J. Phys. B **24**, 3763 (1991).
16. V. D. Irby, S. Datz, P. F. Dittner, N. L. Jones, H. F. Krause, and C. R. Vane, Phys. Rev. A **47**, 2957 (1993).
17. R. D. DuBois, Phys. Rev. A **48**, 1123 (1993).
18. S. Suárez, C. Garibotti, G. Bernardi, P. Focke, and W. Meckbach, Phys. Rev. A **48**, 4339 (1993).
19. R. K. Janev, G. Ivanovski, and E. A. Solov'ev, Phys. Rev. A(RC) **49**, R645 (1994).
20. M. Pieksma, S. Y. Ovchinnikov, J. van Eck, W. B. Westerveld, and A. Niehaus, Phys. Rev. Lett. **73**, 46 (1994).
21. R. D. DuBois, Phys. Rev. A **50**, 364 (1994).
22. R. E. Olson, Phys. Rev. A(RC) **36**, 1519 (1987).
23. R. E. Olson and T. J. Gay, Phys. Rev. Lett. **61**, 302 (1988).
24. T. J. Gay and R. E. Olson, Nucl. Instrum. and Meth. **B40/41**, 104 (1989).
25. D. R. Schultz, R. E. Olson, and C. O. Reinhold, J. Phys. B **24**, 521 (1991).
26. See, e.g., M. E. Rudd, Y.-K. Kim, D. H. Madison, and T. J. Gay, Rev. Mod. Phys. **64**, 441 (1992).
27. T. G. Winter and C. D. Lin, Phys. Rev. A **29**, 3071 (1984).
28. R. E. Olson, Phys. Rev. A **27**, 1871 (1983).
29. N. Stolterfoht, D. Schneider, J. Tanis, H. Altevogt, A. Salin, P. D. Fainstein, R. Rivarola, J. P. Grandin, J. N. Scheurer, S. Andriamonje, D. Bertault, and J. F. Chemin, Europhys. Lett. **4**, 899 (1987).
30. P. van der Straten and R. Morgenstern, J. Phys. B **19**, 1361 (1986).
31. P. Arcuni, Phys. Rev. A **33**, 105 (1986).
32. K. Swenson, C. C. Havener, N. Stolterfoht, K. Sommer, and F. W. Meyer, Phys. Rev. Lett. **63**, 35 (1989).
33. D. K. Gibson and I. D. Reid, J. Phys. B **19**, 3265 (1986).
34. M. W. Gealy, private communication (1993).
35. G. Bernardi and W. Meckbach, Phys. Rev. A **51**, 1709 (1995).
36. H. Eyring, J. Walter, and G. E. Kimball, *Quantum Chemistry*, (Wiley, New York, 1964) pps. 162-163.
37. G. Bernardi, S Suárez, P. Focke, and W. Meckbach, Nucl. Instrum. and Meth. **B33**, 321 (1988).
38. R. A. Phaneuf, R. K. Janev, and M. S. Pindzola, Oak Ridge National Laboratory Report ORNL-6090/V5 (Controlled Fusion Atomic Data Center, ORNL, 1987).
39. G. Basbas, Nucl. Instrum. and Meth. **B4**, 227 (1984).
40. P. D. Fainstein, V. H. Ponce, and R. D. Rivarola, Phys. Rev. A **36**, 3639 (1987).
41. A. Adamo *et al.* (OBELIX Collaboration), Nucl. Phys. **A558**, 665c (1993).

SADDLE ELECTRON EMISSION FROM ION-ATOM COLLISIONS AT INTERMEDIATE PROJECTILE ENERGIES

Sergio G. Suárez [#]
Centro Atómico Bariloche and Instituto Balseiro
Comisión Nacional de Energía Atómica
8400 S.C. de Bariloche - Argentina

Abstract:

Representative experiments reported with regard to the subject of Saddle Emission are reviewed and a critical summary of their evaluation as well as artifacts that could have affected or distorted the experimental results is also given.

I - INTRODUCTION

Electrons may be emitted when a nude projectile with velocity v_p interacts with an atomic gas target under single collision conditions. Their doubly-differential distribution, in angle θ and energy E_e, presents structures with distinguishing features which permit identification of different mechanisms involved in the electron emission process:

♦ The binary electron peak (BE) [1], understood as produced by a binary projectile-electron collision, is observed as a concentration in velocity v_e- space of electrons in a spherical shell of radius $v_e = v_p$, centered at the projectile velocity v_p. The shape of this peak reflects the momentum distribution of the electron in its initial bound state.

♦ The peak of electron capture into the continuum (ECC) of the projectile, produced essentially by pure Coulomb interaction between the projectile and emitted electrons with velocities $v_e \sim v_p$ [2].

♦ The peak of soft electrons (SE), mainly produced by distant collisions and centered around the residual target ion at $v_e \sim 0$ [3].

The sharpness of these two latter electron distributions puts into evidence the expected divergence of electron states around an isolated charge, represented by the projectile and the residual target ion, respectively.

Each one of the mentioned structures has been interpreted as due to a particular mechanism, generally associated with the main interaction of the ejected electron with either the ionized target or the projectile. However, much experimental and theoretical effort has been devoted recently to the understanding of ionization processes in which an emitted electron is subject to the simultaneous and comparable interaction with both ions resulting from the collision.

In this connection, from classical trajectory Monte Carlo (CTMC) calculations, Olson [4] has concluded that, for proton energies of about $E_p \sim 60$ keV a large fraction of the electron emission in $H^+ \to H$ collisions could be found close to half the relative velocity v_p of the two Coulomb centers. The first experimental observation, in this sense, that showed a structure in the doubly differential electron distribution was reported by Meckbach et al [5]. This structure, interpreted as due to the evolution of emitted electrons on the potential saddle of two receding ions, was subsequently shown to be a broad ridge centered at the beam direction ($\theta = 0^0$) [6,7] and stretched between the ECC and the SE peaks.

In the meantime, Olson et al [8] reported new measurements and CTMC-calculations of doubly differential electron distributions (DDD) for $H^+ \to He$ collisions at intermediate projectile energies ($E_p > 60$ keV). According to their arguments, the ECC contribution "washes out" the saddle electrons at small angles and, in order to observe evidences of what they called [9] the "major ionization mechanism", measurements should be performed at nonzero ejection angles. By following this idea, they presented DDD for $\theta = 17^0$ and used the cross section $d^2\sigma/dv_e d\Omega$ to depict their data. They concluded finally that the maximum in $d^2\sigma/dv_e d\Omega$ was an evidence of the presence of a new and dominant ionization mechanism, which was henceforth called "Saddle-Point Electron Emission".

The experimental evidence presented by different authors [7-11] caused then ample controversy, focused on the measurements and their interpretation, and a variety of theoretical treatments [12-14] have shed light on the understanding of the proposed mechanism.

In the present paper a critical review of the existing evidence is presented (Section II) and some suggested artifacts that could have affected the experimental results are discussed (Section III). Finally, preliminary conclusions and perspectives are depicted in Section IV.

II - CONTROVERTED EVIDENCES

a) Interpretation of the experimental information

In their paper, Olson et al [8] interpreted the maximum of the cross section $d^2\sigma/dv_e d\Omega$ as a peak, characteristic of what they called "Saddle-point electron emission". In this same paper is stated the very well-known fact that "low-energy ejected electrons dominate the ionization process", however, in the cross section

$d^2\sigma/dv_e d\Omega$, as a function of the angle θ and electron velocity v_e, the "soft electron" peak is kinematically reduced to zero. In fact, a maximum appears between $v_e=0$ and the projectile velocity v_p when the experimental data are represented by using this cross section. It should be noted, in addition, that such a maximum also appears at larger emission angles θ, where the action of the saddle potential, extended between the residual target and projectile ions, has no meaning. Doubly differential cross sections $d^2\sigma/dv_e d\Omega$ [18] are shown in Fig.1 for 106 keV H^+ on Ne and for $\theta=10^0$, 90^0 and 180^0. Here a maximum is observed for each angle, which is centered near $v_p/2$ for $\theta = 90^0$ and 180^0, however, for small angles (e.g. $\theta=10^0$), the process of charge transfer into the continuum of the projectile is dominant and the maximum is found to be shifted towards higher velocities. This hump localized at a velocity slightly smaller than v_p (v_e/v_p~0.9 for $\theta = 10^0$) and first observed by Rudd and Jorgensen [10] led to the discovery of the ECC peak by Crooks and Rudd [12], but however, was later interpreted by Gay et al [11] as an "unambiguous signature of saddle-point ionization".

Subsequently, Meckbach et al [15] have shown for 52 keV $H^+ \rightarrow$ **He**, the continuous transition, as a function of the emission angle, of the ECC peak from $\theta=0$ up to 10^0, which clearly demonstrates that the hump observed at $\theta=10^0$ is essentially a cut through this peak, as it was interpreted originally. For larger angles, where the action of the electron transfer into the continuum of the projectile is not appreciable, the cross section $d^2\sigma/dv_e d\Omega$ presents a maximum close to $v_p/2$, as shown in Fig.1. Obviously, the presence of a maximum in the cross section $d^2\sigma/dv_e d\Omega$ does not clearly support the existence of a "new" ionization mechanism.

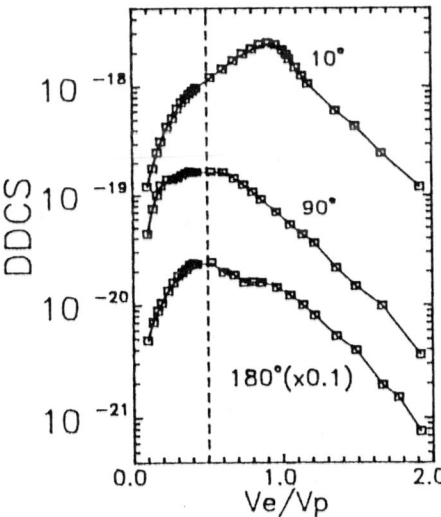

Figure 1: Doubly differential cross sections $d^2\sigma/dv_e d\Omega$ for ionization of Ne by 106 keV H^+ impact and $\theta=10^0$, 90^0 and 180^0.

b) Projectile charge dependence

New reported measurements by Irby et al [16] and Gay et al [17] added new elements to the existing controversy. By investigating the projectile charge dependence of the electron emission near the potential saddle, they found that the maximum in $d^2\sigma/dv_ed\Omega$ shifted as a function of the charge of the impinging projectile. In particular, for He^{2+} on He, Ne and Ar atoms [17], they found a shift of the maximum, with respect to that obtained with H^+, to lower electron velocities. A similar behavior was observed [16] for C^+, C^{2+} and C^{3+}-projectiles in collision with He and Ne targets. The observed shift was understood as resulting from the classical displacement of the potential saddle when the projectile charge is increased and found to follow qualitatively the saddle velocity v_s given by [5]

$$v_s = v_p/[1 + (Z_p/Z_T)^{1/2}], \qquad (1)$$

where Z_p and Z_T are the projectile and target charges, respectively. These results, however, have not been confirmed by experiments performed in our laboratory [7,18] and recently by DuBois [19] who did not find any shift in the electron distributions. Fig.2 shows a comparison between the experimental results obtained by Gay et al [17] and those by Suárez et al [18]. In this figure are presented doubly-differential cross sections $d^2\sigma/dv_ed\Omega$ for the ionization of Ne, for a fixed angle of emission $\theta=10^0$ and as a function of the electron velocity v_e. While the H^+ distributions do not show significant differences in shape, although a discrepancy in the velocity of the maxima is distinguishable, the $^3He^{2+}$ distributions are completely different in both shape and position of their maxima.

It should be mentioned, in addition, that "two center" theoretical calculations performed with the CTMC and CDW-EIS methods do not show any shift between the calculated cross sections $d^2\sigma/dv_ed\Omega$ for different projectile charges [7,9].

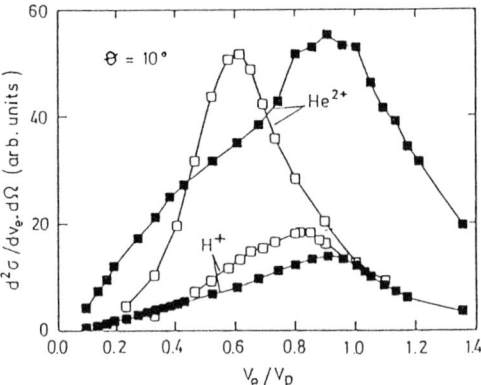

Figure 2: Doubly differential cross sections $d^2\sigma/dv_ed\Omega$ for Ne ionization resulting from 100 keV/amu H^+ and He^{2+} impact. Open symbols: ref.[17]; full symbols: ref. [18].

In order to understand the origin of these experimental discrepancies, mainly produced with ^{3}He^{2+}-projectiles, three experimental artifacts have been suggested [16,17,18,19]. These spurious effects will be evaluated and discussed in the following section.

III - EXPERIMENTAL ARTIFACTS

The experimental discrepancies shown in Fig.2 have been proposed to be caused at least partially by three very different artifacts. The first was related to the extension of the gas target around the collision region [18, 27], which distorts the electron detection for energies E_e<50eV. The second, inherent to electrostatic single-stage electron spectrometers, consists in a distortion of the measured distributions caused by electrons reflected and emitted at the back plate of a parallel plate or cylindrical mirror electron spectrometer [16]. The third was proposed by DuBois as due to the wrong selection of He^{2+}-projectiles or mixture of projectile species in the beam [19].

a) Extended gas target effect

It is known from our previous papers [6,7,18], that the extension of the gas target is increased when the needle tip, source of the gas target, is separated from the beam line by an increasing distance d_{NB}. This procedure has permitted the generation of different gas extensions near the focus of the electron spectrometer, giving rise to distortions in the low-energy range of the measured spectra. The larger d_{NB}, the larger the extension of the gas target. In Fig.3 doubly differential distributions for electron emission at $\theta=10^0$ induced by 106 keV H$^+\rightarrow$Ne collisions are shown for different distances d_{NB}. Since the shape of the spectra was found to be independent, within experimental uncertainties, of the gas target extension in the distance range $d_{NB}\leq 0.5$mm, the distribution corresponding to the smallest distance (d_{NB}=0.1mm) has been interpreted as the doubly differential cross section for $\theta=10^0$. In this figure, it is seen that the largest target extension, reached for the gas target distributed uniformly in the collision chamber, produces the largest distortions of the electron spectra. The spectra obtained for each distance d_{NB} have been normalized at an energy E_e=150eV in the higher energy range, where no dependence of the shape of the electron distributions on the extension of the gas target is observed. The distortions of the low-energy electron spectra have been found to be important even for $\theta=90^0$ for which the beam line is not superimposed on the visual line of the detector. More experimental details concerning the effect produced by extended gas targets have been given in previous papers[6, 18, 27]. An experiment with ^{3}He^{2+}-projectiles gave the same qualitative results.

It is worthwhile to mention, in this connection, that former results reported by Stolterfoht [28] gave qualitatively similar, although smaller, discrepancies between electron distributions obtained with a uniformly distributed gas target and that provided by a nozzle for the case of CH_4-ionization by 300keV protons.

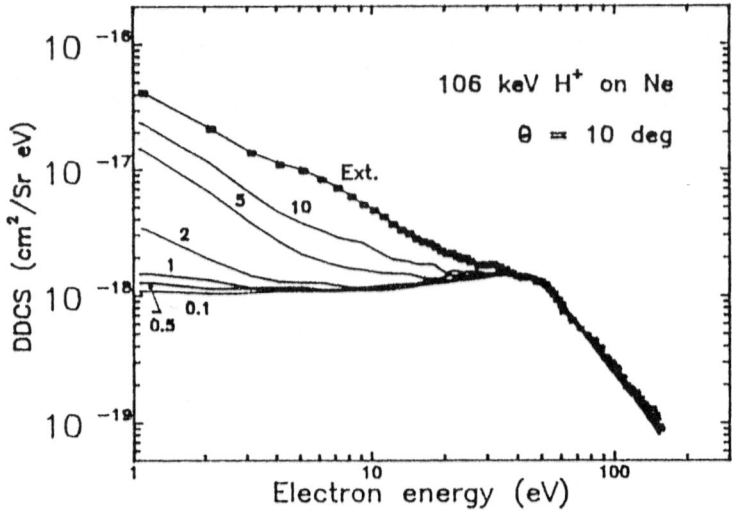

Figure 3: Electron energy distributions at $\theta=10°$ after using a gas target of increasing extension for 106 keV H^+ on Ne (see text).

In order to study the projectile charge dependence of such a target artifact, doubly differential ratios of the electron emissions resulting from projectiles $^3He^{2+}$ and H^+, have been obtained for $\theta=10°$. These ratios are shown in Fig.4. A smooth dependence on the electron energy is found for the ratio corresponding to the best localized target ($d_{NB}=0.1$ mm), whereas, for $d_{NB}=10$ mm, a large maximum centered at about ~5eV dominates the ratio. The ratio calculated from experimental data taken from Gay et al [17] shows qualitatively the same maximum, although there exist some differences in shape and position. It should be noted, in addition, that the experiment of Gay et al [17] has been realized by using an uniformly distributed gas target.

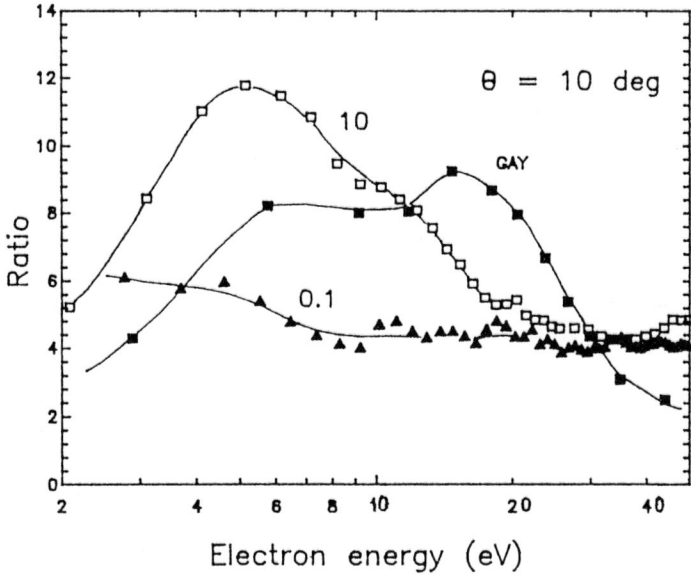

Figure 4: Ratios between detected electron distributions from $^3\text{He}^{2+}$ and H^+ for different target extensions (see Fig.3). Full squares: data of Gay et al [17].

The representation of the experimental data by means of ratios is appropriate in this case since the choice of any particular cross section is avoided. Thus, any shift of the electron distributions with the projectile charge, should be evident in such a representation. Ratios obtained from theoretical calculations furnished by CTMC and CDW-EIS models show only a smooth electron-energy dependence in accordance with the experimental ratio shown in Fig.4 for $d_{NB}=0.1$ mm.

The above discussed experimental artifact could, however, present different features according to the experimental setups used in different laboratories [16,17,18,19,28], therefore a quantitative confirmation of the effect could be casual.

b) Emission and reflection of electrons within the spectrometer.

Here a problem of some discussion that already has led to controversies in the past is again present. In the first place, there is always an "instrumental background". In electron spectroscopy such a background is predominantly caused by low-energy electrons which migrate through the equipment reaching the detector. Of course, if the background is known, one can correct the spectra by subtraction. A simple, although naive, method to know the background is to measure spectra with the target removed. This "target-out" background in some current [9] and early [20] measurements has been certainly quite large. However, this accounts only for a part of a possible instrumental background, because a new source of spurious electrons appears as soon as the target is "in". In this sense, an interesting mechanism of production of spurious electrons, has been suggested by Irby et al [16]. They proposed that secondary electron emission and electron reflection within the spectrometer could be the reason of the experimental discrepancies with other authors [7,19,21]. According to their proposal, electrons entering a parallel plate (or cylindrical mirror) single-stage electron spectrometer, with an energy larger than $E_{max}=e \cdot V_p/\sin^2\phi$ (V_p is the voltage between plates, ϕ the electron entrance angle [16] and e the electron charge) will hit the back plate of the analyzer and then could produce reflected and secondary electrons, part of which may be accelerated into the detector.

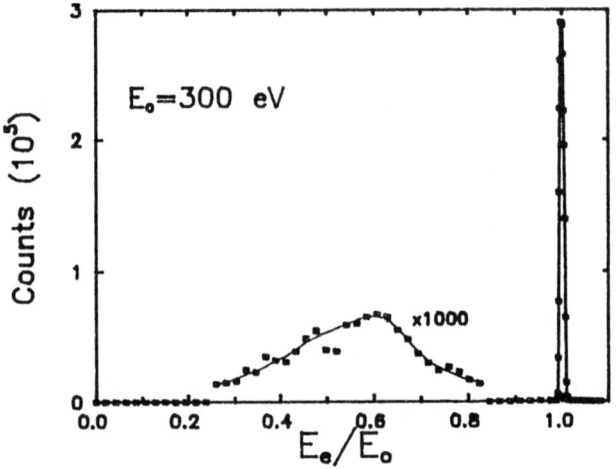

Figure 5: Spectrum of electrons of energy E_0=300 eV resulting from an electron gun with 300 V of acceleration potential.

In former measurements Meckbach et al [22] have shown that such an effect was not observed when using an electron beam with energy E_e=70eV and a cylindrical mirror electron spectrometer [23]. Now, a more extensive study of this effect has been carried out by Bernardi and Meckbach [24] for electron beams with energies within 100 and 500eV. Such a study has shown that, indeed, the effect of reflection and emission of electrons on the outer cylinder of the spectrometer does exist, as can be seen in Fig.5. In this figure, a sharp peak is observed, corresponding to the electron beam of energy E_e=300eV and then, for lower electron energies, a structure with a maximum at about E_e/E_0~0.6 which had to be represented enhanced by a factor 10^3. At the maximum, counts are only 0.02% of those observed at the top of the main peak. For lower energies, the count rates decrease in a continuous manner. It is concluded from this result that, although the effect is observable, it is extremely small and could be considered negligible. Furthermore, a model calculation to study the influence of this effect in the case of broad spectra as obtained in ion-atom collisions, has shown [24] that for energies $E_e<10eV$, the deformation caused by electron emission and reflection reaches not more than 2% of the measured signal, and then decreases quickly for increasing electron energies.

We thus conclude that no reported measurements obtained with electrostatic-single-stage electron spectrometers are significantly affected by the emission of secondaries or reflection of electrons on the back plate (cylinder) of the spectrometer.

c) Beam contamination

In a recent paper [19], DuBois reported spectra for the emission of electrons from helium at $\theta=15^0$ by impact of D^+ and $^4He^{2+}$ with energies ranging from 50 to 250 keV/amu. This study showed qualitative agreement with those of Bernardi et al [7]. It means that no shifts between the cross sections $d^2\sigma/dv_e d\Omega$ for D^+ and $^4He^{2+}$ were observed, in accordance with the result of ref. [18], depicted in Fig.2 for Ne ionization. Beam contamination has been studied carefully by DuBois and he has concluded that the He^{2+} measurements of Irby et al [9] and Gay et al [17] are incorrect and therefore, they should not be considered as an evidence to give support to an independent saddle-point-ionization mechanism.

DuBois considered the possibility that differences with data of references [9,11] and [17] were due to either a wrong selection, or to a mixture, probably with hydrogen ions, of the beam assumed to consist of He^{2+}-projectiles. However, after a systematic study, DuBois did not find evidence for contamination in the proton/alpha experiments. Furthermore, even when the hypothesis of contamination could have been right for the experiments performed with H^+ and He^{2+}, it is difficult to consider it applicable for the new reported data for He ionization by impact of C^+, C^{2+} and C^{3+} [16].

IV - OUTLOOK AND CONCLUSIONS

The proposed Saddle-point-ionization mechanism has found serious criticisms [15,19,24] concerning both the interpretation and reliability of measurements. A critical analysis of the instrumental artifacts discussed in III does not give support to the hypothesis of an independent ionization mechanism, in particular in that the electron distributions obtained with H^+ and He^{2+}-projectiles do not present a shift according to the shift of the saddle-point of the two-center potential. Moreover, an interpretation of such a mechanism based on the cross section $d^2\sigma/dv_e d\Omega$, which has a maximum near $v_p/2$ for all angles of emission, is shown to be incorrect in that the proposed ionization mechanism is not relevant for large emission angles.

Although the hypothesis of DuBois about the possibility of beam contamination in the experiment of Gay et al [17] seems to be plausible for He^{2+}-projectiles, the new results of Irby et al [16] obtained with C^+, C^{2+} and C^{3+}, are not expected to be affected by this same kind of contamination. However, the effect of extended gas targets, mentioned in III, produces a distortion of the low-energy electron distributions which certainly has been present in the experiment of Gay et al [17]. The projectile charge Z_p dependence of such a target artifact could be confused or interpreted as a saddle-point shift when Z_p is increased.

On the other hand, the effect of reflection of high energy electrons and emission of secondaries within the electron spectrometer, claimed by Irby et al [16], has been found to be negligible when a single stage cylindrical mirror spectrometer is used [24].

There is no doubt, as was stated by Olson from CTMC calculations [4], that "a considerable number of small angle scattering events for ejected electron velocities approximately equal to one half the projectile velocity" is present in ion-atom ionization at intermediate energies, but however, these scattering events give rise to the well-known ridge structure centered at $\theta=0^0$ and stretched between the SE and the ECC peaks. Of course, ridge electrons evolve in a two-center potential, in which the saddle-point is a point of equilibrium but, however, unstable. Thus, for intermediate energies, the evolution of electrons in a potential of two receding ions spreads the emitted electron distribution in velocity space [5] and any correlation between such a distribution and the velocity of the saddle point is consequently diluted. In this sense, the existence of an independent saddle-point-ionization mechanism does not imply the observation of a shift between the electron distributions resulting from different projectile charges. To expect a shift, it would be equivalent to consider that the emitted electrons do not evolve in a two-center potential and are only subject to an attractive center at the saddle point. Such a three-center picture has been employed successfully by Winter and Lin [25] to improve calculations of total cross section for projectiles with energies up to 25 keV/amu.

For low enough projectile velocities v_p, the velocity v_b of an electron bound to the target is larger than v_p ($v_b >> v_p$), and the saddle potential remains deeper during a longer time because the two ions (projectile and target) recede slowly giving rise to a larger density of electron states at the saddle point. After the concentration of an electron density at the saddle point, the two ions should recede from each other quickly in a non-adiabatic way, in order to give rise to a structure with a maximum at about the saddle velocity in the forward emission. However, a saddle evidence could be found by obtaining the ratio between the electron emission induced by a multiply charged projectile and that resulting for protons within the low-velocity range. Future experiments should be directed towards this direction in order to clarify the subject.

It should be mentioned, however, that Pieksma and Ovchinnikov [26] have shown from calculations, within the framework of the theory of non-adiabatic transitions, that the major contribution to the total H ionization is found for energies between 5 and 20 keV for H^+, and between 30 and 65 keV/amu for He^{2+}-projectiles. At lower projectile energies the importance of such a mechanism was shown to decrease rapidly.

ACKNOWLEDGMENTS

The author would like to thank Prof. W.Meckbach for useful discussions and a critical reading of the manuscript.

(#) Consejo Nacional de Investigaciones Científicas y Técnicas (CONICET).

REFERENCES

[1] - A.González, P.Dahl, P.Hvelplund and P.Fainstein, J.Phys. B26, L135 (1993), and refs. therein.
[2] - Forward Electron Ejection in Ion Collisions, Vol. 213 of Lectures Notes in Physics, Edited by K.O.Groeneveld, W.Meckbach and I.A.Sellin, (Springer-Verlag, Berlin, 1984).
[3] - S.Suárez et al, Phys. Rev. Lett. 70, 418 (1993).
[4] - R.E.Olson, Phys. Rev. A16, 1871 (1983); Phys. Rev. A33, 4397 (1986).
[5] - W.Meckbach, P.Focke, A.Goñi, S.Suárez, J.Macek and M.Menéndez, Phys. Rev. Lett.57, 1587 (1986).
[6] - G.Bernardi, S.Suárez, P.Focke and W.Meckbach, in High-Energy Ion-Atom Collisions, Vol.294 of Lectures Notes in Physics (Springer-Verlag, Berlin, 1987).
[7] - G.Bernardi, S.Suárez, P.Fainstein, C.Garibotti, W.Meckbach and P.Focke, Phys. Rev. A40, 6863 (1989).
[8] - R.E.Olson, T.J.Gay, H.G.Berry, E.B.Hale and V.D.Irby, Phys. Rev. Lett. 59, 36 (1987).
[9] - V.D.Irby, T.J.Gay, J.Wm.Edwards, E.B.Hale, M.L.McKenzie and R.E.Olson, Phys. Rev. A37, 3612 (1988).
[10] - M.E.Rudd and T.Jorgensen, Phys. Rev.131, 666 (1963).
[11] - T.J.Gay, H.G.Berry, E.B.Hale, V.D.Irby and R.E.Olson, Nucl. Instr. and Meth.B31, 336 (1988).
[12] - S.Ovchinnikov, Phys. Rev. A42, 3865 (1990) and references therein.
[13] - G.Bandarage and R.Parson, Phys. Rev. A41, 5878 (1990).
[14] - J.Wang, J.Burgdörfer and A.Bárány, Phys. Rev. A43, 4036 (1991).
[15] - W.Meckbach, S.Suárez, P.Focke and G.Bernardi, J.Phys. B24, 3763 (1991).
[16] - V.Irby, S.Datz, P.Dittner, N.Jones, H.Krause and C.Vane, Phys. Rev. A47, 2957 (1993).
[17] - T.Gay, M.Gealy and M.Rudd, J.Phys. B23, L823 (1990).

[18] - S.Suárez, C.Garibotti, G.Bernardi, P.Focke and W.Meckbach, Phys. Rev. A48, 4339 (1993).
[19] - R.DuBois, Phys. Rev. A48, 1123 (1993).
[20] - R.Cranage and M.Lucas, J.Phys.B9, 445 (1976).
[21] - L.Toburen, R.DuBois, C.Reinhold, D.Schultz and R.Olson, Phys. Rev. A42, 5338 (1990).
[22] - Meckbach and Nemirovsky, Private communication Unpublished (1981).
[23] - W.Meckbach, I.Nemirovsky and C.Garibotti, Phys Rev. A24, 1793 (1981).
[24] - G.Bernardi and W.Meckbach, Phys. Rev. A51, 1709 (1994).
[25] - Winter and Lin, Phys. Rev. A29, 3071 (1984).
[26] - M.Pieksma and S.Ovchinnikov, J.Phys. B24, 2699 (1991).
[27] - G.Bernardi et al, to be published (1995).
[28] - N.Stolterfoht, Z. Phys. 248, 81 (1971).

Saddle Point Electrons in Slow Ion-Atom Collisions

Marc Pieksma, S.Y. Ovchinnikov*, J. van Eck,
W.B. Westerveld and A. Niehaus

Debye Institute, Department of Atomic and Interface Physics, Utrecht University, P.O. Box 80000, 3508 TA Utrecht, The Netherlands
*Oak Ridge National Laboratory, Oak Ridge, Tennessee 37831-6372

Abstract. Ionization in atomic collisions at adiabatic impact energies is discussed. Special attention is given to the saddle point ionization mechanism. This mechanism can be described by a quantum mechanical theory based on the concept of hidden crossings. The fundamental H^+-H collision system is studied as a test case. For 1–6 keV collisions both the experimental velocity distributions and the relative total ionization cross sections of this system can be explained by the theory in a consistent way. At collision energies of 4 keV and higher there is strong evidence for the existence of a saddle point ionization mechanism. Saddle point effects in the He^{2+}-H (theory) and H^+-He (experiment) collision systems are briefly commented upon.

1 Introduction

During the last decade the mechanism of saddle point ionization has drawn quite some attention. In short, this mechanism acts as follows: on the outgoing part of the collision, as the two nuclei are receding, there is a finite probability that an electron stays localized near or on the saddle point of the internuclear potential barrier, balanced by the attractive Coulomb forces of the nuclei. This will eventually lead to ionization at large internuclear distances, where the electron is no longer bound to the nuclei. In the special case of two nuclei with equal charges one speaks of $v/2$ electrons, since for such systems the saddle point lies exactly between the nuclei, and therefore an electron located on it travels with half the incoming projectile ion velocity. Here the emphasis will be on such a symmetric system, namely H^+-H, for which, recently, we have found evidence for the actual existence of saddle point electrons [1, 2]. Our investigations on this fundamental system were both theoretical and experimental. Total ionization cross sections and – providing more sensitive proof – electron velocity distributions will be discussed.

After a brief review of the work that has been carried out on saddle point electrons so far, the theory of hidden crossings will be outlined in some detail. Next, the experimental set-up is described, and the results obtained are compared with the theoretical predictions. Besides the H^+-H system, other collision systems are considered as well. The theoretical prediction for the electron velocity distribution of the He^{2+}-H system and the measured electron spectrum of the H^+-He system are chosen as examples.

2 Historical overview

The basic idea of saddle point ionization can be traced back to 1953, when Wannier [3] published his theory on ionization of ions and atoms by electron impact near threshold. Wannier found that ionization is most probable when 'all orbits are symmetric in the two electrons'. In other words, ionization occurs if the heavy particle remains at the saddle point of the electronic Coulomb potential. By switching from a system consisting of two electrons and one nucleus to a system consisting of one electron and two nuclei, we obtain the analogue of Wannier ionization in ion-atom collisions. This idea was first exploited by Olson [4, 5] and Olson et al. [6] using classical trajectory Monte Carlo (CTMC) calculations, and also by Winter and Lin [7], who applied a triple center atomic state close coupling method.

A quantum mechanical theory which incorporates the saddle point mechanism, known as the theory of hidden crossings, was formulated by Ovchinnikov and Solov'ev [8]. Based on this theory Pieksma and Ovchinnikov [9] calculated total ionization cross sections of 5−20 keV/amu H^+-H collisions. They found good agreement with already existing experimental data [10], and showed that the saddle point ionization mechanism can contribute up to 50% to the total ionization cross section. This is demonstrated in figure 1, in which also the results of other theoretical approaches are shown, namely two center [11], triple center [7] and low-energy CTMC calculations [12]. In the next section more details concerning the theory of hidden crossings can be found.

It remains unclear whether saddle point electrons have been observed by other groups. Meckbach et al. [13] were the first to report saddle point effects, but later withdrew their conclusions and ascribed their saddle point peak to experimental artifacts [14]. Nowadays this group claims that saddle point electrons do not exist at all [15]. Olson et al. [6], Irby et al. [16, 17] and Gay et al. [18] independently found saddle point shifts in their electron energy spectra, but these results could not be confirmed by DuBois [19]. A possible explanation for this controversy could be that all of these experiments were carried out at relatively high impact energies (typically 100 keV/amu). The saddle point mechanism has been uncovered in adiabatic (i.e. slow, below 25 keV/amu) collisions [9, 12]. It might well be that in the experiments referred to above, the collision process proceeds too fast for any clear saddle point features to be

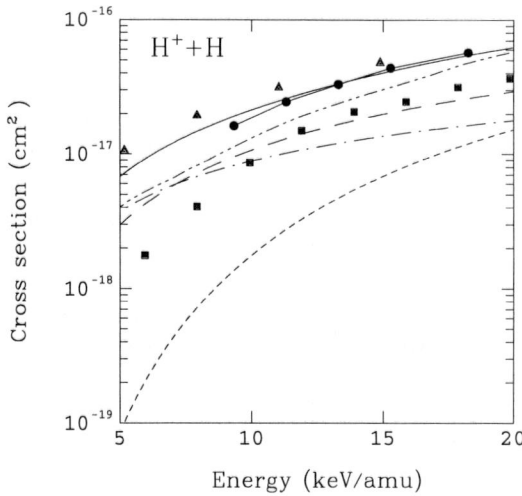

Figure 1: The H$^+$-H total ionization cross section as a function of impact energy. ($\bullet$) = experimental results [10], (——) = theory of hidden crossings [9], ($\blacktriangle$) = triple center calculations [7], ($\blacksquare$) = CTMC calculations [12] and (– ·· –) = two-center close coupling calculations [11]. The contributions of the three dominant ionization channels are also shown: (- - -) = saddle point σ channel, (— —) = saddle point π channel and (— · —) = S promotion.

established. Formulated otherwise, the electron should be allowed sufficient time to build up probability in the saddle point region. The results found by Pieksma et al. [1] at low collision energies (which will be discussed below) seem to supply the only reliable proof for the existence of saddle point electrons at present.

3 Theory

The quantum mechanical theory that treats the saddle point mechanism in a very manageable way is the theory of hidden crossings. Hidden crossings resemble the well-known Landau-Zener avoided crossings, the difference being that complex rather than real values of the internuclear distance are considered [20]. For a extended review of this theory see Ref. [21].

Two types of hidden crossings are relevant for the H$^+$-H system, the S- and the T-type crossings [9]. The T-type crossings are connected with the saddle point mechanism. A chain of successive T-type transitions establishes a saddle point ionization channel. At every internuclear distance which coincides with the real part of a T-type crossing the electron is promoted – on top of the internuclear potential barrier, which is, by definition, the saddle point – to a higher level. The S-type crossings are associated with the transition from a quasi-molecular to an united atom behaviour of the system. In this case the centrifugal barrier, which becomes increasingly dominant at smaller internuclear distances, is of relevance. S-type ionization occurs on the incoming part of the collision, while saddle point ionization occurs on the outgoing part.

The reason why one speaks of crossings that are 'hidden' is clarified in figure 2. In this figure several potential energy curves of the H^+-H system are shown, both for real as well as complex internuclear distances. Note that along the real internuclear axis no crossings can be seen, while in the complex plane the T-type crossings are clearly visible.

In a simple two-level model Hamiltonian the hidden crossings already come out in a natural way. Consider

$$H(R) = \begin{pmatrix} H_{11}(R) & H_{12}(R) \\ H_{21}(R) & H_{22}(R) \end{pmatrix}. \tag{1}$$

The two energy eigenvalues of this Hamiltonian are

$$E_{1,2}(R) = \frac{H_{11}(R) + H_{22}(R)}{2} \pm \frac{1}{2}\sqrt{(H_{11}(R) - H_{22}(R))^2 + 4H_{12}(R)H_{21}(R)}. \tag{2}$$

In general the two energy curves E_1 and E_2 do not cross for real R, but an exact crossing of the levels is possible at a complex distance R_c, which is found by putting the square root in equation (2) equal to zero.

The analytic structure of the Hamiltonian appears to contain the same information as do the matrix elements. Namely, the transition probability between the two energy curves is given by [20, 21]

$$P_{12} = \exp\left(-\frac{2\Delta_{12}}{v_0}\right) \tag{3}$$

where v_0 is the collision velocity. Δ_{12} is a contour integral in the complex plane:

$$\Delta_{12} = \left| \text{Im} \int_{\text{Re}R_c}^{R_c} \frac{(E_2(R) - E_1(R))}{v_R/v_0} dR \right| \tag{4}$$

with v_R being the radial collision velocity.

The theory of hidden crossings is both powerful and elegant, particularly because it makes it possible to obtain *analytic* expressions for ionization cross sections [9]. The restriction on its application is that this theory only works for one-electron systems, because only for those systems the potential energy

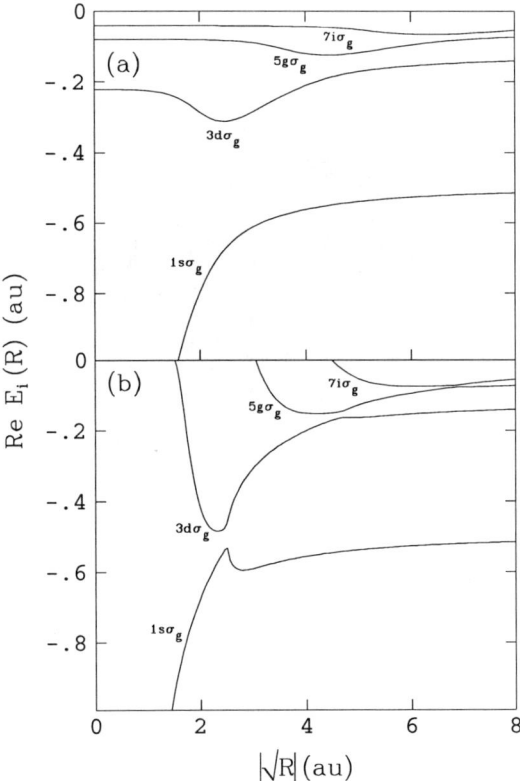

Figure 2: Potential energy curves of the H_2^+ quasi-molecule (a) along the real $\sqrt{R}$ axis and (b) along a straight path in the complex $\sqrt{R}$ plane, which approaches closely the T-type hidden crossings.

curves, and therefore the Massey parameter, can be calculated to arbitrary precision.

Also for the cross sections differential in electron velocity and ejection angles analytic expressions can be derived [2, 22]. For the S-promotion the electron distribution can directly be obtained from the theory of hidden crossings. For the saddle point mechanism the electron distribution is found by solving the Schrödinger equation near the saddle point. However, there is one parameter that critically determines the saddle point electron distribution, but that is difficult to derive in a rigorous way. This is the distance where ionization in the saddle point process occurs. It will appear that a reasonable estimate for this ionization distance R_{ion} is obtained if one assumes that ionization in a saddle point channel effectively occurs when the transition probability be-

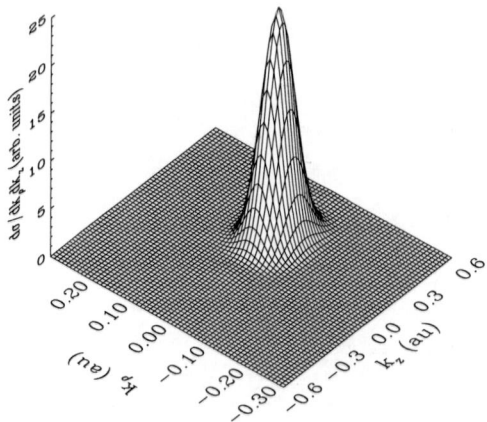

Figure 3: The doubly differential cross section of the saddle point σ channel of the H^+-H system at a collision energy of 4 keV/amu. z is along the internuclear axis, ρ is the perpendicular coordinate.

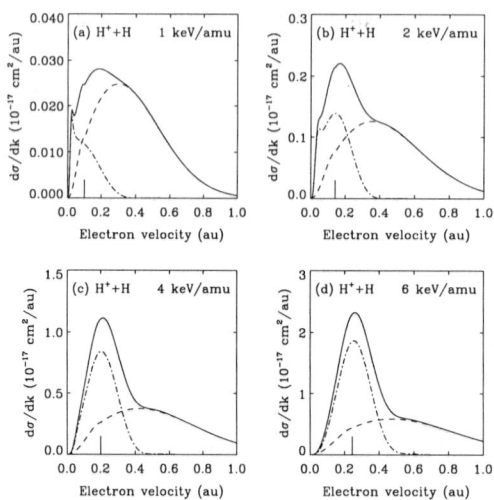

Figure 4: Theoretical velocity distributions of electrons ejected in 1, 2, 4 and 6 keV/amu H^+-H collisions. Both the contributions of the S promotion (- - -) and the saddle point ionization mechanism (- · -) are shown.

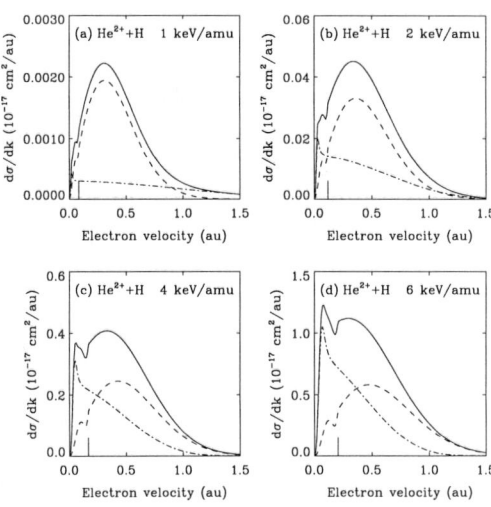

Figure 5: The same as figure 4, but now for 1, 2, 4 and 6 keV/amu He^{2+}-H collisions.

comes larger than 50% [23], i.e. when the system changes from an adiabatic to a diabatic behaviour. In that case the electronic motion is no longer well coupled to that of the nuclei and the electron may considered to be 'quasi-free'. This model predicts an effective ionization distance R_{ion} which equals $50/v_0$ for the H^+-H system and $28/v_0$ for the He^{2+}-H system. At R_{ion} the collisional interaction has not terminated completely. The long-range Coulomb interaction of an electron in the saddle point region from the moment of ionization until infinite internuclear distance can be taken into account by using Macek's saddle point propagators [24].

As an example, in figure 3 the doubly differential cross section of the saddle point σ channel (i.e. the channel with $m=0$) is presented. By integrating this distribution the electron velocity distribution of the saddle point σ channel is obtained. Theoretical electron velocity distributions for 1−6 KeV/amu collisions of H^+-H and He^{2+}-H are shown in figures 4(a) to (d) and 5(a) to (d), respectively. The theoretical calculations include all relevant saddle point and the S promotion channels. Note that for the H^+-H system at impact energies above 2 keV/amu a clear saddle point peak is predicted. For the He^{2+}-H system this is not the case. Here the saddle point electrons are expected to be smeared out over the entire low electron energy range.

At very low collision energies (typically below 2 keV/amu) the recently discovered radial decoupling mechanism [25] should also be taken into account. This mechanism is not incorporated in the theory of hidden crossings. It occurs at very small internuclear distances where the classically accessible region of

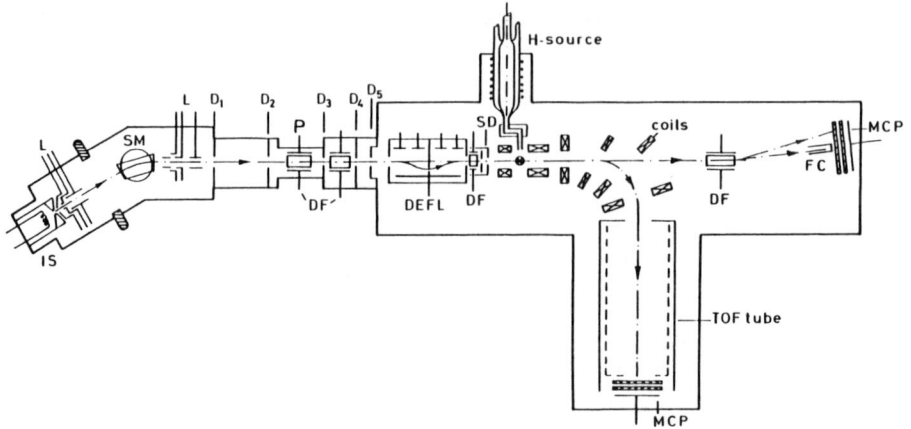

Figure 6: The experimental set-up (not to scale). IS = ion source, L = electrostatic lenses, SM = 60° sector magnet, $D_{1,2,3,4,5}$ = diaphragms, DF = deflection plates, P = ion beam pulser, DEFL = deflector, SD = electron suppression device, FC = Faraday cup, MCP = micro channel plates (at the bottom for electron, at the right for proton detection).

the electron changes from a configuration of two centers to that of one center. This causes an effective radial decoupling of the electronic and nuclear motions, simply because in an united atom the behaviour of the system is independent of the motion of the nucleus. The radial decoupling ionization mechanism is not shown in figures 4 and 5, because at the moment there exists no theory which describes this mechanism in enough detail that reliable theoretical differential cross sections can be calculated.

4 Experimental

The experimental set-up used for measuring these electron velocity distributions is shown in figure 6 (see also [2, 26]). A pulsed, stabilized proton beam with a width of about 50 ns (FWHM) and a repetition rate of 15 kHz is crossed with either a partially dissociated thermal hydrogen beam or a thermal helium beam. The hydrogen beam is produced by a RF discharge source [27]. A magnetic time-of-flight (TOF) spectrometer was especially designed to measure the velocity distribution of the slow saddle point (and other slow) electrons. A photograph of this spectrometer is shown in figure 7. The spectrometer consists of a configuration of separate circular coils, which allows exact, numerical simulations of the electron trajectories. The principle of operation is that slow electrons are guided along the magnetic field lines towards the detector. The pattern of the field lines is shown in figure 8. Immediately after the reaction

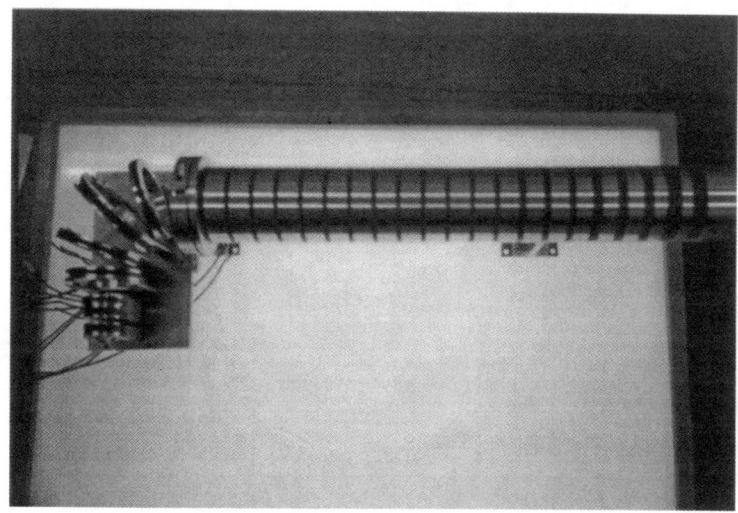

Figure 7: The magnetic time-of-flight electron spectrometer. The diameter of the TOF tube is 62 mm.

region it has the shape of a 'bottleneck' to realize a large angle of acceptance: at the end of the bottleneck region the paths of slow electrons have become (almost) parallel. After the bottleneck region the magnetic field is bent in order to guide the electrons out of the proton beam direction and into a time-of-flight tube. The spectrometer can collect essentially all electrons ejected in the forward hemisphere as long as they satisfy the approximate transmission condition $k\sin\theta \leq 0.42$ (au), where k is the electron velocity and θ is the ejection angle with respect to the symmetry axis of the spectrometer.

The RF source caused a severe (uncorrelated) background of slow electrons appearing in the recorded TOF distributions, which could not be completely suppressed, even by applying suitable electric fields. A strong cooling of the source wall, which would increase the dissociation degree, also appeared to increase the number of background electrons. As a compromise the temperature of the source wall was kept at room temperature. This resulted in a stable dissociation degree of 70±6%.

Electron TOF distributions were recorded by using the proton beam pulser as a start and the electron signal as a stop for a time-to-amplitude converter (TAC). A pulse height analyzer was used to analyze the TAC signals and to accumulate the TOF distributions.

The contribution of signal of electrons produced outside the reaction region has to be corrected for. In the case of a helium beam this is realized by measuring two spectra, one with a gas inlet directly in the reaction region,

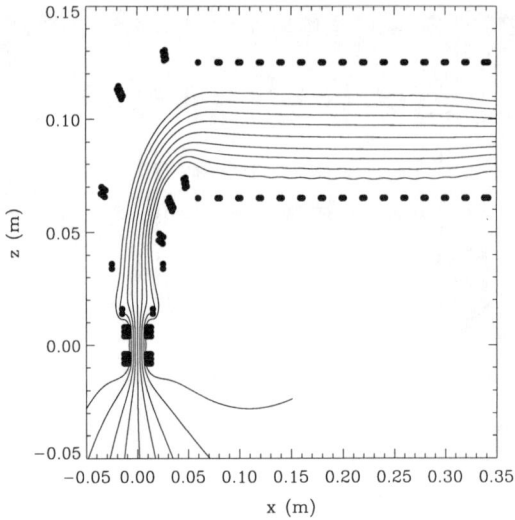

Figure 8: The pattern of magnetic field lines of the electron spectrometer.

and one with a gas inlet elsewhere, but with the same diffuse background pressure. The true electron velocity spectrum, normalized to ion current and target density is given by

$$\Sigma_{He} = \Sigma_{He,direct} - \Sigma_{He,diffuse} \ . \tag{5}$$

For the measurements with an atomic hydrogen beam a 'diffuse' measurement is not possible. However, corrected, normalized spectra can still be obtained, namely by subtracting the spectra measured with the hydrogen source switched 'on' and 'off', and by correcting for the partial contribution of H_2 to the signal:

$$\Sigma_H = \frac{1-D}{\beta D}\left(\Sigma_{on}^{H_2} - \Sigma_{H/H_2}^{off} + (1-\beta)\Sigma_{H_2}\right) \tag{6}$$

with $\beta = (1-D)/(1-D+D/\sqrt{2})$, D being the dissociation fraction, and where Σ_H is normalized to the density of H atoms, while Σ_{H/H_2}^{on}, $\Sigma_{H_2}^{off}$ and Σ_{H_2} are normalized to the H_2 density.

The electron velocity distributions of 1 to 6 keV collisions of proton projectiles with atomic hydrogen and helium targets are shown in figures 9(a) to (d) and 10(a) to (d), respectively.

5 Theory versus experiment

In figures 9(c) and (d) the theoretical electron velocity distributions are compared with the experimental ones. The comparison shows that there is a

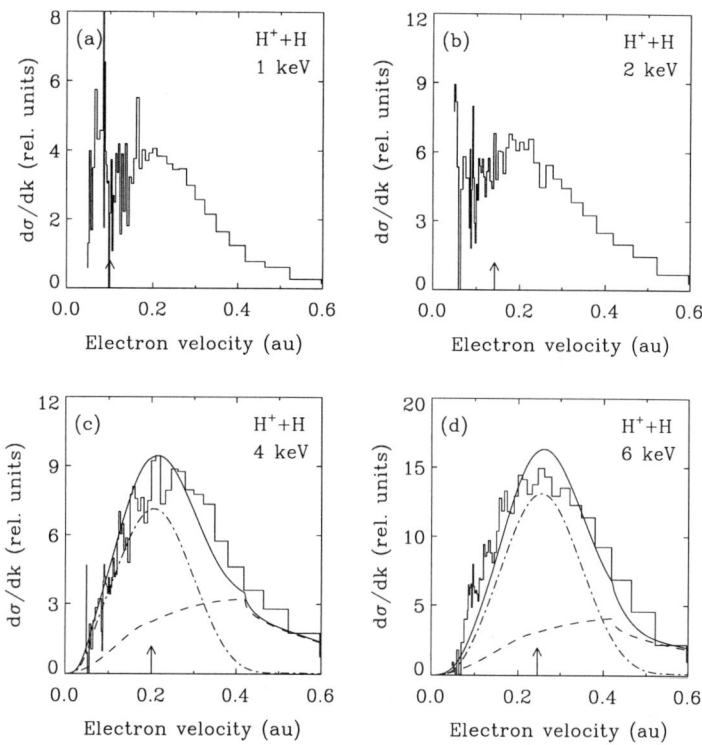

Figure 9: The velocity distributions of electrons ejected in (a) 1, (b) 2, (c) 4 and (d) 6 keV collisions of protons with hydrogen atoms. The solid curves (—) in the figures (c) and (d) are least squares fits of the theoretical velocity distributions to the experimental distributions, obtained by using only the relative height as a fitting parameter. Both the contributions of the S promotion (- - -) and the saddle point mechanism (- · -) are shown. The change in slope in the theoretical distributions at an electron velocity of 0.42 au is the result of taking the approximate transmission function of the electron spectrometer $k \sin\theta \leq 0.42$ (au) [26] into account. The markers at the bottom indicate the saddle point velocity.

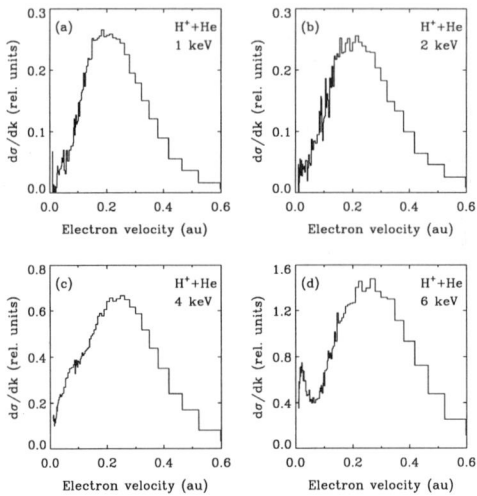

Figure 10: Electron velocity distributions of the H^+-He system at (a) 1, (b) 2, (c) 4 and (d) 6 keV impact energy.

good agreement between theory and experiment at 4 and 6 keV impact energies. Relative total ionization cross sections of the H^+-H system can be obtained by integrating the normalized electron velocity distributions. In figure 11 these cross sections, scaled to the theoretical result at a collision energy of 4 keV/amu, are presented. Again, good agreement is found with the theoretical prediction. We have not tried to fit the experimental results at 1 and 2 keV in figures 9(a) and (b) with the theoretical spectra, because, as figure 11 demonstrates, at these low collision energies the radial decoupling ionization mechanism gives the major contribution to the total ionization. Moreover, this mechanism also produces enough low-energy electrons to explain why the maximum of the velocity distributions no longer shifts with the collision energy below 2 keV [22].

Note that true saddle point electrons are ejected in a forward cone only. Although the presented set-up does not distinguish between different ejection angles, so we can not check this feature of the saddle point electrons, the observed shift in the electron velocity distributions still strongly suggests the presence of a saddle point ionization mechanism: no other low-impact energy ionization mechanism is predicted to exhibit such shifts.

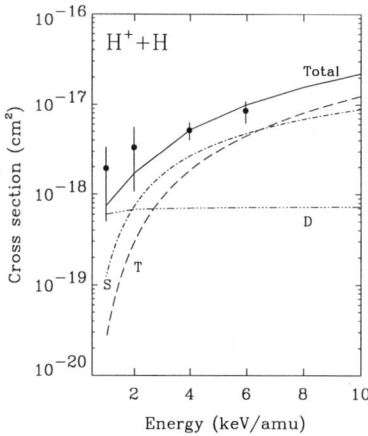

Figure 11: Ionization cross sections of the H^+-H system: ($\bullet$) = experimental results obtained with our set-up, (—) = total theoretical ionization cross section. The contribution to the the cross section of the S (– · –), the saddle point (T) (– – –) and the radial decoupling (D) (– · · · –) ionization mechanisms, are shown as well. The experimental cross sections were scaled to the theoretical result at 4 keV.

6 Conclusions

Strong evidence has been found for the actual existence of a saddle point ionization mechanism at adiabatic energies. The theory consistently explains both the total and differential cross sections. Above a collision energy of 4 keV/amu the saddle point mechanism is the dominant ionization process, especially regarding the emission of low-energy electrons. Below 4 keV/amu another ionization mechanism becomes of increasing importance, the radial decoupling mechanism. At high collision energies, i.e. above 25 keV/amu it is very uncertain whether any saddle point features may be expected.

For the H^+-He collision system the shift of the maximum of the electron spectra with impact energy resembles that of the H^+-H system, as figures 9 and 10 demonstrate. Unfortunately, the electron velocity distributions of this two-electron system can not be compared with any existing theory, but the observed shift indicates that also in this system the saddle point ionization mechanism plays an important role. There are, however, also unexpected features. For example, the shape of the low velocity tail of the spectra, particularly at 4 and 6 keV, is as yet not understood.

As far as future experiments are concerned, the measurement of the velocity distributions of electrons ejected in highly charged ion-atom collision systems, particularly one-electron systems such as He^{2+}-H and C^{6+}-H, which allow a direct comparison with theory, are planned.

Acknowledgment

Professor J. Macek is gratefully acknowledged for his help with the calculation of the propagated saddle point wave functions. This work is part of the research program of the 'Stichting voor Fundamenteel Onderzoek der Materie' (FOM), which is financially supported by the 'Nederlandse Organisatie voor Wetenschappelijk Onderzoek' (NWO).

References

[1] M. Pieksma, S.Y. Ovchinnikov, J. van Eck, W.B. Westerveld and A. Niehaus, Phys. Rev. Lett. **73**, 46 (1994)

[2] M. Pieksma, Ph.D. thesis, Utrecht University (1993)

[3] G.H. Wannier, Phys. Rev. A **90**, 817 (1953)

[4] R.E. Olson, Phys. Rev. A **27**, 1871 (1983)

[5] R.E. Olson, Phys. Rev. A **33**, 4397 (1986)

[6] R.E. Olson, T.J. Gay, H.G. Berry, E.B. Hale and V.B. Irby, Phys. Rev. Lett. **59**, 36 (1987)

[7] T.G. Winter and C.D. Lin, Phys. Rev. A **29**, 3071 (1984)

[8] S.Y. Ovchinnikov and E.A. Solov'ev, Comments on Atomic and Molecular Physics **XXII**, 69 (1988)

[9] M. Pieksma and S.Y. Ovchinnikov, J. Phys. B **24**, 2699 (1991)

[10] M.B. Shah, D.S. Elliott and H.B. Gilbody, J. Phys. B: At. Mol. Phys. **20**, 2481 (1987)

[11] W. Fritsch and C.D. Lin, Phys. Rev. A **26**, 762 (1983)

[12] G. Bandarage and R. Parson, Phys. Rev. A **41**, 5878 (1990)

[13] W. Meckbach, P.J. Focke, A.R. Goñi, S. Suárez, J. Macek and M.G. Menendez, Phys. Rev. Lett. **57**, 1587 (1986)

[14] G. Benardi, S. Suárez, P. Focke and W. Meckbach, Nucl. Instrum. Meth. Phys. Res. B **33**, 321 (1988)

[15] W. Meckbach, S. Suárez, P. Focke and G. Bernardi, J. Phys. B: At. Mol. Opt. Phys. **24**, 3763 (1991)

[16] V.D. Irby, T.J. Gay, J.W. Edwards, E.B. Hale, M.L. McKenzie and R.E. Olson, Phys. Rev. A **37**, 3612 (1988)

[17] V.D. Irby, S. Datz, P.F. Dittner, N.L. Jones, H.F. Krause and C.R. Vane, Phys. Rev. A **47**, 2957 (1993)

[18] T.J. Gay, M.W. Gealy and M.E. Rudd, J. Phys. B **23**, L823 (1990)

[19] R.D. DuBois, Phys. Rev. A **48**, 1123 (1993)

[20] J.-T. Hwang and P. Pechukas, J. Chem. Phys. **67**, 4640 (1977)

[21] E.A. Solov'ev, Sov. Phys. Usp. **32**, 228 (1989)

[22] M. Pieksma and S.Y. Ovchinnikov, J. Phys. B: At. Mol. Opt. Phys. **27**, 4573 (1994)

[23] E.A. Solov'ev, Phys. Rev. A **42**, 1331 (1990)

[24] J. Macek, Phys. Rev. A **41**, 1361 (1990)

[25] S.Y. Ovchinnikov and J.H. Macek, *in Proceedings of the Eighteenth ICPEAC, Aarhus 1993*, Abstracts p. 676

[26] M. Pieksma, H.J. van der Meiden, J. van Eck, W.B. Westerveld and A. Niehaus, Rev. Sci. Instrum. **66**, 72 (1995)

[27] J. Slevin and W. Stirling, Rev. Sci. Instrum. **52**, 1780 (1981)

II. TARGET AND PROJECTILE ELECTRON INTERACTIONS

Target and Projectile Ionization in Ion-Atom Collisions: Theoretical Aspects

Steven T. Manson

Department of Physics and Astronomy, Georgia State University, Atlanta, Georgia 30303

A theoretical overview of target and projectile ionization in ion-atom collisions is presented within the framework of the First Born Approximation. The dual role of the electrons in screening and electron-electron interactions is emphasized, along with the observable consequences of each role. Examples from theory and experiment are given.

INTRODUCTION

The study of electron emission is an invaluable tool for the investigation of the dynamics of target and projectile electrons in fast ion-atom and atom-atom collisions. The interplay between theory and experiment has proved to be of great value in developing our understanding of these processes. This has led to a fairly deep comprehension of electron emission in *bare*-ion impact ionization of target atoms (1,2), although some questions still exist, e.g., the existence of saddle-point electrons (3). The situation for structured projectile ions, projectiles which bring one or more of their own bound electrons into the collision, is otherwise. On the experimental side, there is simply less data than for bare projectile collisions. From the theoretical point of view, the introduction of projectile electrons complicates the problem considerably (4,5).

In this review, we focus upon the basic physics of the dual role of the projectile electrons in screening the projectile nucleus and in direct projectile electron - target electron interactions in target ionization, along with the analogous role of target electrons in projectile ionization. In addition, the experimentally realizable "fingerprint" of each of the various pathways for electron emission is emphasized. The First Born Approximation (FBA) provides an excellent framework to elucidate many of the salient features of the electron emission spectrum, along with the crucial role of the projectile electrons in the addition of electron emission channels and the dramatic alteration of the electron emission spectrum from the bare projectile case.

In the next section, a presentation of the FBA is made, with emphasis on the modifications to the theory engendered by the existence of projectile electrons. The final section presents several illustrative examples which exemplify the physics of the situation.

THEORY

Bare Projectiles

The FBA theory applied to bare projectiles has been employed for some time to treat ionization in fast collisions (1,2). Consider a bare ion of charge Z_P incident on a neutral atomic target of nuclear charge Z_T. Denoting the reduced mass of the system as μ, the Hamiltonian for this collision system is given by

$$H = H_0 + H_1 \qquad (1)$$

in which

$$H_0 = \frac{P_P^2}{2\mu} + \sum_{j=1}^{Z_T} \left[\frac{p_j^2}{2m} - \frac{Z_T e^2}{r_j} + \sum_{j'<j}^{Z_T} \frac{e^2}{r_{jj'}} \right] \qquad (2)$$

and the perturbation

$$H_1 = \frac{Z_P Z_T e^2}{R} - \sum_{j=1}^{Z_T} \frac{Z_P e^2}{|\bm{R}-\bm{r}_j|} \qquad (3)$$

where e and m are the electron charge and mass respectively, $\bm{R}$ is the position vector of the projectile with respect to the target nucleus, $\bm{r}_j$ is the position vector of the j-th target electron, and $\bm{P}_P$ and $\bm{p}_j$ are the corresponding momentum operators. In the FBA theory, the total Hamiltonian of the system is broken up as indicated in Eqs. 1-3, and H_1 is treated as a perturbation. The wave functions employed in the perturbation theory are solutions of H_0; a product of the plane wave describing the relative motion and the target wave function. In actual fact, except for a one-electron target, the target wave function is not known exactly, but it has been found that FBA works quite well with reasonably good approximations to the target wave function (6).

For any inelastic collision, the initial and final state wave functions of the target are orthogonal; thus the matrix element of the first term in the perturbation, $Z_P Z_T e^2/R$, vanishes since it is independent of the $\bm{r}_j$'s, and only the second term in the perturbation contributes. Physically this means that the interaction between projectile and target nucleus does not have any effect on the collision cross section, in first order; the process is carried by the interaction of the target electrons with the bare projectile, the second term in the perturbation, Eq. 3.

For ionization processes, the most basic cross section is the triple differential cross section (TDCS), differential in the ejected electron energy, the ejection angle, and the scattering angle of the projectile. While the TDCS is often investigated experimentally for incident electrons (7), for protons and other heavy projectiles the TDCS is not generally measured owing to the very small scattering angles of such projectiles. Thus, it is the double differential cross section (DDCS), differential in ejected electron energy and angle, that is the most basic cross section usually studied for heavy projectiles. Applying FBA then yields for the DDCS (5)

$$\frac{d\sigma}{d\varepsilon d\Omega} = \int Z_P^2 A(K) dK \qquad (4)$$

where ε is the energy of the ejected electron, Ω the solid angle of ejection, $\hbar K$ is the momentum transferred from the projectile to the target in the collision, and $A(K)$ is an extremely complicated function of the target wave functions which also includes a myriad of angular momentum coupling coefficients (5,6). The details of $A(K)$ are not required for our purposes; thus, for the sake of simplicity, they shall be omitted. In any case, it is clear from Eq. 4 that the DDCS is proportional to Z_P^2. It therefore follows that the single differential cross section (SDCS) and total cross section (TCS), which are integrals of the DDCS over solid angle, and ejected electron energy and solid angle, respectively, also go as Z_P^2.

Structured Projectiles

For structured charged particle impact ionization, where the projectile brings in its own N_P electrons, the Hamiltonian is modified to

$$H_0 = \frac{P_P^2}{2\mu} + \sum_{j=1}^{Z_T} \left[\frac{p_j^2}{2m} - \frac{Z_T e^2}{r_j} + \sum_{j'<j}^{Z_T} \frac{e^2}{r_{jj'}} \right] + \sum_{k=1}^{N_P} \left[\frac{p_k^2}{2m} - \frac{Z_P e^2}{r_k} + \sum_{k'<k}^{N_P} \frac{e^2}{r_{kk'}} \right] \qquad (5)$$

and

$$H_1 = \frac{Z_P Z_T e^2}{R} - \sum_{j=1}^{Z_T} \frac{Z_P e^2}{|R-r_j|} - \sum_{k=1}^{N_P} \frac{Z_T e^2}{|R-r_k|} + \sum_{j=1}^{Z_T}\sum_{k=1}^{N_P} \frac{e^2}{|R+r_k-r_j|} \qquad (6)$$

where p_k and r_k refer to projectile electrons. The unperturbed Hamiltonian, H_0, now has solutions which include the internal wave function of the projectile electrons in the antisymmetric product. The perturbing Hamiltonian is now a sum of four terms, and this modification, the last two terms in Eq. 6, which are the result of the existence of projectile electrons, has profound implications for the electron ejection process; these last two terms describe, respectively, the interaction of the projectile electrons with the target nucleus and the interaction of target electrons with the projectile electrons. Of importance here is that the projectile electrons open up several physically distinct alternative electron emission channels which are characterized as:

    a: Target ionization, projectile remains in initial state;
    b: Target ionization, projectile excited (including ionized);
    c: Projectile ionization, target remains in initial state; and
    d: Projectile ionization, target excited (including ionized).

Clearly, only process a is possible for a bare projectile. Note also that the totality of processes a through d include all of the processes leading to ionization and electron emission, i.e., the complete electron emission cross section is the sum of the cross sections for each of these four pathways.

To understand the origin of each of these electron emission pathways, it is necessary to consider the matrix elements of the perturbation, Eq. 6, in some detail. As in the case of bare projectiles, for any inelastic collision the first term in the perturbation, which describes the interaction between projectile and target nuclei, vanishes since it is independent of any of the electron coordinates. For process a, the second term in the perturbation which describes the interaction of the target electrons with the projectile nucleus contributes, along with the last term which is the interaction between projectile electrons and target electrons; the third term vanishes for process a, where the initial and final state wave functions of the target electrons are orthogonal, because it involves only nuclear and projectile electron coordinates. Evaluation of the FBA cross section for process a yields (5)

$$\frac{d\sigma}{d\varepsilon d\Omega} = \int |Z_p - F_{ii}^P(K)|^2 A(K) dK \quad (7)$$

where the elastic scattering form factor of the projectile

$$F_{ii}^P(K) = \langle \psi_i^P | \sum_{k=1}^{N_p} \exp(i\boldsymbol{K} \cdot \boldsymbol{r}_k) | \psi_i^P \rangle \quad (8)$$

in which ψ_i^P is the wave function for the initial state of the projectile electrons. Comparing Eq. 4 with Eq. 7, it is seen that the effect of projectile electrons on process a is to introduce electron-electron interactions which cause a dynamic screening of the projectile nucleus, a screening that depends upon the momentum transfer K. Further, it is evident from Eq. 8 that $F_{ii}^P(0) = N_p$ and $F_{ii}^P(\infty) = 0$. Thus, for small energy transfer collisions, which implies small K

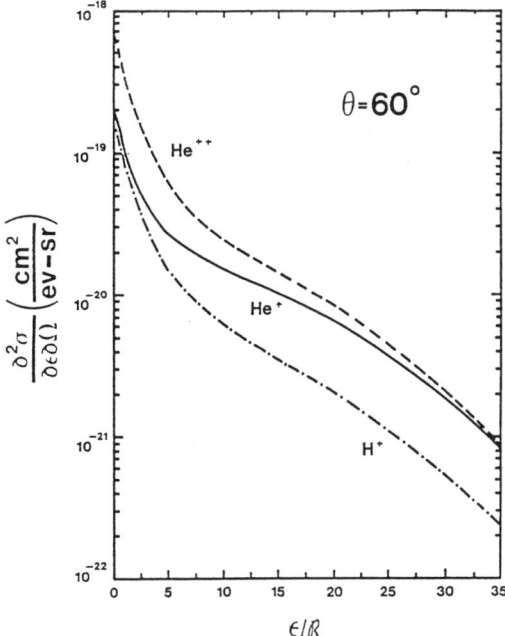

FIGURE 1. Theoretical double differential cross section (DDCS) for the ionization of He by 0.5 MeV/u H^+, He^{++}, and He^+ (projectile left in ground state) at an electron emission angle of 60 degrees as a function of ejected electron energy in Rydbergs (13.6 eV).

(large impact parameter), the screening function approaches $Z_P - N_P$, i.e., full screening of the projectile by its electrons. On the other hand, for large energy transfer, which implies large K (small impact parameter), the screening function approaches the full Z_P, i.e., no screening at all.

As an illustration of this effect, consider the DDCS for target electron emission in He^++He collisions. By the above arguments, for low-energy electron ejection, the He^+ nucleus should be almost completely screened by its electron, thus behaving like a proton of the same velocity. Similarly, for high-energy electron ejection, the arguments suggest that the nucleus would be virtually unscreened and behave like an alpha particle. These conclusions are validated by the results of a calculation (8) for the DDCS (process *a*) for equal velocity 0.5 MeV/u He^+, H^+, and He^{++} collisions with He shown in Fig. 1 where the transition of the screening from the small K to the large K limit is clear.

Turning our attention to process *b*, target ionization with simultaneous excitation of the projectile, the situation is rather different. This is a doubly inelastic collision in which both the target and the projectile undergo transitions, thereby rendering the final state wave function of each orthogonal to the initial state. Because of this orthogonality, the matrix element between initial and final states of any transition operator that does not depend upon both target electron and projectile electron coordinates must vanish. Thus, doubly inelastic collisions can arise only from the last term in the perturbing Hamiltonian, Eq. 6, the projectile electron - target electron interaction term. Within the framework of FBA, then, the DDCS for electron emission from the target with simultaneous excitation of the projectile from initial state ψ_i^P to final state ψ_f^P is given by (5)

$$\frac{d\sigma}{d\varepsilon d\Omega} = \int |F_{if}^P(K)|^2 A(K) dK \qquad (9)$$

where the inelastic scattering form factor of the projectile

$$F_{if}^P(K) = \langle \psi_i^P | \sum_{k=1}^{N_P} \exp(i\boldsymbol{K} \cdot \boldsymbol{r}_k) | \psi_f^P \rangle. \qquad (10)$$

Because of the orthogonality of initial and final state projectile wave functions, this inelastic form factor vanishes in the $K = 0$ limit; at $K = \infty$ it vanishes due to the oscillations in the exponential. Note that the last term in the perturbation, the projectile electron - target electron interaction, plays two differing roles. When the projectile remains in the initial state, it acts to screen the projectile nucleus, while when the projectile changes state, it is this interaction alone that carries the process.

Formally, the cross sections of Eq. 7 and Eq. 9 appear quite similar except for the "screening" functions; the function $A(K)$ is exactly the same in both cases, being only a function of target properties. But there is another difference that is important, the lower limit on the integration over K. To an excellent approximation, this limit is given by (1)

$$(Ka_0)_{min} = \frac{\Delta E/v}{\sqrt{2m}} \qquad (11)$$

where ΔE is the kinetic energy of the projectile transferred to the target and projectile electrons, v is the projectile velocity, and m is electron mass. Thus, to produce electrons of a given energy, ΔE is larger if the projectile is also excited and, in fact, is clearly different for different excitations of the projectile. In Eq. 11, then, the lower limit on the integration over K differs for each of these situations. For process *b* we are generally interested in the DDCS summed over all possible projectile excitations. Since there are an infinity of possible projectile excitations, a technique for summing over all of them is especially useful. The closure relation provides a way to perform such a sum approximately since (5)

$$\sum_{f \neq i} |F_{if}^P(K)|^2 = N_P - |F_{ii}^P(K)|^2 + \langle \psi_i^P | \sum_{k,k',k \neq k'}^{N_P} \exp[i\mathbf{K} \cdot (\mathbf{r}_k - \mathbf{r}_{k'})] | \psi_i^P \rangle \quad (12)$$

is an exact sum rule which has the added simplicity of depending only on the initial (generally ground) state projectile wave function. However, in order to use this sum rule to perform the sum over final states of the projectile electrons in Eq. 9, it must be realized that the lower limit on the integration changes for each of the projectile final states. Various ways of choosing an approximate lower limit of integration have been introduced (9). The total target electron emission DDCS is, then, a sum over process a and all possible processes b.

The analogous projectile electron emission cross sections, processes c and d, are handled exactly like the target ionization cases except that the projectile and the target reverse roles. All of the above discussion then, applies to electron ejection by the projectile as well, with one very important proviso; the results for the projectile electron emission DDCS is obtained in the projectile reference frame, a frame that is moving at a velocity **v** with respect to the laboratory (and target) frame. Thus, to compare with experiment, the projectile frame DDCS must be transformed to the laboratory frame. To do this, we note that $d\sigma/d\mathbf{k}$ is a Galilean invariant, where $\hbar\mathbf{k}$ is the momentum of the ejected electron. In terms of the Galilean invariant cross section, the DDCS is given by (10)

$$\frac{d\sigma}{d\varepsilon d\Omega} = \left(\frac{m}{\hbar^2}\right) k \frac{d\sigma}{d\mathbf{k}} \quad (13)$$

so that it follows that the DDCS in the laboratory frame is related to the DDCS in the projectile frame by

$$\left(\frac{d\sigma}{d\varepsilon d\Omega}\right)_{lab} = \left(\frac{\varepsilon_L}{\varepsilon_P}\right)^{1/2} \left(\frac{d\sigma}{d\varepsilon d\Omega}\right)_{proj} \quad (14)$$

where ε_L and ε_P are the ejected electron energies is laboratory and projectile frames respectively.

ILLUSTRATIVE EXAMPLES

To give some idea of the adequacy of FBA for electron emission in fast atomic collisions, the first example is the DDCS for the simple $H^+ + He$ system where the atomic wave functions, while approximate, are rather accurate. A comparison of experimental results and FBA calculation (6) for the DDCS as a

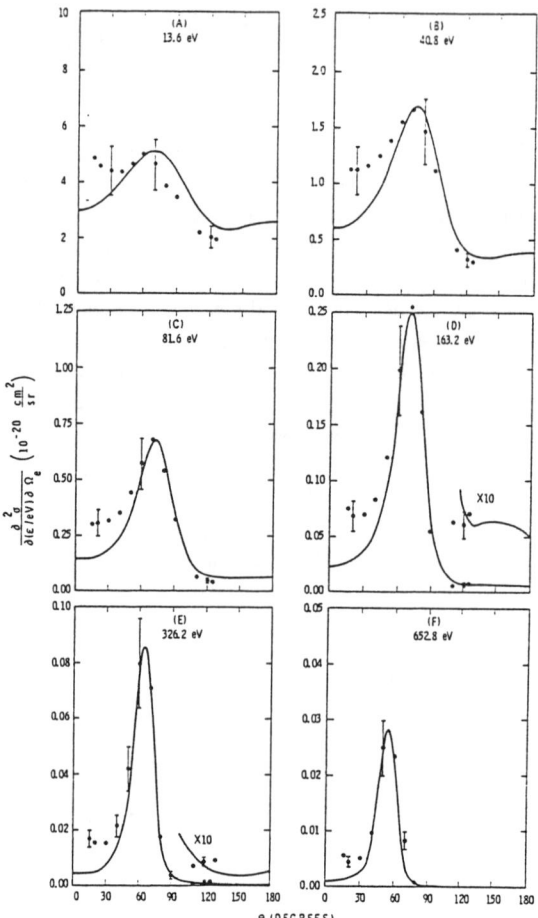

FIGURE 2. Comparison of experiment and First Born Approximation (FBA) theory for the double differential cross section (DDCS) for the ionization of He by 1.0 MeV H^+ as a function of electron emission angle for a range of ejected electron energies.

function of ejection angle for a range of electron emission energies produced by 1.0 MeV protons is presented in Fig. 2. From this comparison the agreement between FBA theory and experiment over an electron energy range from fairly slow electrons (13.6 eV) to rather fast electrons (652.8 eV) moving at a velocity larger than the projectile velocity is seen. Except for very small angles, where the charge-transfer-to-the-continuum process (11), which is not included in the FBA, is important, agreement is quite good.

As a second example, scrutiny of the DDCS for the same He target, but

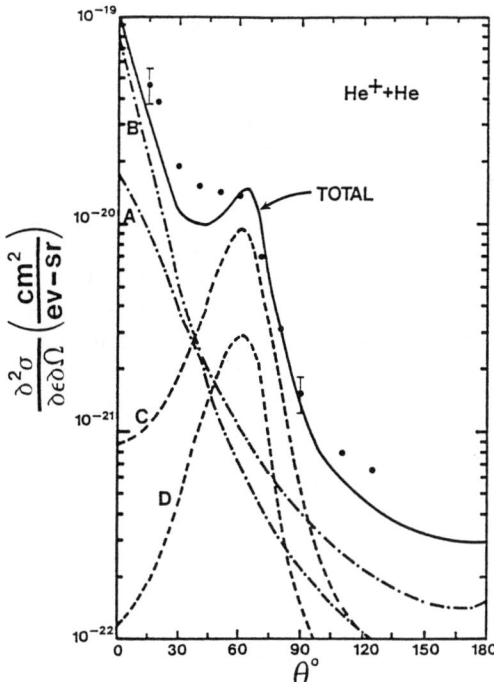

FIGURE 3. Comparison of experiment and First Born Approximation (FBA) theory for the double differential cross section (DDCS) for the ionization of He by 2.0 MeV He$^+$ as a function of electron emission angle for 217.68 eV ejected electrons. The solid curve is the total DDCS and the dash, C and D, and dash-dot (A and B) curves are processes *a-d* respectively, as described in text.

for 2.0 MeV (0.5 MeV/u) He$^+$ projectiles, illustrates the effects of the projectile electrons. A comparison of experiment and FBA calculation of the DDCS for total electron emission for this process (8) is shown in Fig. 3. The DDCS looks rather different from the case of a bare projectile as can be seen from a comparison of Fig. 3 to Fig. 2. The major difference is the sharp rise at small angle. In addition, it is clear that the FBA calculation gives excellent qualitative and reasonably good quantitative agreement with experiment. Also shown in Fig. 3 are the contributions of each of the processes *a-d* detailed above. The forward peak is seen to be the result of projectile ionization with both ionization without target excitation and ionization with target excitation making important contributions; this occurs owing to the transformation from projectile to target frame, Eq. 14. The largest disagreement between theory and experiment, in the region of 30 degrees, appears to arise principally from the use of the approximate sum rule. The calculation shown in Fig. 3 employed a rather simple approximation; more sophisticated treatments of the sum rule shrink the discrepancy considerably (9). Note also that although the projectile ionization plus target excitation is a very important process, target ionization plus projectile excitation is not. This is basically because the target is ionized by an ionic projectile while the projectile is ionized by a neutral atom; for distant collisons, the target "sees" a charge of unity (the projectile nucleus fully screened by its electron), while the projectile "sees" a charge of zero (the target nucleus fully screened by its two electrons). Thus, it is evident that the doubly inelastic process will be much more important when the ionization is caused by a neutral particle. This notion of the dominance of doubly inelastic processes has been known in connection with total cross sections for light atoms for some time (12). Recent and current work has focussed on atom-atom collisions (13) and projectile ionization by the neutral target (14) to test these ideas.

ACKNOWLEDGEMENT

This work was supported by the National Science Foundation.

REFERENCES

1. Inokuti, M., *Rev. Mod. Phys.* **43**, 297 (1971).
2. Fano, U., *Ann. Rev. Nuc. Sci.* **13**, 1 (1963).
3. DuBois, R.D., *Phys. Rev. A* **48**, 1123 (1993).
4. Briggs, J.S. and Taulbjerg, K., in *Structure and Collisions of Ions and Atoms*, ed. by I.A. Sellin (Springer-Verlag, Berlin, 1978), pp.105-153.
5. DuBois, R.D. and Manson, S.T., *Phys. Rev. A* **42**, 1222 (1990).
6. Manson, S.T., Toburen, L.H., Madison, D., and Stolterfoht, N., *Phys. Rev. A* **12**, 60 (1975).
7. Ehrhardt, H., Hesselbacher, K.H., Jung, K., and Willmann, K., *Case Studies in Atomic Collision Physics*, **2**, 159 (1972).
8. Manson, S.T., and Toburen, L.H., *Phys. Rev. Letters* **46**, 529 (1981).
9. Hartley, H.M., and Walters, H.R.J., *J. Phys. B* **20**, 1983 (1987).
10. Drepper, F., and Briggs, J.S., *J. Phys. B* **9**, 2063 (1976).
11. Briggs, J.S., and Macek, J.H., *Advances in Atomic Molecular and Optical Physics* **28**, 1 (1991).
12. Gillespie, G.H., and Inokuti, M., *Phys. Rev. A* **22**, 2430 (1980).
13. Manson, S.T., and DuBois, R.D., *Phys. Rev A.* **46**, R6773 (1993).
14. Montenegro, E.C., Melo, W.S., Meyerhof, W.E., and de Pinho, A.G., *Phys. Rev. Letters* **69**, 3033 (1992).

Two-Center Effects in Electron Spectra from Ion-Atom Collisions

Pat Richard

J.R. Macdonald Laboratory, Department of Physics
Kansas State University, Manhattan, Kansas 66506 USA

Abstract. This lecture will focus on one aspect of electron spectroscopy from ion-atom collisions. This topic is the ionization of target atoms by highly charged projectile ions. It deals with the overall broad features of electron spectra created by direct ionization and only requires low energy resolution electron spectroscopy.

TARGET IONIZATION

Mechanisms

The subject of electron spectroscopy of target ionization has been covered in several review articles: Rudd and Macek (1), Sellin (2), Stolterfoht (3-5), Matthews (6), Mehlhorn (7), and Rudd *et al.* (8). These papers form an important background for the more recent work to be discussed here.

In order to begin our discussion of target ionization, I would like to outline some of the simple processes one may expect to be important. I will, for the sake of discussion, consider a bare ion incident on atomic hydrogen.

The first mechanism, see Fig. 1, is a binary collision (1) where the projectile ion interacts with the quasi-free electron of atomic hydrogen in a hard collision. If we assume the electron is free, the velocity of the emitted electron by two body elastic scattering kinematics is $v_e = 2 v_p \cos\theta_L$, where v_p is the projectile velocity and θ_L is the laboratory scattering angle of the electron. The emitted electrons are broadened for any laboratory observation angle by the momentum distribution of the target electrons, but they are however localized in a peak near $2 v_p \cos\theta_L$ (referred to as 2v electrons). In this process there is a large momentum and energy transfer to the electron in the laboratory frame. At zero degrees in the lab they are $\sim 2 m_e v_p$ and $\sim 4t$, respectively, where t is the cusp energy $1/2 m_e v_p^2 = m_e E_p/M_p$. E_p and M_p are the projectile energy and mass, respectively. As viewed in the ion-electron center of mass, the zero degree laboratory frame scattering corresponds to the electron approaching the ion with the velocity v_p and elastically scattering with a velocity -v_p (i.e., 180 degree scattering in the center of mass system).

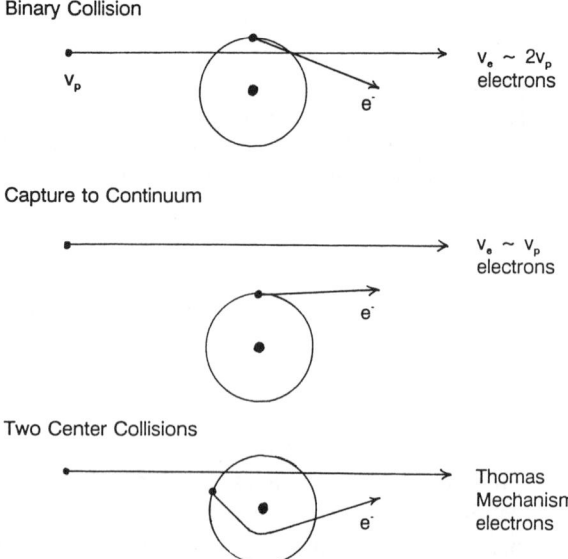

FIGURE 1. Mechanisms for projectile ionization in bare-ion atomic-hydrogen collisions.

The second mechanism is the capture of the target electron to the continuum of the projectile. This process gives rise to electrons moving with a velocity near that of the incident projectile ($v_e = v_p$) and leads to an observed peak near the cusp energy, t, in the forward direction (1-2).

The third mechanism is a second order scattering process where the electron scatters from the projectile ion and then rescatters from the target nucleus. This process can proceed by two hard collisions, two soft collisions and everything in between leading to a broad continuous electron distribution at all observation angles. This process is known to be important in describing the double differential cross section, DDCS, in the region of the cusp electrons, Shakeshaft and Spruch (9). This process is similar to the Thomas scattering mechanism (10) for electron capture. Most of the two center scattering events do not satisfy the Thomas mechanism for capture which occurs for a small range of angles and energies of the outgoing captured electron.

All three of these mechanisms should be in effect in producing ionization electrons and should produce spectral features which overlap in electron energy. The electron yield in the binary encounter region of the spectra can be influenced by electrons from the capture to the continuum and vice versa. In addition, the amplitudes for the one-center and the two-center mechanisms add coherently in a quantum mechanical treatment.

Bare Ions on H_2 and He

The electron DDCS from Lee et al. (11) for $H^+ + H_2$ and He, and F^{9+} + He given in Fig. 2 exemplifies the case of a bare heavy ion of a few MeV/u on a light target at zero degrees in the lab. The binary encounter peak at $v_e \sim 2 v_p$ ($E_e \sim 4$ t) is very evident as is the cusp electron peak at $v_e = v_p$ ($E_e = t$). The cross sections in the other energy regions of the spectra are monotonically decreasing with electron energy and come from all three of the mechanisms mentioned in the previous section.

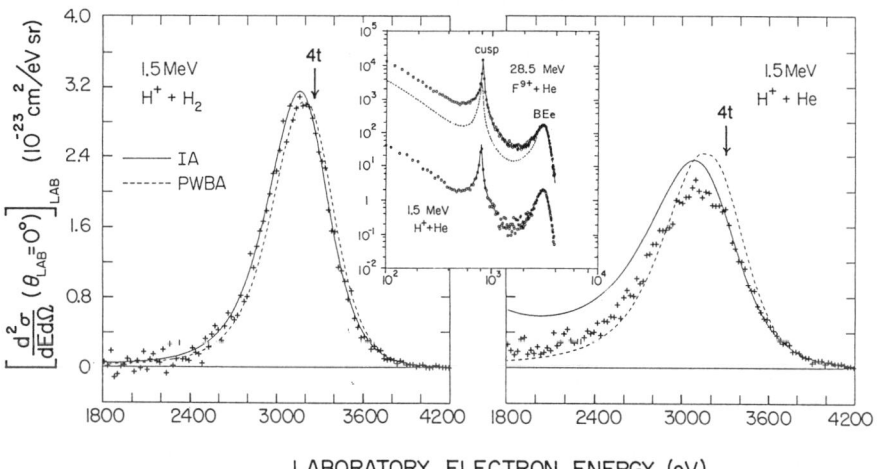

FIGURE 2. BEe DDCS for 1.5 MeV/u $H^+ + H_2$ and H^+ + He collisions. Solid line IA or ESM model and dashed line PWBA. Inset: DDCS for 1.5 MeV/u H^+ and F^{9+} + He, where dot-dashed line is the H^+ data multiplied by Z_p^2 (=81), from Lee et al. (11).

Lee et al. (11), analyzed these data in the region of the BEe peak using the impulse approximation, IA, modeled after Brandt (12). A similar analysis was previously discussed by Burch et al. (13) and by Böckl and Bell (14). In the IA model the DDCS is given in terms of the elastic scattering cross section, $(d\sigma/d\Omega)_{el}$, of the target electron-projectile ion system and the Compton profile of the target electrons $(J_T(p_z))$ by the relation

$$\frac{d^2\sigma}{d\varepsilon\, d\Omega} = \frac{d\sigma}{d\Omega}\Big)_{el} \frac{J_T(P_z)}{v_p + P_{z/m}} \quad (1)$$

It is assumed in deriving this relation that the electron energy is given by

$$E_e = (\vec{s} + \vec{p})^2/2m - E_I. \quad (2)$$

E_I is the target ionization potential, $\vec{p}$ is the momentum of the target electron in the target frame, and $\vec{s}$ is the cusp momentum. The IA model in the remainder of the text will be referred to as the elastic scattering model, ESM. Fig. 2 shows the result of the comparison of the BEe region of the zero degree electron spectra to the IA or ESM and to the PWBA calculations, Lee et al. (11). Both calculations reproduce the experiment very well.

Several additional calculations recently have been presented and compared to the data of Lee et al. (11): Miraglia and Macek (15) - a quantum mechanical impulse approximation, Brauner and Macek (16) - a Brinkman-Kramers model and a distorted wave strong potential Born model, Madsen and Taulbjerg (17) - a channel distorted Brinkman-Kramers model, Jakubassa-Amundsen (18) - ECC-impulse approximation, and Schultz and Reinhold (19) -

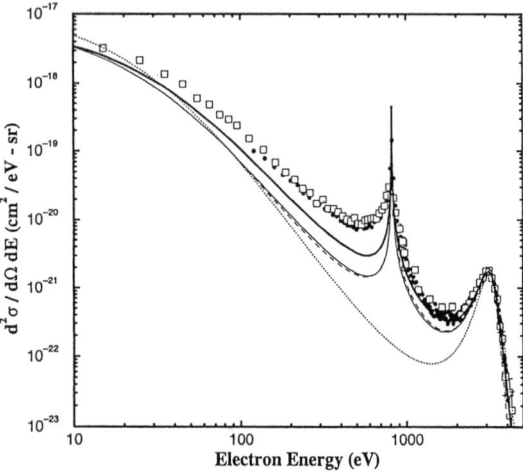

FIGURE 3. 1.5 MeV/u F^{9+}+He; ••• data Lee et al. (11), ☐☐☐ CTMC, ——— CDW-EIS, – – – model potential CDW-EIS, --- DSPB, ··· FBA from Schultz and Reinhold (19).

an extensive classical trajectory Monte-Carlo, CTMC, simulation and a model potential continuum distorted wave-eikonal initial state calculation, CDW-EIS. Fig. 3 is from Schultz and Reinhold (19) and gives a comparison of the 1.5

MeV/u F^{9+} + He data of Lee *et al.* (11) to the various calculations listed in the caption. The CTMC calculation which includes the two center effects fits the data extremely well over the entire electron energy range. Both the DSPB and the CDW-EIS calculations fit the F^{9+} + He spectrum near and below the binary peak down to the cusp but not so well below the cusp. The single-particle channel-distorted Brinkman-Kramers, DBK, calculation by Madsen and Taulbjerg (17), not shown here, also fits the spectrum very well over the entire energy range even though the model does not contain the two step mechanism deemed necessary to describe the cusp region. The BEe peak depends on the central part of the Compton profile whereas the ECC cusp in the DBK approximation depends on the high-momentum components of the Compton profile.

Lee *et al.* (11) have shown that the BEe cross section scales as Z^2 in the range of projectile Z between 1 and 10. One interesting feature of the bare ion scaling in going from H^+ + He to F^{9+} + He in the electron energy region from above the BEe peak down to the cusp peak is shown in Fig. 4. The scaling varies from Z^2 near the BEe peak to almost Z^3 near the cusp peak as was suggested to be the case by Brauner and Macek (16).

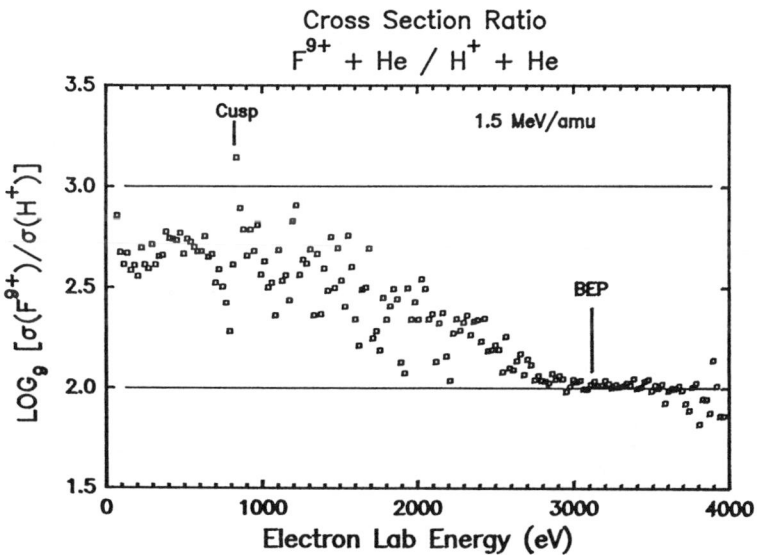

FIGURE 4. The log base 9 of the ratio of the experimental DDCS for F^{9+} + He to that of H^+ + He. The solid lines at 2 and 3 are Z^2 and Z^3 scaling, respectively, for Z=9. From Lee *et al.* (11).

Non-bare Ions on H_2 and He

As reported by Richard et al. (20) the cross section for BEe production by non-bare projectiles shows an inverted q-scaling at $\theta_L=0°$, i.e., the BEe cross section increases as the projectile q decreases. Reinhold, Schultz and Olson (21), Shingal et al. (22) and Schultz and Olson (23) demonstrated that this was due to the static short range potential of the projectile ion. In ionization processes occurring in distant collisions, the cross section should show the normal Z_{eff}^2 behavior where $Z_{eff} \sim q$, whereas in the BEe case the close collision, θ_L = zero degrees, Z_{eff} is not only greater than the q of the projectile but it is greater than the Z of the projectile. Taulbjerg (24) and Bhalla and Shingal (25) demonstrated that the cross section at zero degrees in the lab is further enhanced by the effects of electron exchange. Fig. 5 shows a comparison of the measured DDCS for $C^{2+} + H_2$ to the ESM model calculation of Bhalla and Shingal (25). The calculation uses Eq. 1 with the Rutherford cross section being replaced by the elastic scattering differential cross section for an electron of energy E_e colliding with C^{q+} ions. This cross section can be calculated considering only the static potential, as well as when the exchange is included.

DuBois et al. (27) recently presented results for 0.5 MeV/u B, C, O, and F on He where they showed that the screening for the distant collisions, impact parameters between 1.5 to 3 au (low energy electrons), varied from a Z_{eff} slightly greater than one for the low q ions to a Z_{eff} slightly less than one for the high q ions.

Hidmi et al. (28) extended the studies of zero degree BEe spectra to Z=29 projectiles. For all the data, the inverted q-scaling was observed. Fig. 6 shows the result of the cross section ratio $\sigma(q)/\sigma(1)$ where $\sigma(q)$ is the BEe peak cross section for an ion of charge q and $\sigma(1)$ is for H^+ projectiles.

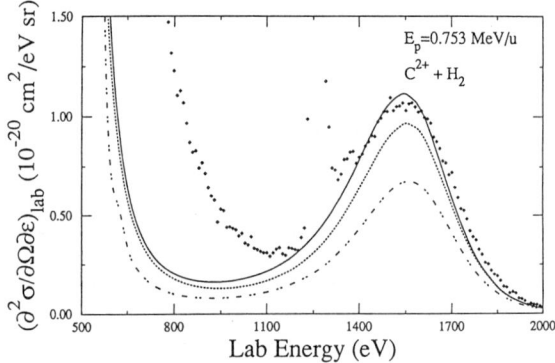

FIGURE 5. Comparison of ESM calculation of DDCS with experiment: ♦♦♦, data; theory: ·· — ··, pure Coulomb potential; ···, static potential; ———, static plus exchange. From Hidmi et al. (26).

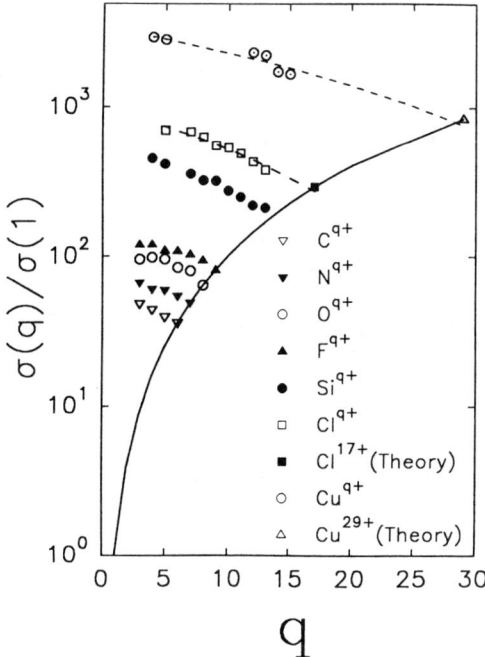

FIGURE 6. Ratio of DDCS for ions of charge state q on H_2 relative to the DDCS for protons of H_2. Cu projectile ions have an energy equal to 0.5 MeV/u and all other ions have a beam energy of 1 MeV/u. The solid line is the Z_p^2 prediction for $q=Z_p$, and the dashed lines are the best fits to the Cl and Cu data extrapolated to the bare ions for each case. From Hidmi et al. (28).

Each projectile specie shows an inverted q scaling. For the Cu case the enhancement ratio of Cu^{4+} to the bare ion cross section is 3.5 compared to the first reported case of F^{3+} (20) where the enhancement ratio is 1.48.

Recent zero degree observations of Sataka et al. (29) for Au^{q+} + He, q = 12, 30, 35, and 37, at 1 MeV/u show the normal q-scaling as opposed to the inverted q-scaling. Even though the normal q-scaling was observed, the Z_{eff} was greater than Z, varying from Z_{eff} = 1.0 × Z for q=12 to 2.1 × Z for q=37. The case with the most projectile electrons shown in Fig. 6 is for Cu^{4+} which has 25 electrons, whereas for the Au case, there are between 42 and 67 electrons on the projectile which presents quite a challenge for the theory.

BEe Energy Shifts

The classical two-body collision of a free electron with a projectile ion produces an electron with energy 4t in the laboratory frame at zero degrees as discussed in the first section. As can be seen in Fig. 2 the position of the

BEe peak occurs at an energy less than the free electron model value of 4t. The ESM model, Eqs. 1 and 2, predicts a BEe peak shifted to lower energy due to the binding energy of the target electron, E_I, and due to the $1/E^2$ behavior of the elastic scattering cross section. For low Z ions (Z<10) the peak shift, $E = 4t - E_{BEe}$, where E_{BEe} is the peak energy of the BEe, is independent of q and observed to be ~ 95 eV for 1 MeV/u ions and is in good agreement with the ESM calculation (25). However, for Z>10 projectiles the BEe peak definitely exhibits a shift which is increasing with q. A q-dependent shift is predicted by several published models; the Bohr-Lindhart (BL) model (30), the resonant tunneling model (31), the CTMC (21), and the CDW-EIS (15). Fig. 7 shows the data of Hidmi et al. (28) for 1 MeV/u ions between C and Cu.

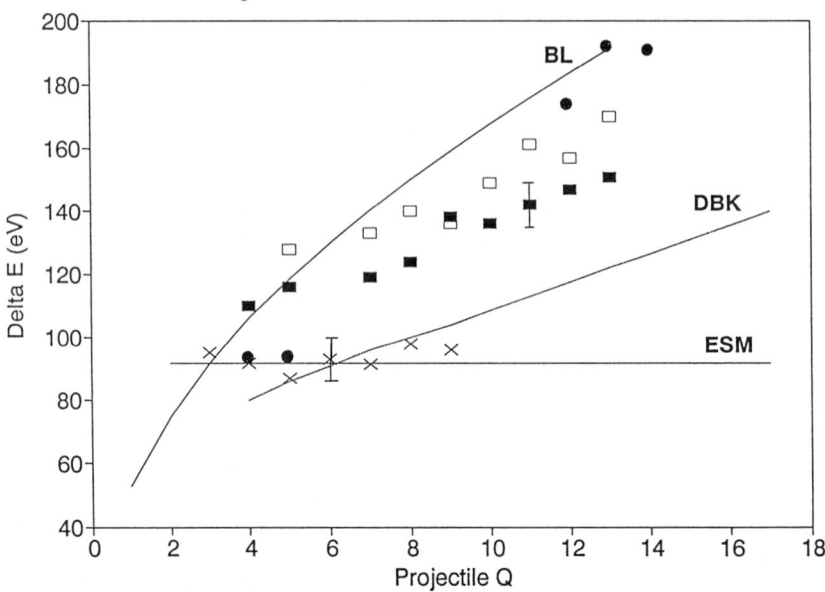

FIGURE 7. The systematics of the BEe peak shift $\Delta E = 4t-E_{BEe}$: × low Z data, ■ Si, □ Cℓ, and ● Cu; theory Bohr-Lindhart, BL (30), distorted Brinkman-Kramers, DBK (17), and the elastic scattering model, ESM (25), see text.

The X's are the average energy shifts for the C, N, O, and F vs q, the closed and open squares are respectively the energy shifts for Si and Cℓ, and the open circles are for Cu. The results of BL ($E = 53\sqrt{q}$ eV) (30), the DBK of Madsen and Taulbjerg (17), and the ESM (25) are given for comparison.

The DBK result of Madsen and Taulbjerg (17) are particularly interesting in that the predicted energy shift comes from the single step capture-to-the-continuum process. They find that this process reproduces the ionization amplitude derived from the ESM model for electrons near the BEe peak, but shifts the peak and changes its shape, and that this calculated shift varies with projectile q. This scenario of the q-dependent shift being due to the tail of cusp electron distribution (i.e., from the high components of the Compton profile of the target) seems particularly plausible. Fig. 8 shows the comparison between the ESM prediction and the experiment for the cases of $C\ell^{7+}$ and $C\ell^{13+}$, Bhalla and Grabbe (32). It is noteworthy that the ESM model fits the magnitude of the BEe cross section very well but does not reproduce the energy shift or the shape of the peak.

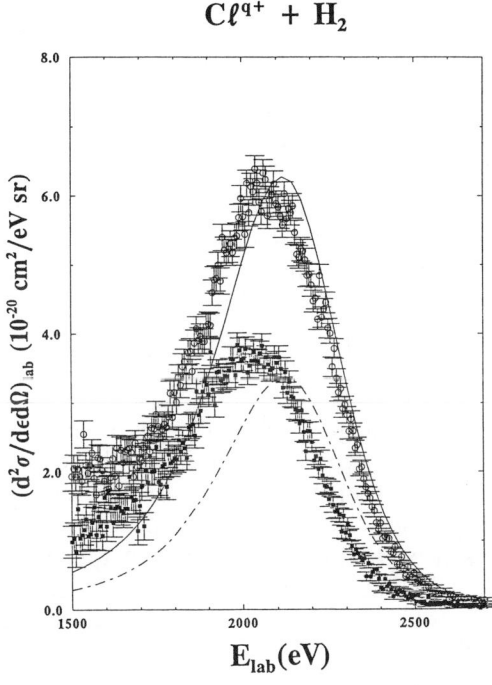

FIGURE 8. The systematics of the 1 MeV/u $C\ell^{q+}$ + H_2 system for q=7 open circles, and q=13 closed circles. The solid and the dot-dashed lines are the results of the ESM calculations. Data are from Hidmi et al. (28); calculations from Bhalla and Grabbe (32).

Multiple Shell Targets

An experiment to study the target shell and subshell effects on the BEe peak for the low Z ions was recently reported by Zouros et al. (33). The study reported on the inverted q-dependence of the cross section for O^{q+} on O_2 targets. It was found that the q-dependence was similar to that observed for H_2 and He targets at zero degrees. The BEe peaks are in fair agreement with the ESM model when the 1s shell electrons are neglected and the Compton profile is taken as

$$J_O^2 = 1.9 \, [2J_{2s}(P_z) + 4J_{2p}(P_z)] \qquad (3)$$

where J_{2s} and J_{2p} are the Compton profiles of atomic oxygen calculated from a Hartree-Fock model (33). Studies of Ne are still in progress and show some small disagreement between ESM calculations and experiment for O^{q+} ions, but is in overall fair agreement.

Angular Distributions and Diffraction Effects

The angular distribution of electrons by fast high-q ions, $Z<10$, has been reported recently by several authors (27, 30, 34, 35). Fig. 9 shows the data of Liao et al. (35) for 1.0 MeV/u F^{q+} + H_2 for $\theta_L = 0°$, 10°, 40°, and 50°. The BEe peaks for the different charge states are connected by the dashed line for each observation angle. The change in the slope of these lines indicates that the q-scaling changes from an inverted q-scaling for 0° to a normal q-scaling for $\theta \geq 20°$. The angular dependence of the BEe DDCS is fairly well reproduced by the ESM model as shown in Fig. 10 for the case of 19 MeV F^{4+} + H_2 (35).

Several important studies of the angular dependence of electron emission using very heavy ion-beams have progressed in parallel with the work on light ions (36-40). Kelbch et al. (36) observed an anomalous splitting of the BEe peak at certain laboratory observation angles for the case of 1.4 MeV/u U^{21+} ions on various targets in 1989. It has been shown that the same phenomena leading to the inverted q-scaling, i.e., the non-Coulomb nature of the screened projectile field, also leads to an electron diffraction spectrum (39, 41) as reported by Kelbch et al. (36).

Fig. 11 shows the data of Hagmann et al. (39) and the ESM calculation of Bhalla, Shingal and Grabbe (41) for 0.62 MeV/u Au^{11+} on He. At $\theta_L = 17.5°$ and 35.0° the BEe structure is a single peak whereas the $\theta_L = 27.5°$ spectrum shows a double peak. This double peak is approximately reproduced by the ESM calculation. The diffraction structure is a general quantum mechanical scattering feature of a short range potential giving rise to the interference of a few active partial waves. These diffraction structures

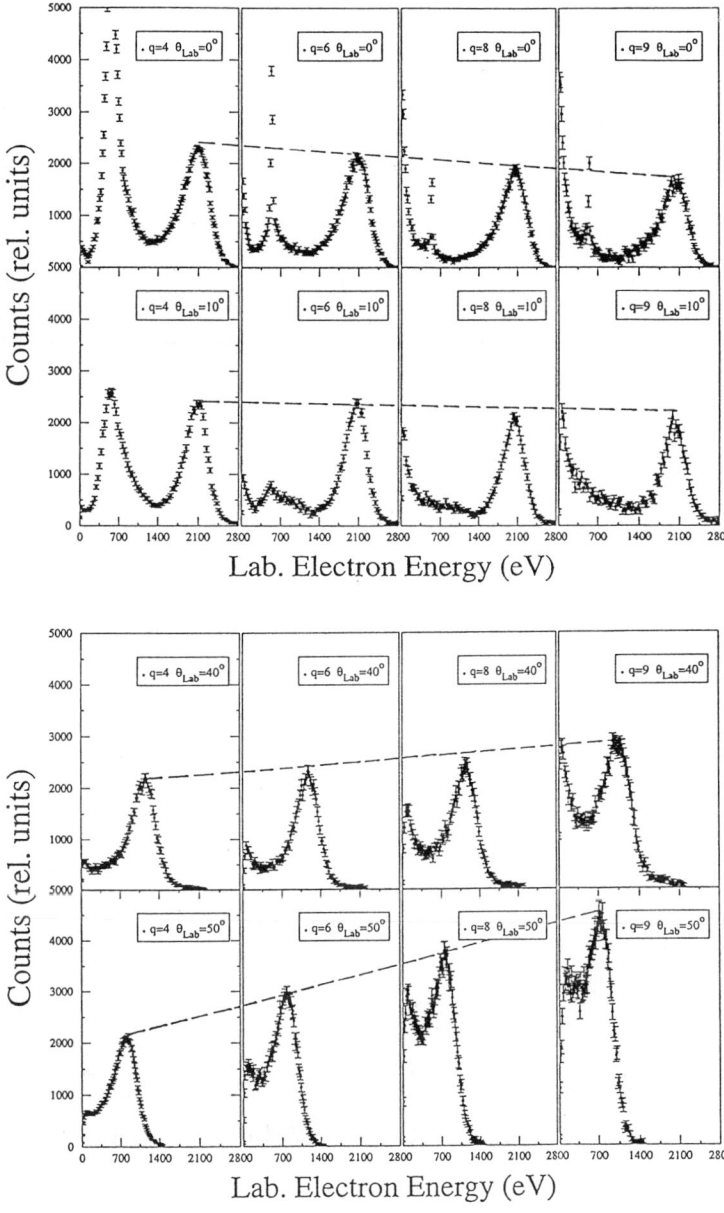

FIGURE 9. The measured DDCS for 1.0 MeV/u F^{q+} + H$_2$ collisions. From Liao et al. (35).

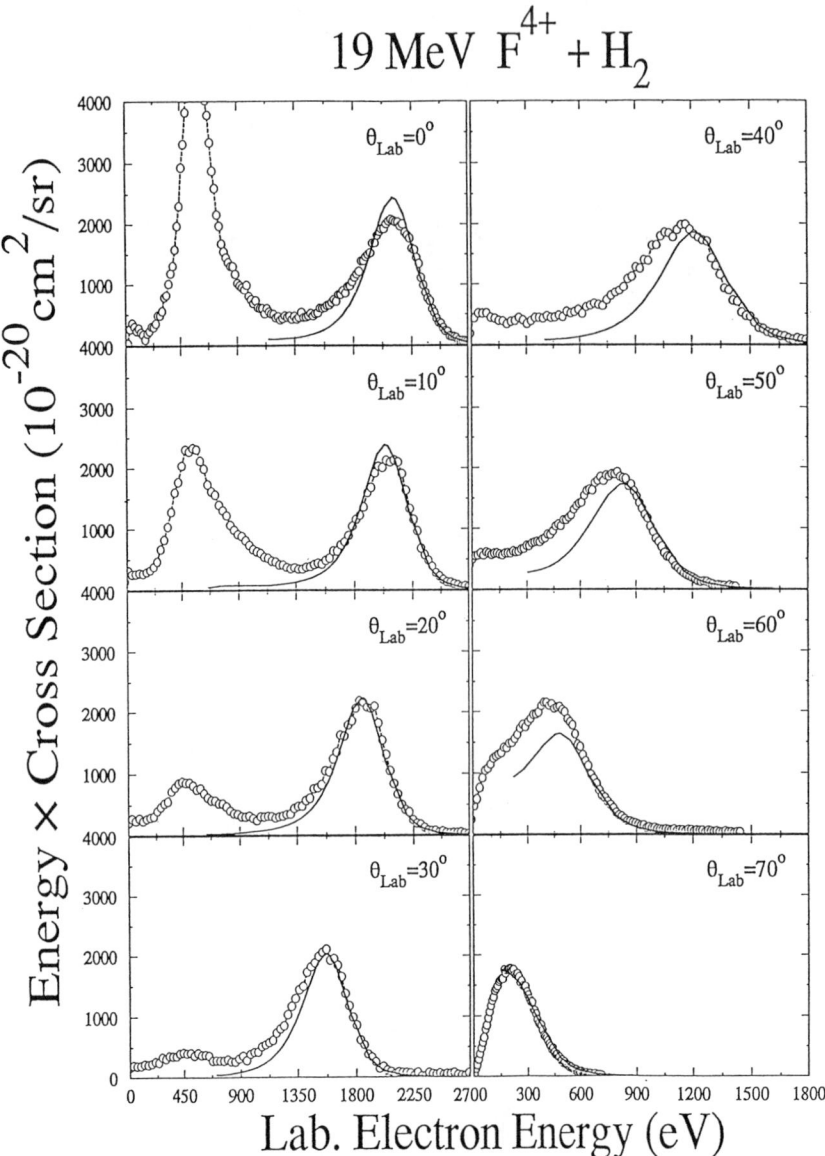

FIGURE 10. Comparison of the measured DDCS to the ESM calculation for 19 MeV F^{4+} + H_2. From Liao et al. (35).

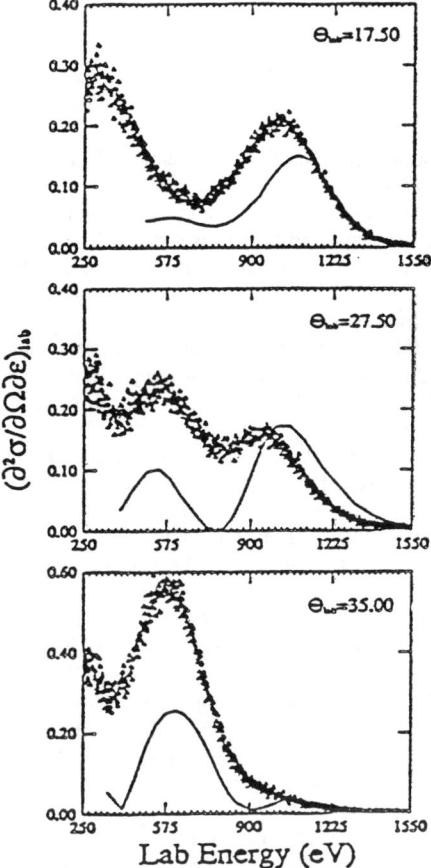

FIGURE 11. The DDCS for 0.62 MeV/u Au^{11+} + He: ••• data of Hagmann et al. (39), —— ESM calculation of Bhalla et al. (41). A diffraction pattern is observed and calculated for $\theta_L = 27.5°$ but not for the nearby angles of 17.5° and 35.0°.

are observed in electron-neutral atom scattering and can be used as a guide to map the region of energy and angular scattering where one may expect to see this phenomenon in ion-atom collisions (39).

ACKNOWLEDGEMENTS

The work reported in this paper was done in close collaboration with Profs. C.P. Bhalla and S. Hagmann of Kansas State University, and Prof. T.J.M. Zouros, University of Crete and Institute of Electronic Structure and Laser, Heraklion, Crete, Greece. The help of Shon Grabbe, Chunlei Liao, and Gabor Toth of Kansas State University is also acknowledged. This work was supported by the Division of Chemical Sciences, Office of Basic Energy Sciences, Office of Energy Research, U.S. Department of Energy.

REFERENCES

1. Rudd, M.E. and Macek, J.H., *Case Stud. At. Phys.* **3**, 47 (1972).
2. Sellin, I.A., "Physics of Electron and Atomic Collisions," ed. by S. Datz (Amsterdam, 1982) p. 199.
3. Stolterfoht, N., *Topics in Current Phys.* **5**, ed. by I.A. Sellin (Springer-Verlag, Berlin, 1978) p. 155.
4. Stolterfoht, N., "Fundamental Processes in Energetic Atomic Collisions," Eds. H.O. Lutz, J.S. Briggs, and H. Kleinpoppen (Plenum Press, New York, 1983) p. 295.
5. Stolterfoht, N., *Physics Reports* **146**, 315 (1987).
6. Matthews, D.L., *Methods of Experimental Physics* **17**, ed. by P. Richard (Academic Press, New York, 1980) p. 433.
7. Mehlhorn, W., "Electron Spectroscopy of Auger and Autoionizing States," Lecture Notes University of Aarhus (1978) unpublished.
8. Rudd, M.E., Toburen, L.H., and Stolterfoht, N., *At. Data Nucl. Data Tables* **23**, 405 (1979).
9. Shakeshaft, R., and Spruch, L., *Rev. Mod. Phys.* **51**, 369 (1979).
10. Thomas, L.H., *Proc. Roy. Sec.* **114**, 561 (1927).
11. Lee, D.H., Richard, P., Zouros, T.J.M., Sanders, J.M., Shinpaugh, J.L., and Hidmi, H., *Phys. Rev. A* **41**, 4816 (1990).
12. Brandt, D., *Phys. Rev. A* **27**, 1314 (1983).
13. Burch, D., Wieman, H., and Ingalls, W.B., *Phys. Rev. Lett.* **30**, 823 (1973).
14. Böckl, H., and Bell, F., *Phys. Rev. A* **28**, 3207 (1983).
15. Miraglia, J.E., and Macek, J.H., *Phys. Rev. A* **43**, 5919 (1991).
16. Brauner, M., and Macek, J.H., *Phys. Rev. A* **46**, 2519 (1992).
17. Madsen, J.N., and Taulbjerg, K., 1994, to be published.
18. Jakubassa-Amundsen, D.H., 1994, to be published.
19. Schultz, D.R., and Reinhold, C.O., 1994, to be published.
20. Richard, P., Lee, D.H., Zouros, T.J.M., Sanders, J.M., and Shinpaugh, J.L., *J. Phys. B* **23**, L213 (1990).
21. Reinhold, C.O., Schultz, D.R., and Olson, R.E., *J. Phys. B* **23**, L591 (1990).
22. Shingal, R., Chen, Z., Karim, K.R., Lin, C.D., and Bhalla, C.P., *J. Phys. B* **23**, L637 (1990).
23. Schultz, D.R., and Olson, R.E., *J. Phys. B* **24**, 3409 (1991).
24. Taulbjerg, K., *J. Phys. B* **23**, L761 (1990).
25. Bhalla, C.P., and Shingal, R., *J. Phys. B* **24**, 3187 (1991).
26. Hidmi, H.I., Bhalla, C.P., Grabbe, S.R., Sanders, J.M., Richard, P., and Shingal, R., *Phys. Rev. A* **47**, 2398 (1993).
27. DuBois, R.D., Toburen, L.H., and Middendorf, M.E., *Phys. Rev. A* **49**, 350 (1994).
28. Hidmi, H.I., Richard, P., Sanders, J.M., Schöne, H., Giese, J.P., Lee, D.H., Zouros, T.J.M., and Varghese, S.L., *Phys. Rev. A* **48**, 4421 (1993).
29. Sataka, M., Imai, M., Yamazaki, Y., Komahi, K., Kawatsura, K., Kanai, Y., Tawara, H., Schultz, D.R., and Reinhold, C.O., 1994, to be published.
30. Pedersen, J.O., Hvelplund, P., Petersen, A.G., and Fainstein, P.D., *J. Phys. B* **24**, 4001 (1991).
31. Fainstein, P.D., Ponce, V.H., and Rivarola, R.D., *Phys. Rev. A* **45**, 6417 (1992).
32. Bhalla, C.P., and Grabbe, S., 1994, private communication.
33. Zouros, T.J.M., Richard, P., Wong, K.L., Hidmi, H.I., Sanders, J.M., Liao, C., Grabbe, S., and Bhalla, C.P., *Phys. Rev. A* **49**, 3155 (1994).
34. Gonzalez, A.D., Dahl, P., Hvelplund, P., and Fainstein, P.D., *J. Phys. B* **26**, L135 (1993).

35. Liao, C., Richard, P., Grabbe, S.R., Bhalla, C.P., Zouros, T.J.M., and Hagmann, S., (1994) to be published.
36. Kelbch, C., Hagmann, S., Kelbch, S., Mann, R., Olson, R.E., Schmidt, S., and Schmidt-Böcking, H., *Phys. Lett.* **139A**, 304 (1989).
37. Kelbch, C., Koch, R., Hagmann, S., Ullmann, K., Schmidt-Böcking, H., Reinhold, C.O., Schultz, D.R., Olson, R.E., and Kraft, G., *Z. Physik D* **22**, 713 (1992).
38. Wolff, W., Shinpaugh, J.L., Wolf, H.E., Olson, R.E., Wang, J., Lencinas, S., Piscevic, D., Herrmann, R., and Schmidt-Böcking, H., *J. Phys. B: At. Mol. Opt. Phys.* **25**, 3683 (1992).
39. Hagmann, S., Wolff, W., Shinpaugh, J.L., Wolf, H.E., Olson, R.E., Bhalla, C.P., Shingal, R., Kelbch, C., Herrmann, R., Jagutzki, O., Dörner, R., Koch, R., Euler, J., Ramm, U., Lencinas, S., Dangendorf, V., Unverzagt, M., Mann, R., Mokler, P., Ullrich, J., Schmidt-Böcking, H., and Cocke, C.L., *J. Phys. B: At. Mol. Opt. Phys.* **25**, L287 (1992).
40. Shinpaugh, J.L., Wolff, W., Wolf, H.E., Ramm, U., Jagutzki, O., Schmidt-Böcking, H., Wang, J., Olson, R.E., (1994) to be published.
41. Bhalla, C.P., Shingal, R., and Grabbe, S., *Nucl. Instrum. Meth. in Phys. Res. B* **79**, 170 (1993).

Two-Center Effects in the Ejected-Electron Spectra in Ion-Atom Collisions

David R. Schultz[1], Carlos O. Reinhold[2], and Ronald E. Olson[3]

[1] *Physics Division, Oak Ridge National Laboratory, Oak Ridge, TN 37831-6373*
[2] *Department of Physics, University of Tennessee, Knoxville, TN 37996-1200*
[3] *Department of Physics, University of Missouri-Rolla, Rolla, MO 65401*

Features of the spectra of electrons ejected in ion-atom collisions which arise due to "two-center" effects are described. After a brief survey of these effects, they are illustrated in more detail through examples of studies involving intermediate energy collisions.

I. INTRODUCTION

The ionization of atoms by ions is a ubiquitous phenomenon. On one hand, such collisions provide a very basic testing ground for atomic physics. Study of the yield of electrons, especially the singly and doubly differential cross sections as a function of ejection angle and energy, is a sensitive probe of ionization, testing our basic understanding of atomic collisions. On the other hand, this understanding is also critical to a variety of practical concerns. Ion-induced electrons produce damage to biological material when, for example, humans are exposed to radiation, a situation which can be either harmful if uncontrolled, or beneficial, as in the radiotreatment of cancer. In addition, a wide range of applications exist in which ions are used to probe, prepare, or modify surfaces in semiconductor and materials research and manufacturing. Indeed, fields of study as richly diverse as astrophysics, radiation biology, and plasma physics all rely heavily on our knowledge of the details of the process of ionization.

Within the last decade it has become customary to describe the ejected electron spectrum by partitioning it into two components reflecting the principal mechanisms of emission. The first represents a contribution describable within the context of "one-center" theories like the Born approximation, while the other is the contribution from "two-center" effects. The term "one-center" refers to describing theoretically the electrons as being ejected into the continuum of, say, the target or the projectile alone, while "two-center" refers to the fact that electrons escape in the combined field of both the residual target and projectile ions. Several fairly recent works have made a strong case for this distinction (1–5). However, the experimental data of Rudd, To-

© 1996 American Institute of Physics

buren, Stolterfoht and others (see e.g. (6-9)) have shown clear evidence for almost thirty years of the manifestations of both one- and two-center effects. Most recently, the number of papers reporting or describing two-center effects has become extremely large, testifying to the success of experimentally derived insight and theoretical approaches which have been able to develop a fuller understanding of both the seminal and recent measurements of ejected electron spectra.

Here, we will provide an overview of two-center effects, describing the mechanism of "saddle point" ionization, the formation of the electron-capture-to-the-continuum (ECC) peak, enhancement, oscillation and shifting of the binary peak, and the combination of projectile and target electron emission. We will illustrate some of these effects with specific reference to a few case studies. It is also worth noting here that a number of works have reviewed the extensive literature regarding cross sections for electron emission (8-13) caused by light ion impact.

II. SURVEY OF "TWO-CENTER" EFFECTS

To orient our view of two-center effects, we display in Figure 1 a typical forward-ejected (i.e. $\theta < 90°$) electron spectrum, at a collision energy at the high end of the intermediate energy range where the features are well separated. We define intermediate collision energies as the range in which the projectile velocity, v_p, is on the order of the orbital electron velocity, up to several times this speed. In this figure, the height of the surface indicates the magnitude of the cross section, with the radii from the $v_e = 0$ (v_e is the ejected electron velocity) peak giving the electron energy. A projectile ion impinges upon an atom initially at rest and travels away towards the right of the page, indicating the extreme forward (i.e. $\theta = 0°$) direction. In addition, we display several arcs over the spectrum indicating energies corresponding to electron velocities equal to zero, $v_p/2$, v_p, and $2v_p$.

The peak surrounding $v_e = 0$ is the soft-electron peak which results from collisions in which only a very small momentum transfer between the projectile and electron occurs. At high collision velocities or for small projectile charge states, q, these electrons are essentially well described as being emitted in just the field of the target ion, since relative to their slow velocity, the projectile recedes very rapidly. Clearly, at intermediate energies where the projectile recedes relatively more slowly, a different situation is encounterd. Another prominent feature of the forward spectrum is the peak near an emission angle of zero degrees for $v_e \approx v_p$. This is the so-called electron-capture-to-the-continuum peak, and consists of electrons traveling with velocity nearly that of the projectile (thus "capture") but not bound to it (thus "to-the-continuum"). The existence of this feature is predicted by one-center theories in which the emitted electron is considered to be in the continuum of the projectile. However, a complete description of this feature requires a two-center theory.

Lying between these two regions is the range of ejection velocities which is termed the "saddle-point" region. For simplicity, if we consider the case of

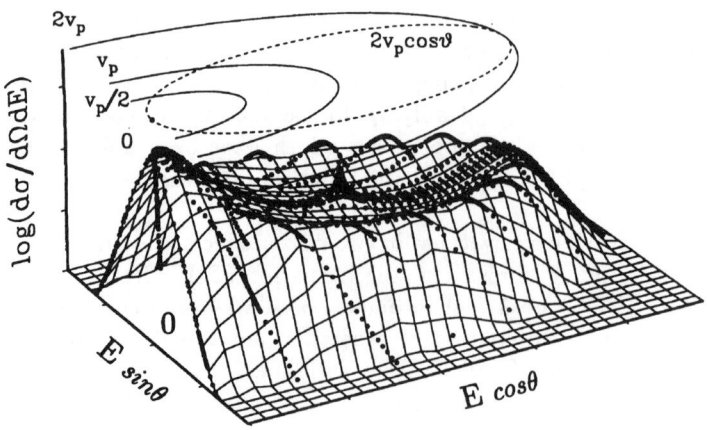

FIG. 1. A typical spectrum of electrons ejected into forward ($\theta < 90°$) angles in the laboratory frame (depicted is the spectrum for the collision of 1.5 MeV/u F^{9+} with helium). The arcs indicate regions in which the ejected electron has a velocity near 0, $v_p/2$, v_p, and $2v_p$.

proton impact of atomic hydrogen, one can see that halfway between the projectile and target protons, the electron experiences an extreme (saddle point) of the potential energy. For other projectile and target charges, the position of the saddle, or equiforce, point is changed in accordance with the appropriate balancing of attractive Coulomb fields. If the electron finds itself in this region in which it is bound to neither target nor projectile, the probability of ionization is enhanced. For $H^+ + H$, the position of the saddle point moves through the collision with a velocity $v_p/2$, and the electron experiences no force at that point. Therefore, it has been speculated that it is likely that the electron can be ejected with just this velocity. Description of electrons ejected from the target through this mechanism clearly requires that the electron be treated as being simultaneously in the field of both the target and projectile.

The remaining prominent feature is the binary peak, or ridge, which is mainly a one-center feature since it results from the two-body (binary) collision of the electron with the projectile. In such a collision, energy and momentum conservation indicates that the electron should leave with an energy $E = 2v_p^2 cos^2\theta - E_i$, where θ is the ejection angle and E_i the ionization potential. The width of the ridge reflects the fact that the electron has a momentum distribution in the atom. One-center theories such as the Born or the impulse (or binary encounter) approximations represent the binary peak reasonably well for $q/v_p \ll 1$. However, for $q/v_p \geq 1$ the peak shifts to lower energy and broadens, compared to the shape predicted by one-center theories. Thus, even the binary peak requires, in many instances, a proper two-center treatment. We also note that anomalous behaviors of the binary peak have recently been found for impact by partially stripped projectile ions and have been explained by models based on one-center theories.

In what follows we focus on each of these two-center effects in somewhat more detail, primarily from the point of view of the theoretical approaches which have led to their explication.

A. "Saddle point" electrons

Before the widespread recognition of the need for two-center theories of ionization, electron emission was described primarily by considering the ejected electrons as being either target-centered or projectile-centered. Then in 1983, Olson (14) noted that the classical trajectory Monte Carlo (CTMC) method (15,16) produced a large number of electrons ejected at small angles associated neither with the target nor projectile, but emitted with velocities about $v_p/2$, in proton-hydrogen collisions at intermediate energies. That work attributed the enhancement of the probability for ejection near this velocity to the stranding of the electrons on the saddle point region.

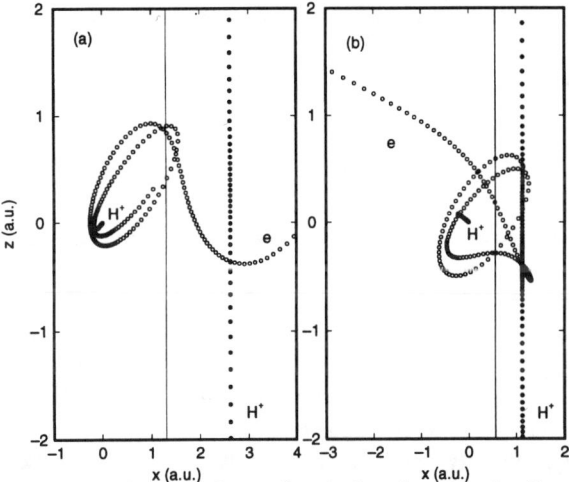

FIG. 2. Projection onto the x–z plane of typical trajectories leading to asymptotic ejected electron velocities $v_e \approx v_p/2$ for a 20 keV proton colliding with atomic hydrogen. The proton is incident from the bottom and the track of the "saddle" or equiforce point through the collision is given by the solid line. A small circle representing each particle is plotted at each time interval through the computation. The time step varys in size with the relative accelerations of the particles thus accounting for the spacing of the circles seen in the figure.

Calculations performed by Winter and Lin (17) implied that by including basis functions on the midpoint between the projectile and target in a coupled channels calculation, a dramatic improvement of the agreement of the total cross section with the available experimental measurements occurred, giving further support to the idea that a significant amount of electron density existed near the saddle point energy.

Further early CTMC calculations in 1986 (2) displayed a peak around $v_p/2$ at intermediate impact energies and indicated that many more electrons were associated with this ejection energy than that corresponding to electron-capture-to-the-continuum. In addition, experimental work by Meckbach and coworkers (18,19) and theory by Macek (18) seemed to show that a ridge existed stretching between $v_e = 0$ and $v_e = v_p$ at small forward angles. Even though the conclusion that such a ridge existed has been invalidated by an experimental error (20), that work and the CTMC predictions sparked a number of additional experimental investigations seeking to identify a peak in the distribution of electrons ejected in association with the saddle point.

In particular, in 1987 experimental measurements (21) found a peak near the expected saddle point energy for 60 keV proton collisions with helium, at an observation angle of $17°$. This work cited early measurments of Rudd and coworkers (6,7,22) which also seemed to possess a peak in the doubly differential ionization cross section at energies corresponding to the saddle point velocity. Next, Irby et al. (23) showed that the peak shifted to lower energies when the projectile charge was changed. This provided another indication that the saddle point ionization "model" was valid since the shift would be expected due to the new position of the equiforce point when the projectile (Z_p) and target (Z_t) charges were different. In this case, the saddle point velocity scales as

$$v_{saddle} = \frac{v_p}{1 + (\frac{Z_p}{Z_t})^{\frac{1}{2}}} \tag{1}$$

and therefore, for larger projectile charges, the saddle point moves more slowly. Later measurements by two other groups (24–26) did not confirm the shift, while new measurements by the first group with Rudd's apparatus in Nebraska once again found the shift (27). To this point, no two-center theory has predicted such a shift. More recent measurements of a peak, and of its shift, in partially stripped ion impact at intermediate collision energies have been put forth (28); however, experiments contradicting them also have been published (29). Finally, we note that Pieksma has recently presented evidence of the saddle point peak in very low energy collisions of protons with atomic hydrogen, and supports his observations with calculations based on an adiabatic collision theory (30).

Thus, the existence of a saddle point peak arising in intermediate energy ion-atom collisions, along with its behavior with varying projectile and target charge, still remains a rather contentious issue. In retrospect, these attempts to find a saddle point peak, especially one which exists for a wide range of forward angles, seems to have been hampered by three key points which now seem clear. First, the experimental measurements by the Argentine group which indicated that a ridge might exist connecting the soft electron and electron-capture-to-the-continuum peaks, and thus manifesting an enhancement due to saddle point electron production, have been retracted. Secondly, from the point of view made possible by contemporary computational resources, the

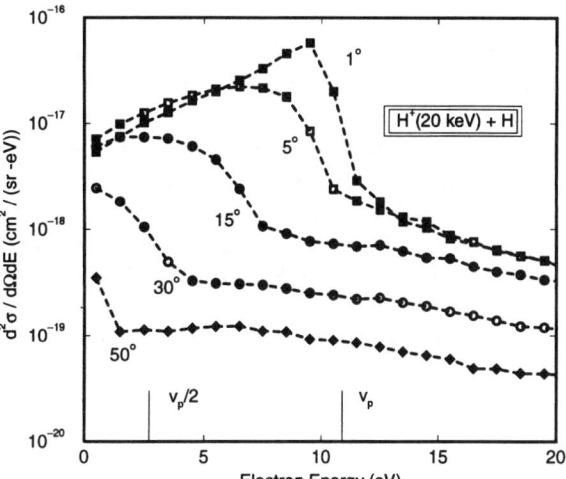

FIG. 3. The doubly differential cross section for ionization in collisions of 20 keV protons with atomic hydrogen for various forward angles of ejection. The ejected electron energies corresponding to $v_e = v_p/2$ and $v_e = v_p$ are indicated.

number of trajectories which could be utilized in the early CTMC calculations, and the distance to which they could afford to be integrated, hindered a complete view of the small forward-angle, doubly differential cross section. We will return to this point in more detail below in our discussion of the cusp. And thirdly, the experiments have looked only at a limited range of angles, so that they too have not taken into account the variation of the "peak" position with angle.

So, what is our view of the situation? To begin with, if one examines the types of trajectories which the CTMC technique gives for electrons which have asymptotic velocities near $v_p/2$ in intermediate energy collisions of protons with atomic hydrogen, one sees that they come from a wide range of three-body collisions. Figure 2 displays two trajectories, typical of these $v_p/2$ electrons in 20 keV $H^+ + H$. For example, for intermediate collision energies, the effect of the saddle point is to liberate the electron because, for a time, it feels little or no force binding it to the target ion. Because the electron emerges with a rather low velocity characterized by the initial electronic momentum distribution, it is attracted by the receding projectile into the forward direction. Also shown is a trajectory in which the electron has a rather close collision with the projectile, but still obtains an asymptotic velocity close to $v_p/2$ due to its interaction with both centers. This type of trajectory is also characteristic of electrons ejected with velocities near v_p. Thus, at intermediate impact energies, we infer that there are a number of possible collision geometries contributing to $v_p/2$ electrons.

Further evidence of this interplay of mechanisms is shown in Figure 3 where we display the doubly differential ionization cross section for the same collision

(20 keV H$^+$ + H) as a function of electron energy for several forward ejection angles. One sees that near zero degrees, the ECC cusp is a prominent feature of the low energy spectrum. However, as one goes to larger angles, the peak near zero velocity, the soft-electron peak, becomes dominant. Thus, as a function of increasing ejection angle, the peak in the spectrum shifts from $v_e = v_p$ at zero degrees, to $v_e = 0$ at larger angles. Somewhere in between, say at around 15°, the peak lies at approximately $v_e = v_p/2$. Therefore, to observe a "saddle point peak" one has to make a measurement at the correct ejection angle. In conclusion, the mechanism whereby electrons are liberated because of the presence of a saddle or equiforce point is clearly a valid and essential point to describing the low electron energy distribution, but the presence of a peak at intermediate impact energies is a result of an interplay of the $v_e = 0, v_p/2$, and v_p ejection processes.

B. The v_p cusp

A body of work too large to summarize here has sought to explore the behavior of the ECC cusp for a wide variety of projectiles (fully stripped, clothed, and neutral), and in particular, to explore the degree of asymmetry that the cusp displays for these projectiles. From a historical point of view, the cusp was first definitely shown in about 1970, experimentally by Crooks and Rudd (31) and Harrison and Lucas (32), and theoretically by Salin (33) and Macek (34). It is composed of ejected electrons which have their velocity vectors approximately equal to that of the projectile. In this volume Macek describes some of the early history of the description of the ECC cusp, and here we only touch on a few points, particularly with respect to its exploration with the CTMC technique.

It was not immediately clear that this phenomenon was classical in origin. In fact, it was only recently that Ovchinnikov and Khrebtukov (35) predicted that the classical post-collisional electron-projectile interaction would produce the correct cusp in the doubly differential ionization cross section. Independently, Reinhold and Olson (5) showed that this feature can be obtained using the CTMC method if the trajectories with nearly a zero binding energy in the projectile frame are integrated out to very large distances. Even though the binding energy of the cusp electrons is determined soon after the collision, it requires a very large distance for the kinetic energies to converge to their asymptotic value. The corresponding quantal picture is that of a wavepacket of projectile-centered states which is created during the collision and subsequently propagates radially very slowly. Classically, each free electron comprising the wavepacket is on its way out in a hyperbolic trajectory and is still very close to the projectile right after the collision. We also note that the cusp is a direct consequence of threshold laws associated with the two-body final state interaction and its existence is independent of the particular collision process (see e.g. (36) for a more general discussion).

Thus, stopping the time integration too soon results in the complete absence of the cusp, and in fact, a hole or anti-cusp in the spectrum. This observation

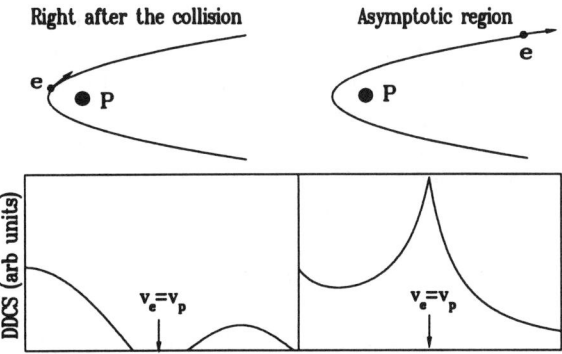

FIG. 4. Schematic illustration of the doubly differential cross section (DDCS) near the ECC peak position for two different electron-projectile separations. In the projectile frame, the ECC electron is still bound right after the collision, and thus the DDCS for ionization reflects the absence of free electrons with $v_e = v_p$. In the asymptotic region, the ECC electron following the hyperbolic trajectory is free and the DDCS displays a cusp at $v_e = v_p$.

sheds light on the earlier studies which showed no cusp, and a peak near $v_p/2$. In that case, since the trajectories are stopped too soon to accurately determine the asymptotic kinetic energies of cusp electrons, a peak appears near $v_p/2$, a peak which would vanish as the trajectories are continued to very large distances. Upon extending the integration of trajectories to much larger distances, not only did the CTMC method yield an ECC cusp, but the symmetry of the cusp was found to be in excellent agreement with that which was observed experimentally (5,37). This method has also been used to successfully predict and explain the variation of cusp asymmetry depending on whether the projectile is bare or paritally stripped (38).

In addition, application of the CTMC technique has played a role regarding the possibility and nature of an ECC cusp for light-particle impact (i.e. for positrons). Only fairly recently have suggestions been made that the cusp might also exist for such a projectile (39–42), and early calculations based on quantum mechanical approximations indicated that the positron-impact cusp should look very much like the cusp produced by isotachic proton-impact (42,43). However, the CTMC method predicted (44) that the cusp should be much smaller and broader, forming essentially a ridge rather than a cusp. This behavior arises owing to the fact that the significantly lighter positron will be deflected to a greater extent than the proton, and that its energy loss will be larger. Recent experiments (45) have confirmed the view that the ECC feature forms only a small ridge.

Also of interest is the fact that when a negatively charged particle impinges upon an atom, a feature which is essentially the opposite of the ECC cusp is formed. Whereas the positively charged ion focuses the ejected electrons into trajectories converging with its path, the negatively charged particles repel the electrons. The result has been termed an "anti-cusp" since there is essentially a hole in the doubly differential cross section near zero degrees. This effect has been observed experimentally for antiproton impact by Yamazaki et al. (46) and for electron impact by Guang-yan et al. (47). The anticusp has also been described theoretically using two-center treatments such as the continuum-distorted-wave–eikonal-initial-state (CDW-EIS) approximation by Fainstein et al. (48) and using the CTMC method (49).

C. The binary peak

In energetic ion-atom collisions, the binary peak arises from a hard collision between the projectile and a target electron. Since the process which leads to its formation is essentially elastic scattering of the electron from the projectile, it is expected that its magnitude will increase when the projectile charge is increased. In particular, for fully stripped ions the magnitude of the peak should be proportional to the square of the projectile charge, in accordance with Rutherford scattering. This has indeed been demonstrated for a variety of fully stripped ions (50–52).

For partially stripped ions, the situation has until recently not been quite as clear. Early work established that the magnitude of the peak did not scale as q^2 for such clothed ions (50,51,53), and a semiempirical effective charge model was proposed (11,54). However, a strong anomaly was observed in 1990 by Richard et al. (55) in the binary peak observed near an ejection angle of zero degrees. They found for a particular range of projectile charge states of light ions, that instead of possessing a smaller binary peak, as would be expected on the basis of the effective charge model, partially stripped ions produced a larger binary peak than that arising from equivelocity bare ions. Just prior to this observation made at Kansas State University, a group from Frankfurt (Kelbch et al. (56)) was exploring the behavior of the fast ejected electrons from much heavier ion-impact. They found that at certain ejection angles, instead of the usual binary peak, an unexpected double peak structure appeared. A schematic view of these two effects is given in Figure 5.

As with many anomalies, their origin has a quite simple explanation which was only at first not apparent. The common link was that the target electrons were scattered by a partially stripped ion which presents to them a strongly non-Coulombic potential. The enhancement of the binary peak for partially stripped ions has been explained by including in the binary encounter approximation (57) a screened potential rather than a pure Coulombic interaction between the projectile and the target electron (58). A number of authors found (59–61) at essentially the same time the same explanation by performing elastic scattering calculations to estimate the degree of enhancement. The oscillation in the binary peak observed by the Frankfurt group was similarly explained using the binary encounter model with a screened potential shortly thereafter (62). Subsequently, a number of works have sought to describe these behaviors for a wider range of projectile ions and observation angles

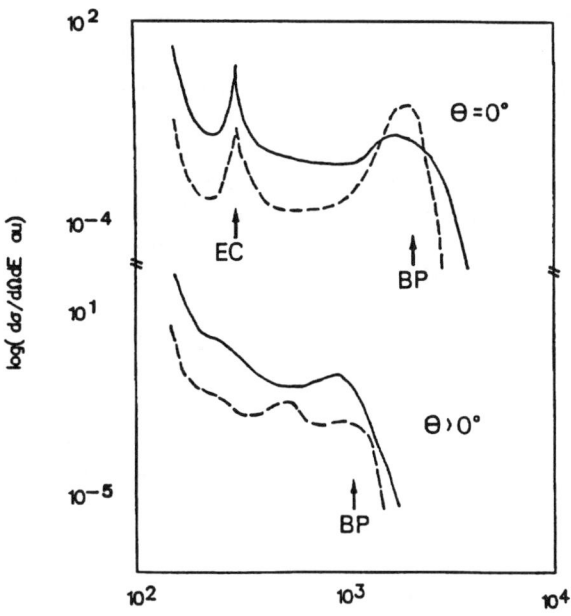

FIG. 5. Schematic illustration of the enhancement and oscillation of the binary peak in the ejected electron spectrum resulting from partially stripped ion impact (broken curves) compared to the yield of electrons from fully stripped ion impact (full curve). The arrows indicate the position of the electron (capture or loss) cusp, and the binary peak predicted from conservation of energy and momentum in a two-body collision. The scales represent units typical for impact by an ion such as Xe^{11+} at 1 MeV/u and for ejection angles of $0°$ and $30°$. (This figure is reprinted from Ref. (63)).

(see e.g. (63–65)). Even though the nature of these anomalies is explicable with just a one-center treatement, we briefly summarize here the mechanism responsible for their appearance.

In the binary encounter or impulse model, the ejected electron spectrum is given by the convolution of the elastic cross section for scattering of the electron by the impinging ion and its initial momentum distribution. For bare ion impact, the result of the convolution of the elastic (Rutherford) cross section with the electronic momentum distribution gives a peak near $v_e = 2v_p \cos\theta$ with a width proportional to the Compton profile. We note that backscattering of the electron in the projectile frame results in forward emission in the laboratory frame. Also, the scattering of an electron from a screened potential results in an elastic cross section which is larger than for scattering in a Coulomb potential at backward angles. Thus, the convolution of this larger elastic cross section at backward angles in the projectile frame with the momentum distribution yields an enhanced forward binary peak.

In addition, if the projectile ion has a very small ratio of its charge state q to its nuclear charge Z_p, meaning that the scattered electron experiences a very low charge state before the collision, but penetrates to expose a very large charge during the binary collision, then oscillations in the elastic cross section may be observed. That is, diffractive electron scattering occurs. In other words, if the oscillations in the elastic scattering cross section are deep and the target momentum profile narrow, the diffraction pattern may survive the convolution and manifest itself as the split or double binary peak observed for heavy ion impact. Thus, the difference in scattering from a bare ion (Coulomb, Rutherford scattering), which scales as the square of the projectile charge, is quite different from the scattering from a clothed ion which has a much more complicated scaling and variation with the ionic and nuclear charges of the projectile.

We emphasize that though the enhancement and oscillation of the binary peak can thus be explained by a one-center treatment, in order to reproduce the full spectrum of ejected electrons which underlies the peak requires a two-center theory. Thus, much of the spectrum at intemediate collision energies results from a subtle interplay of effects which can be described by one-center theories, but have two-center modifications, and effects which may only be described by two-center approaches.

Finally, we might also note that the latest trend in the study of the anomalous behavior of the binary peak is to ascertain the degree to which its position is shifted to lower energies compared to that given by the one-center, binary encounter or Born approximations. Two-center theories such as the CTMC method, the CDW-EIS approximation (1,66,67), or the strong-potential-Born (SPB) approximation (68,69) and a number of experiments (see e.g. (70,71)) show such a shift. It has been proposed that this effect might be due to a post-collisional retarding of the ejected electrons, or that it is described by a simplified model of ionization (Bohr-Lindhard (72)) or its slight elaboration (73), or that it is simply due to the addition of the correct two-center background of electrons to those produced in binary collisions.

III. RECENT CASE STUDIES

A. Collisions of H^+ with H

The most fundamental testing ground for theoretical approaches seeking to describe the ejected electron spectrum is the study of protons colliding with atomic hydrogen. This is the case since in the absence of many electrons the exact unperturbed initial wavefunction is known and its evolution does not involve electron-electron interactions. Essentially no well-founded approximations exist for treating the doubly differential cross section for ionization in $H^+ + H$ collisions at very low collision energies. At very high energy the projectile-electron interaction may be treated as a perturbation, and one-center approaches based on the Born or distorted-wave approach are valid. However, at intermediate energies, other approximations are necessary which can account for the strong coupling among all reaction channels, as well as the combined influence of the projectile and target ions.

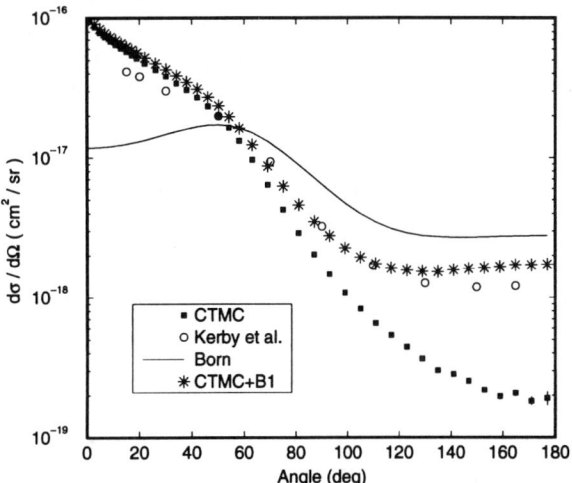

FIG. 6. The singly differential cross section as a function of electron ejection angle for collisions of 114 keV protons with atomic hydrogen, comparing the measurments of Kerby et al. (76) with various theoretical approximations.

Recent measurements of the doubly differential ionization cross section for collisions of 70 keV protons with atomic hydrogen by Rudd's group (74) provide, therefore, an important benchmark against which to judge various theoretical approaches. In fact, comparison with results of the CTMC method, the Born approximation, and the CDW-EIS approximation yielded (74) considerable insight into the electron energy and angle regimes in which each of these models best reflects the actual collision dynamics. In particular, it was found that the best fit of the data could be found if the CTMC calculation were followed except for slow, backward-ejected electrons where the CDW-EIS model should be used. Other work (75) explored in great detail the classical limit of ionization and led to the understanding that, in essence, the classical model of the atom is too stable to small perturbations, and therefore the soft electron component which should arise in this way is not completely reproduced. These ideas also led to a suggestion as to how to approximate that portion of the cross section not well represented classically by including a contribution from the Born, or other quantal perturbation approximation, for small momentum transfers.

To illustrate this proposed correction of the CTMC results, we display in Figure 6 the singly differential ionization cross section for 114 keV protons colliding with atomic hydrogen, comparing the very recent measurements of Kerby et al. (76) with the Born approximation, CTMC results, and the Born-corrected CTMC approximation. The figure shows the well-known underestimation of the forward spectrum of electrons by the Born approximation, due to the lack of saddle-point and ECC-like electron emission. In addition, the longstanding underestimation of results of the CTMC method for backward angles compared to experiment is seen to have just been due to this lack of electrons produced in very small momentum transfer collisions. This effect also accounts for the difference between the quantum mechanical rate of drop of the total ionization cross section ($\ln E/E$) as a function of impact energy, E, as compared to the classical result ($1/E$).

B. Collisions of F^{9+} with He

Another strenuous test of theoretical approaches was recently posed by the detailed experimental measurement of the ejected electron spectrum in collisions of 1.5 MeV/u F^{9+} with helium made at zero degrees in the laboratory by the group at Kansas State University (52). In Figure 7 we show a comparison of these data with various theoretical approaches. An extensive CTMC calculation (77) for this system gave excellent agreement with the experimental measurements, much better in fact than any quantum mechanical perturbation theory result. This theoretical work (77) also emphasized the sensitivity of the CDW-EIS approximation, and by implication any perturbation theory, to the choice of initial and final effective charges. Certain common choices of these parameters which are not readily justifiable because they give non-orthogonal initial and final states, fit the experiment better than more justifiable choices. Inclusion of a model potential, which circumvents the need to make these choices of effective charges, was shown to substantially improve the CDW-EIS result. Further, the conventional effective charge CDW-EIS and distorted-wave strong potential Born (DSPB) approximation of Brauner and Macek (78) were shown to yield essentially identical results, even though the starting points of the two approaches are significantly different.

We note the excellent agreement of the CTMC calculations with experiment as to the asymmetry of the ECC cusp, evidence of the good representation of the distribution of electronic angular momenta in the low-lying continuum of the projectile. The model potential CDW-EIS improves over the conventional CDW-EIS regarding the cusp asymmetry and yields excellent agreement with the experiment around the binary peak. All the two-center treatments predict the shift of the binary peak's position to lower energies with respect to one-center treatments, the model potential CDW-EIS and CTMC methods giving the best agreement with the experiment.

C. Collisions of C^+ with He

When the electrons carried by a partially stripped ion are only weakly bound, or bound by an amount comparable to that of the target electrons, a strong contribution from projectile electron loss is found in the ejected electron spectrum, as well as in the total cross section. Early work concentrated on separating the total cross sections for projectile and target electron

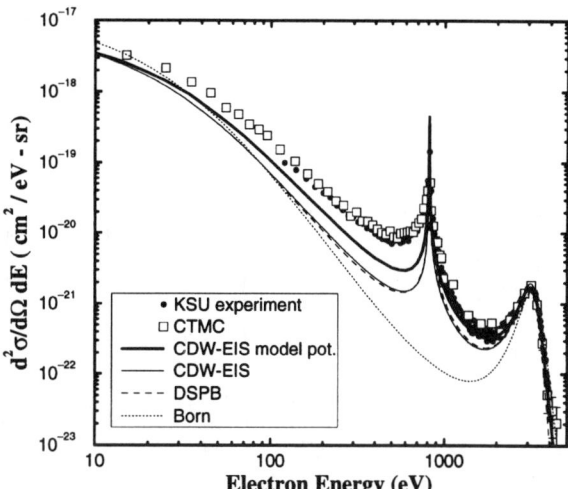

FIG. 7. The ejected electron spectrum in collisions of 1.5 MeV/u F^{9+} ions with helium at an ejection angle of zero degrees. The experimental measurements of the Kansas State Group (52) are compared with results of several theoretical approaches. (This figure is adapted from Ref. (77)).

loss (79–81) by comparing the yield of electrons in partially stripped ion impact to those found when bare projectiles were used. Important progress was made with the advent of coincident measurements of the charge states of the projectile and target so as to determine the loss from each simultaneously (82–85). Considerable insight has been gained through these works as to the relative contribution of the electron-capture-to-the-continuum and electron-loss-to-the-continuum mechanisms.

Despite these important experimental measurements, only few theoretical works on the subject had appeared by 1990, such as the first Born calculations of Manson and coworkers (82,86) (see also the contributions by Manson and by Montenegro in this volume). In light of the fact that the studies we have described above demonstrated the inability of one-center treatments to give an overall good description of the ejected electron spectra at intermediate energies, we tested in a joint theory-experiment collaboration the ability of the CTMC technique to describe the evolution of electrons ejected from both the target and projectile.

We have assumed that the role of the electron-electron interaction is small compared with the interaction of the electrons with the ionic cores, and therefore have added incoherently two three-body calculations. In the first, the three bodies were the projectile ion, the target electron, and the target core. The interaction of the electron with each ion was given by a model potential, accounting for the screening provided by the other (spectator) electrons. In the second, the role of target and projectile were reversed, in that the electron was initially bound to the projectile.

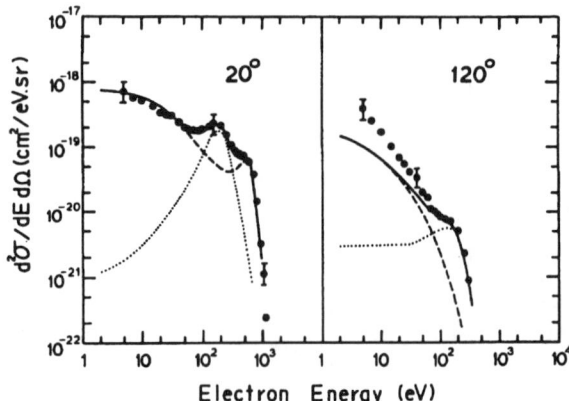

FIG. 8. The ejected electron spectrum in collisions of C^+ ions with helium at a collision energy of 350 keV/u for ejection angles of 20° and 120°. The theoretical total yield of electrons (full curve) is divided into electron emission from He (broken curve) and C^+ (dotted curve), while the filled circles represent the experimental measurements. (This figure reprinted from Ref. (87)).

Such a calculation was performed for 66.7 to 350 keV/u C^+ ions colliding with He (87,88) and compared with experiments performed by Toburen and DuBois. Very good overall agreement between theory and experiment was found regarding both singly and doubly differential cross sections for several impact energies in the stated range. In Figure 8 we illustrate how the experimental ejected electron spectrum could be separated into the yield of electrons from either the target or projectile. In general terms, the spectrum arising from projectile ionization displays a peak at electron velocities close to v_p. At forward angles, it forms a relatively sharp cusp and is termed the electron-loss-to-the-continuum (ELC). The ELC peak is evident in Figure 8 for an ejection angle of 20° as the peak in the experimental measurements. Also evident is a contribution at backward angles (e.g., 120° in Figure 8) which arises from binary collisions between projectile electrons and the target core (backward projectile electron binary peak (89,90)). Thus, the features observed in the experimental measurements may be explained as arising from a combination of projectile and target electron emission. We also note the underestimation of the experiment by the CTMC result for low electron energies and large ejection angles, explanation of which was described above.

IV. CONCLUSIONS

Thus, study of the spectrum of electrons ejected in intermediate energy ion-atom collisions is a sensitive probe of the interactions governing atomic collisions. It constitutes a fundamental exploration of physics on the atomic level, and of the basic information needed for modeling and diagnosis of a wide variety of practical applications. Prof. Rudd's contributions to this field have been numerous and pioneering. His work, and the work of others, has led to a fuller understanding of ionization in general, and of the two-center effects which dominate much of the spectrum of ejected electrons in particular.

ACKNOWLEDGEMENTS

Support for the theoretical works described here has been provided by grants from the Office of Fusion Energy, U.S. Department of Energy, to the University of Missouri-Rolla, and by that Office and the Office of Basic Energy Sciences, U.S. Department of Energy, under Contract No. DE-AC05-84OR21400 with Oak Ridge National Laboratory, managed by Martin Marietta Energy Systems, Inc.

REFERENCES

1. D.S.F. Crothers and J.F. McCann, J. Phys. B **16**, 3229 (1983).
2. R.E. Olson, Phys. Rev. A **33**, 4397 (1986).
3. N. Stolterfoht, D. Schneider, J. Tanis, H. Altervogt, A. Salin, P.D. Fainstein, R. Rivarola, J.P. Grandin, J.N. Scheurer, S. Andriamonje, D. Bertault, and J.F. Chemin, Europhys. Lett **4**, 899 (1987).
4. P.D. Fainstein, V.H. Ponce, and R.D. Rivarola, J. Phys. B **21**, 287 (1988).
5. C.O. Reinhold and R.E. Olson, Phys. Rev. A **39**, 3861 (1989).
6. M.E. Rudd and T. Jorgensen, Phys. Rev. **131**, 666 (1963).
7. M.E. Rudd, C.A. Sautter, and C.L. Bailey, Phys. Rev. **151**, 20 (1966).
8. M.E. Rudd, L.H. Toburen, and N. Stolterfoht, At. Data Nucl. Data Tables **18**, 413 (1976).
9. M.E. Rudd, L.H. Toburen, and N. Stolterfoht, At. Data Nucl. Data Tables **23**, 405 (1979).
10. M.E. Rudd, Radiat. Res. **64**, 153 (1971).
11. N. Stolterfoht, in *Structure and Collisions of Ions and Atoms*, Vol. 5 of *Topics in Current Physics*, edited by I.A. Sellin (Springer-Verlag, New York, 1978), pp. 155-199.
12. D. Berényi, Vacuum **37**, 53 (1987).
13. M.E. Rudd, Y.-K. Kim, D.H. Madison, and T.J. Gay, Rev. Mod. Phys. **64**, 441 (1992).
14. R.E. Olson, Phys. Rev. A **27**, 1871 (1983).
15. R. Abrines and I.C. Percival, Proc. Phys. Soc. **88**, 861 (1966).
16. R.E. Olson and A. Salop, Phys. Rev. A **16**, 531 (1977).
17. T.G. Winter and C.D. Lin, Phys. Rev. A **29**, 3071 (1984).
18. W. Meckbach, P.J. Focke, A.R. Goni, S. Suarez, J. Macek, and M.G. Menendez, Phys. Rev. Lett. **57**, 1587.
19. W. Meckbach, I.B. Nemirovsky, and C.R. Garribotti, Phys. Rev. A **24**, 1793 (1981).
20. G.C. Bernardi, P. Focke, S. Suarez, and W. Meckbach, *Lecture Notes in Physics*, ed. by D. Berényi and G. Hock (Spinger-Verlag, Berlin, 1988), pp. 295-9.
21. R.E. Olson, T.J. Gay, H.G. Berry, E.B. Hale, and V.D. Irby, Phys. Rev. Lett. **59**, 36 (1987).
22. M.E. Rudd and D.H. Madison, Phys. Rev. A **14**, 128 (1976).
23. V.D. Irby, T.J. Gay, J.W. Edwards, E.B. Hale, M.L. McKenzie, and R.E. Olson, Phys. Rev. A **37**, 3612 (1988).

24. G.C. Bernardi, S. Suarez, P.D. Fainstein, C.R. Garibotti, W. Meckbach, and P. Focke, Phys. Rev. A **40**, 6863 (1989).
25. G.C. Bernardi, P.D. Fainstein, C.R. Garibotti, and S. Suarez, J. Phys. B **23**, L139 (1990).
26. R.D. DuBois, Phys. Rev. A **48**, 1123 (1993).
27. T.J. Gay, M.W. Gealy, and M.E. Rudd, J. Phys. B **23**, L823 (1990).
28. V.D. Irby, S. Datz, P.F. Dittner, N.L. Jones, H.F. Krause, and C.R. Vane, Phys. Rev. A **47**, 2957 (1993).
29. R.D. DuBois, Phys. Rev. A **50**, 364 (1994).
30. M. Pieksma, *Saddle Point Electrons in Slow Ion-Atom Collisions*, Ph.D. Thesis University of Utrecht (1993); M. Pieksma, S.Yu. Ovchinnikov, J. van Eck, W.B. Westerveld, Phys. Rev. Lett. **73**, 46 (1994).
31. G.B. Crooks and M.E. Rudd, Phys. Rev. Lett. **25**, 1599 (1970).
32. K.G. Harrison and M.W. Lucas, Phys. Lett. **33A**, 142 (1970).
33. A. Salin, J. Phys. B **2**, 1225 (1969).
34. J.H. Macek, Phys. Rev. A **1**, 235 (1970).
35. S. Yu. Ovchinnikov and D.B. Khrebtukov, *Abstracts of Contributed Papers of the 15th International Conference on the Physics of Electronic and Atomic Collisions, Brighton, United Kingdom, 1987*, edited by H.B. Gilbody et al. (North-Holland, Amsterdam, 1987), p. 596.
36. J. Burgdörfer, in *Interaction of Charged Particles with Solids and Surfaces*, edited by Gras-Marti et al. (Plenum Press, New York, 1991), p. 459.
37. C.O. Reinhold and R.E. Olson, J. Phys. B **22**, L39 (1989).
38. C.O. Reinhold and D.R. Schultz, J. Phys. B **22**, L565 (1989).
39. J.S. Briggs, Nucl. Instrum. Methods A **240**, 577 (1985).
40. J.S. Briggs, J. Phys. B **19**, 2703 (1986).
41. M. Brauner and J.S. Briggs, J. Phys. B **19**, L325 (1986).
42. P. Mandal, K. Roy, and N.C. Sil, Phys. Rev. A **33**, 756 (1986).
43. N.C. Sil, K. Roy, and P. Mandal, *Abstracts of Contributed Papers of the 17th International Conference on the Physics of Electronic and Atomic Collisions, Brisbane, 1991*, edited by I.E. McCarthy, W.R. MacGillivray, and M.C. Standage (Griffith University, Brisbane, 1991), p. 354.
44. D.R. Schultz and C.O. Reinhold, J. Phys. B **23**, L9 (1990).
45. J. Moxom, G. Larrichia, M. Charlton, G.O. Jones, and Á. Kövér, J. Phys. B **25**, L613 (1992).
46. Y. Yamazaki, K.-I. Kuroki, L.H. Andersen, E. Horsdal-Pedersen, P. Hvelplund, H. Knudsen, S.P. Moller, E. Uggerhoj, and K. Elsener, J. Phys. Soc. Japan **59**, 2643 (1990).
47. P. Guang-yan, P. Hvelplund, H. Knudsen, Y. Yamazaki, M. Brauner, and J.S. Briggs, Phys. Rev. A **47**, 1531 (1993).
48. P.D. Fainstein, V.H. Ponce, and R.D. Rivarola, J. Phys. B **21**, 2989 (1988).
49. C.O. Reinhold and D.R. Schultz, Phys. Rev. A **40**, 7373 (1989).
50. L.H. Toburen and W.E. Wilson, Phys. Rev. A **19**, 2214 (1979).
51. L.H. Toburen, W.E. Wilson, and R.J. Popowich, Rad. Res. **82**, 27 (1980).
52. D.H. Lee, P. Richard, T.J.M. Zouros, J.M. Sanders, J.L. Shinpaugh, and H. Hidmi, Phys. Rev. A **41**, 4816 (1990).
53. N. Stolterfoht, D. Schneider, D. Bruch, H. Wieman, and J.S. Risley, Phys. Rev. Lett. **33**, 59 (1974).
54. L.H. Toburen, N. Stolterfoht, P. Ziem, and D. Schneider, Phys. Rev. A **24**, 1741

(1981).
55. P. Richard, D.H. Lee, T.J.M. Zouros, J.M. Sanders, and J.L. Shinpaugh, J. Phys. B **23**, L213 (1990).
56. C. Kelbch, S. Hagmann, S. Kelbch, R. Mann, R.E. Olson, S. Schmidt, and H. Schmidt-Böcking, Phys. Lett. **139A**, 304 (1989).
57. T.F.M. Bonsen and L. Vriens, Physica **47**, 307 (1970).
58. C.O. Reinhold, D.R. Schultz, and R.E. Olson, J. Phys. B **23**, L591 (1990): see also R.E. Olson, C.O. Reinhold, and D.R. Schultz, J. Phys. B **23**, L455 (1990).
59. T. Quinteros and J.F. Reading, private communication (1991).
60. R. Shingal, Z. Chen, K.R. Karim, C.D. Lin, and C.P. Bhalla, J. Phys. B **23**, L637 (1990).
61. K. Taulbjerg, J. Phys. B **23**, L761 (1990).
62. C.O. Reinhold, D.R. Schultz, R.E. Olson, C. Kelbch, R. Koch, and H. Schmidt-Böcking, Phys. Rev. Lett **66**, 1842 (1991).
63. D.R. Schultz and R.E. Olson, J. Phys. B **24**, 3409 (1991).
64. O. Jagutzki, S. Hagmann, H. Schmidt-Böcking, R.E. Olson, D.R. Schultz, R. Dörner, R. Koch, A. Skutlartz, A. González, T.B. Quinteros, C. Kelbch, and P. Richard, J. Phys. B **24**, 2579 (1991).
65. W. Wolff, J.L. Shinpaugh, H.E. Wolf, R.E. Olson, U. Bechtold, and H. Schmidt-Böcking, J. Phys. B **26**, L65 (1993).
66. Dž. Belkić, J. Phys. B **11**, 3529 (1978).
67. P.D. Fainstein, V.H. Ponce, and R.D. Rivarola, J. Phys. B **24**, 3091 (1991).
68. J.H. Macek, J. Phys. B **24**, 5121 (1991) and references therein.
69. K. Taulbjerg, R.O. Barrachina, and J.H. Macek, Phys. Rev. A **41**, 4816 (1990).
70. W. Wolff, H.E. Wolf, J.L. Shinpaugh, J. Wang, R.E. Olson, P.D. Fainstein, S. Lencinas, U. Bechthold, R. Herrmann, and H. Schmidt-Böcking, J. Phys. B **26**, 4169 (1993).
71. J.O.P. Pederson, P. Hvelplund, A.G. Petersen, and P.D. Fainstein, J. Phys. B **23**, L597 (1990): J. Phys. B **24**, 4001 (1991).
72. N. Bohr and J. Lindhard, K. Dan. Vidensk. Selsk. Mat. Fys. Medd. **28**, No. 7 (1954).
73. P.D. Fainstein, V.H. Ponce, and R.D. Rivarola, Phys. Rev. A **45**, 6417 (1992).
74. D.R. Schultz, R.E. Olson, C.O. Reinhold, M.W. Gealy, G.W. Kerby, Y.-Y. Hsu, and M.E. Rudd, J. Phys. B **24**, L599 (1991).
75. C.O. Reinhold and J. Burgdörfer, J. Phys. B **26**, 3101 (1993).
76. G.W. Kerby, M.W. Gealy, Y.-Y. Hsu, M.E. Rudd, D.R. Schultz, and C.O. Reinhold, Phys. Rev. A (1995) to be published.
77. D.R. Schultz and C.O. Reinhold, Phys. Rev. A **50**, 2390 (1994).
78. M. Brauner and J.H. Macek, Phys. Rev. A **46**, 2519 (1992).
79. Á. Kövér, D. Varga, Gy. Szabó, D. Berényi, I. Kadar, S. Ricz, J. Vegh, and G. Hock, J. Phys. B **16**, 1017 (1983).
80. J. Schader, R. Latz, M. Burkhard, H.J. Frischkorn, D. Hoffman, P. Koschar, K.O. Groeneveld, D. Berényi, Á. Kövér, and Gy. Szabó, J. Physique Lett. **45**, L249 (1984).
81. J. Schader, R. Latz, M. Burkhard, H.J. Frischkorn, D. Hoffman, P. Koschar, K.O. Groeneveld, D. Berényi, Á. Kövér, and Gy. Szabó, Nuovo Cimento **7**, 219 (1986).
82. R.D. DuBois and S.T. Manson, Phys. Rev. Lett **57**, 1130 (1986).
83. O. Heil, J. Kemmler, K. Kroneberg, Á. Kövér, Gy. Szabó, L. Gulyas, R. De-

Serio, S. Lencinas, N. Keller, D. Hoffman, H. Rothard, D. Berényi, and K.O. Groeneveld, J. Phys. B **16**, 1017 (1983).
84. L. Sarkadi, J. Pálinkás, Á. Kövér, D. Berényi, and T. Vajnai, Phys. Rev. Lett. **62**, 527 (1989).
85. Á. Kövér, L. Sarkadi, L. Gulyas, Gy. Szabó, T. Vajnai, D. Berényi, O. Heil, K.O. Groeneveld, J. Gibbons, and I.A. Sellin, J. Phys. B **22**, 1595 (1989).
86. S.T. Manson and L.H. Toburen, Phys. Rev. Lett. **46**, 529 (1981).
87. C.O. Reinhold, D.R. Schultz, R.E. Olson, L.H. Toburen, and R.D. DuBois, J. Phys. B **23**, L297 (1990).
88. L.H. Toburen, R.D. DuBois, C.O. Reinhold, D.R. Schultz, and R.E. Olson, Phys. Rev. A **42**, 5338 (1990).
89. W.E. Wilson and L.H. Toburen, Phys. Rev. A **7**, 1535 (1973).
90. D. Burch, H. Wieman, and W.B. Ingalls, Phys. Rev. Lett. **30**, 823 (1973).

Passive and Active Electrons in the Electron Loss Process

E.C.Montenegro[1]
Departamento de Física, Pontifícia Universidade Católica do Rio de Janeiro
Caixa Postal 38071, Rio de Janeiro, RJ 22452, Brazil

W.E.Meyerhof
Department of Physics, Stanford University, Stanford, CA 94305

Abstract. The passive (screening) and active (antiscreening) roles of the target electrons in the electron loss process is reviewed. The semiclassical approximation is used to show that the screening and antiscreening modes are related to the form in which the coherence of the various parts of the target electron cloud is considered in the electron loss process. These concepts are applied to projectile electron loss in 2.0-MeV/u C^{5+} + He collisions.

Introduction

Energetic atomic collisions involving multielectron atoms or ions quite frequently result in the excitation or ionization of several electrons of the two collision partners. In general, the mechanisms leading to these multielectronic excitations can be described in simple terms only if the independent-electron approximation can be used. If correlation effects are present and the independent-particle model fails, the theoretical analysis of these mechanisms can be very difficult and, in some cases, even a simple conceptual physical picture of the problem is not available.

If an ion carrying one or more electrons collides with a neutral, light atom, the projectile ionization (electron loss) is due to the joint action of the target nucleus and of the target electrons. The interaction between the projectile active electron and the target electrons can result in the simultaneous ionization of the projectile and the target due to a correlated process (antiscreening). This two-center electron-electron contribution to the electron loss is one of the few (if not a unique) correlated multielectron processes which can be successfully understood within first-order theories. Because of that, it is useful

---
[1]On leave at the J. R. Macdonald Laboratory, Department of Physics, Kansas State University, Manhattan, KS 66506.

for the understanding of more complex cases that alternative pictures for this mechanism be explored.

A theoretical treatment for two-center processes in atom-atom collisions was given originally by Bates and Griffing about forty years ago (1-3). They used the plane-wave Born approximation (PWBA) to study the H + H system, providing a basic methodology which has been followed by many authors in an effort to generalize the application of the PWBA to multielectron atoms (see Ref. 4 for a recent review). Only recently, the electron loss processes have been studied within an alternative scenario, using the semiclassical approximation (SCA) (5-7). In this paper, some new features of the loss process which emerge from the SCA approach are discussed and applied to the impact parameter dependence of projectile electron loss in 2.0-MeV/u C^{5+} + He collisions.

Basic Ideas

A collision between a charged projectile and a neutral atom (with nuclear charges Z_1 and Z_2, respectively) resulting in projectile electron loss, is governed by two different and competing mechanisms which are directly related to the state of the target after the collision. In the first mechanism, the projectile electron loss is induced by a (target) nucleus–(projectile) electron interaction, with the target electrons assuming the *passive* role of screening the coulomb field of the target nucleus. This is called the *screening mode*, and in this mode the target remains in its ground state after the collision. In the second mechanism, the loss process is induced by a (target) electron–(projectile) electron interaction, and the target atom ends the collision in an excited state which usually is not observed and which is most likely to be the target continuum. This is called *antiscreening mode*, and in this mode the target electrons are the *active* agents responsible for the electron loss. Figure 1 illustrates these two modes.

The above description of the electron loss process reflects the way it is considered in the PWBA. Although the calculations can be performed with relative simplicity, the PWBA scheme does not allow the main dynamic features of the screening and antiscreening processes to be obtained from a simple physical reasoning. An alternative scenario of the antiscreening mode can be obtained through the Impulse Approximation (IA) (8,9), in which the loss collision is observed in the *projectile frame* and the target electrons are consid-

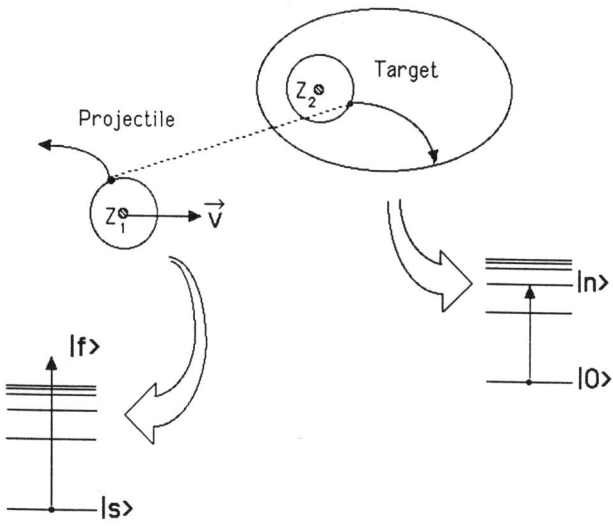

Fig. 1. Diagrammatic representation of first-order electron loss mechanisms. While the active projectile electron goes from a state $|s\rangle$ to a state $|f\rangle$, the target atom can stay in the ground state $|0\rangle$ (screening mode) or be excited into a state $|n\rangle$ (antiscreening mode).

ered to be a beam of particles participating, in the same manner as the target nucleus, as ionizing agents of the projectile. The bound nature of the target electrons is taken into account through the Compton profile of the target atom, which gives a realistic velocity broadening of the electron beam impinging on the projectile.

In the present paper a different point of view is adopted. We also consider the loss process in the projectile frame, but the target electrons are now described in *configuration space* instead of momentum space, as it is done in the IA. In other words, the target cloud is subdivided into volume elements with different densities $|\langle\boldsymbol{\xi}|0\rangle|^2 = |\Phi(\boldsymbol{\xi})|^2$, where $\boldsymbol{\xi}$ is the target electron coordinate with respect to the target nucleus, but with the *same velocity* as the target nucleus. Comparing the two approaches, we can say that in the present case there is a "density broadening", instead of a momentum broadening as in the IA.

This situation is pictured in Fig.2, which defines the electron coordinates

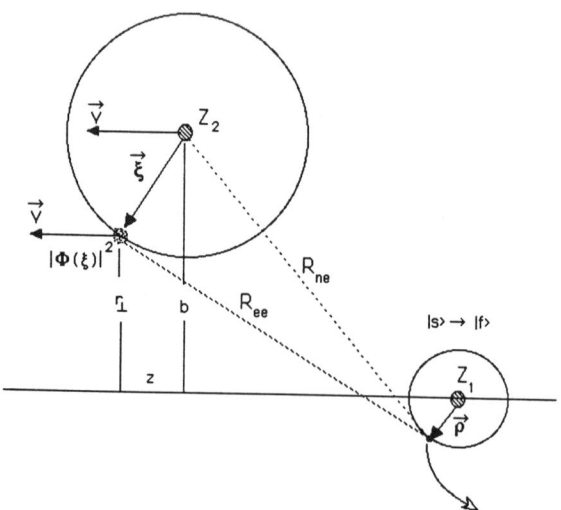

Fig. 2. Diagrammatic representation of the electron loss process in the projectile frame. The impact parameter for the nucleus-electron interaction (R_{ne}) is b, and $r_\perp$ is the impact parameter associated with a particular volume element of the target electron cloud for the electron-electron interaction (R_{ee}). The coordinates of the target and projectile electrons are given by ξ and ρ, respectively.

as well the interactions involving the projectile electron, represented by dashed lines. For simplicity, the target and the projectile are considered one-electron systems. In this picture, each volume element $|\Phi(\xi)|^2$ behaves as an independent particle with the same relative velocity v as the target nucleus ($v=$ projectile velocity) and follows a straight line trajectory. The transition amplitude, as a function of a generic impact parameter $\mathbf{x}$, for the projectile being excited from a state $|s\rangle$ to a state $|f\rangle$ by a unit charge is given in the SCA by (4-6):

$$a_{\text{sca}}(\mathbf{x}) = \frac{ie^2}{\hbar} \int_{-\infty}^{+\infty} dt\, e^{i\omega t} \langle f|\frac{1}{R}|s\rangle \tag{1}$$

where, in our case, the impact parameter $\mathbf{x}$ (implicitly considered in R) can be $\mathbf{b}$ or $\mathbf{r_\perp}$ and R can be R_{ne} or R_{ee}, respectively, depending whether the nucleus-electron or the electron-electron interaction is considered.

The question now arises how the sum over the volume elements of the cloud should be performed. The various volume elements can be added *coherently* with respect to each other as well as with respect to the target nucleus, or can be added *incoherently* with respect to each other. We perform the sum in both ways and show that the first case corresponds to the screening mode and the second to the antiscreening.

Screening

If the volume elements displayed in Fig. 2 are added coherently, the relative phases between them are preserved during the collision process. This assumption is equivalent to supposing that the target electrons remain in the ground state, i.e., we are considering the screening mode. Furthermore, momentum is transferred to the projectile electron as if the target nucleus *and* the target electrons have an effectively infinite mass. This large inertia associated with the target electron cloud is related to its bound nature: the momentum received during the projectile excitation process is shared with the target nucleus, to which the cloud is and remains bound.

To keep the coherence between the elements of the cloud and the target nucleus, a phase $e^{i\omega z/v}$ must be included in the probability amplitudes, to account for the shift $(= z)$ in their relative positions along the beam direction (see Fig. 2). The phase shift due to the differences between the impact parameters $\mathbf{b}$ and $\mathbf{r}_\perp$ is already taken into account in the amplitudes $a_{\text{sca}}(\mathbf{x})$. Within this scenario, the (screening) loss probability becomes:

$$P_{\text{S}} = |Z_2 a_{\text{sca}}(\mathbf{b}) - \int d^3\xi \, |\Phi(\boldsymbol{\xi})|^2 \, e^{i\omega z/v} a_{\text{sca}}(\mathbf{r}_\perp)|^2, \qquad (2)$$

which can be rewritten as:

$$P_{\text{S}} = |Z_2 a_{\text{sca}}(\mathbf{b}) - \langle 0|e^{i\omega z/v} a_{\text{sca}}(\mathbf{r}_\perp)|0\rangle|^2. \qquad (3)$$

Using Eq. (1), the above equation involving the probability amplitudes can be written in terms of the interaction potentials as (5,6):

$$P_{\text{S}} = \left| \frac{ie^2}{\hbar} \int_{-\infty}^{+\infty} dt \, e^{i\omega t} \langle f| \frac{Z_2}{R_{\text{ne}}} - \langle 0|\frac{1}{R_{\text{ee}}}|0\rangle |s\rangle \right|^2 \qquad (4)$$

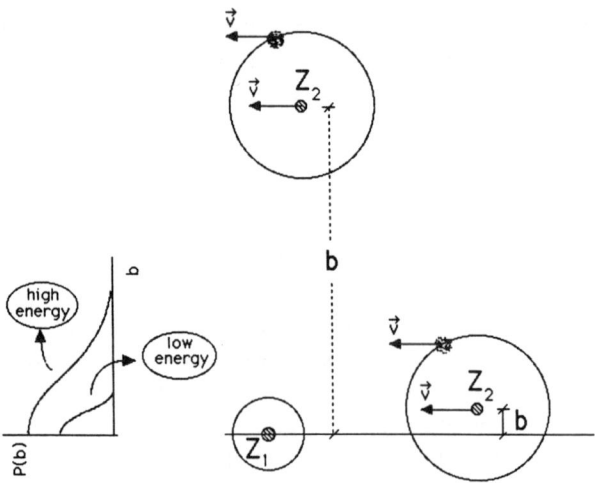

Fig. 3. Diagrammatic representation of the electron loss process, in the projectile frame, for two different impact parameters. The loss probability $P(b) = |a_{\text{sca}}|^2$ for a point-like particle is sketched for high- and low-energy collisions.

$$= \left| \frac{ie^2}{\hbar} \int_{-\infty}^{+\infty} dt\, e^{i\omega t} \langle f|V_{\text{eff}}|s\rangle \right|^2. \tag{5}$$

Here, $V_{\text{eff}} = Z_2/R_{\text{ne}} - \langle 0|1/R_{\text{ee}}|0\rangle$, i.e., V_{eff} is the screened Coulomb potential of the target atom as seen by the projectile electron, and Eq. (5) is the expression one expects for the probability associated with the screening mode. The picture which emerges from Eq. (5) is that of an electron cloud assuming the passive role of screening the nuclear Coulomb field of the target. This apparently *passive* behavior, however, can be viewed as an *active and coherent* action of the target cloud and nucleus on the projectile electron.

The usefulness of having these two pictures available can be illustrated with the help of Fig. 3, in which the collision is displayed in the projectile frame for two different impact parameters. For distant collisions, we can easily see the effect of the target screening decreasing the probability of loss when compared with a "bare target", either by considering the decrease of the net Coulomb field of the target nucleus in the vicinity of the projectile electron, or by considering that the interference between the target nucleus and the target cloud is almost completely destructive, because these two components have practically the same impact parameter.

For close collisions, the dynamic point of view is particularly useful to examine under which conditions screening is important. For slow collisions, the range of impact parameters which gives significant contributions to the projectile electron loss is restricted to the adiabatic radius $\hbar v/\Delta E$ (10), where in the present case ΔE is the energy transfer to the projectile electron. Since this radius becomes smaller as the velocity v decreases, only collisions in which the ionizing particle is close enough to the projectile nucleus contribute to projectile electron loss. It can be seen from Fig. 3, that under these conditions the interference with the target nucleus is significant only for a small part of the cloud and, as a consequence, the effective screening is small. On the other hand, for fast collisions, the range of impact parameters in which the ionizing collision is effective is broader and a larger portion of the cloud participates in electron loss. This enhances the destructive interference with the target nucleus and increases its effective screening, even at impact parameters close to zero.

Antiscreening

Let us return to Fig. 2 and consider the case in which the various volume elements of the cloud are added *incoherently*. Now, each element does not need to keep the relative phase with respect to the others as the collision proceeds, and the total contribution of the cloud is obtained by adding the ionization *probabilities* instead of the ionization amplitudes. Furthermore, the active target electron is now allowed to recoil as a result of its interaction with the projectile electron. This means not only that it will not end the collision in its ground state, but also that the excitation dynamics must be evaluated using the electron rest mass instead of an infinite (nuclear) mass as was done in the screening case. The threshold associated with the antiscreening mode (4,10) then appears naturally as a result of this fact. Also, there is no restriction on the final state of the target electron: the momentum it receives is adjusted to the dynamics of the ejected projectile electron. This situation is equivalent to considering that the final target state is not observed and that a sum over all target states must be carried out.

With these assumptions, the loss probability can be written as:

$$P_{A1} = \int d^3\xi \, |\Phi(\xi)|^2 \, |a_{sca}(\mathbf{r}_\perp)|^2 \tag{6}$$
$$= \langle 0| \, |a_{sca}(\mathbf{r}_\perp)|^2 |0\rangle. \tag{7}$$

Equation (7) gives the loss probability for all possible target final states, including the ground state. In order to obtain the antiscreening probability we must subtract the contribution from the ground state, i.e., the coherent sum over the cloud. Then, we have:

$$P_A = \langle 0| \, |a_{sca}(\mathbf{r}_\perp)|^2 |0\rangle - |\langle 0|a_{sca}(\mathbf{r}_\perp)e^{i\omega z/v}|0\rangle|^2 \tag{8}$$

$$= \sum_{n\neq 0} |\langle n|a_{sca}(\mathbf{r}_\perp)e^{i\omega z/v}|0\rangle|^2. \tag{9}$$

Here, the relation $\sum_{n\neq 0} |\langle n|D|0\rangle|^2 = \langle 0||D|^2|0\rangle - |\langle 0|D|0\rangle|^2$ has been used, which is valid for any sum D of single-particle operators (6).

We are now in position to analyze the two situations displayed in Fig. 3 with respect to the antiscreening mode. For this mode, the target nucleus does not contribute and the electron loss of the projectile is due only to the action of the target electron cloud. However, it is convenient to keep in mind what would be the effect of the nucleus alone, for purpose of comparison. For large impact parameters, the scattering probability is a decreasing function of b. Because the cloud is spread out, there is a part of it which has a smaller impact parameter than its center (where the target nucleus is placed) and this part increases the probability for antiscreening at large impact parameters when compared with that given by a point charge.

At small impact parameters, the semiclassical ionization probability $|a_{sca}|^2$ is maximum. The situation is then reversed with respect to the large impact parameter case. Because the cloud is spread out, there is a part of the cloud which is not located at small impact parameters, giving for the cloud a net contribution which is smaller than that given by a point particle.

Figure 4 gives a quantitative summary of the concepts described in this paper. The probability of electron loss of C^{5+} on He at 2.0 MeV/u is shown as a function of the impact parameter. It can be clearly seen, at large impact parameters, the effect of the destructive interference decreasing the screening mode probability when compared with that of a bare nucleus. One also recog-

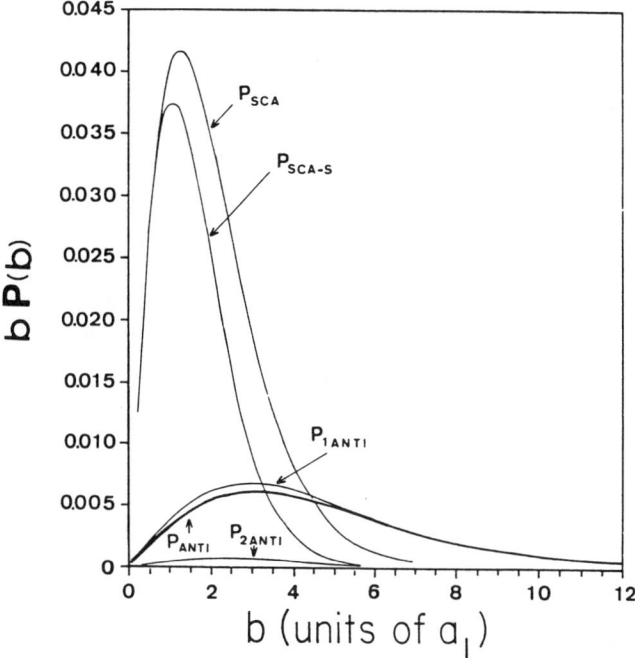

Fig. 4. Probability of loss for C^{5+} on He. P_{SCA} is the probability of loss by a bare target nucleus; P_{SCA-S} is the loss probability for the screening mode; P_{ANTI} is the loss probability for the antiscreening mode; P_{1ANTI} and P_{2ANTI} correspond to the first and second terms of Eq. (8), respectively. From Ref. 6.

nizes the effect of incoherent sum over the target electron cloud, broadening the probability distribution, and resulting in a predominance of the antiscreening mode at large impact parameters.

Acknowledgments

This work was supported in part by the Division of Chemical Sciences, Office of Basic Energy Sciences, Office of Energy Research, U.S. Department of Energy (Kansas State), by National Science Foundation Grants Nos. PHY-9019293 and INT-9101057 (Stanford), and by the CNPq (Brazil). E.C.M. gratefully acknowledges the specially kind hospitality granted during his stay at the J. R. Macdonald Laboratory.

References

1. Bates, D.R., and Griffing, G., *Proc. Phys. Soc. London* A **66**, 961 (1953).
2. Bates, D.R., and Griffing, G., *Proc. Phys. Soc. London* A **67**, 663 (1954).
3. Bates, D.R., and Griffing, G., *Proc. Phys. Soc. London* A **68**, 90 (1955).
4. Montenegro, E.C., Meyerhof, W.E., and McGuire, J.H., *Adv. At. Mol. Opt. Phys.* **33**, (1994), to be published.
5. Montenegro, E.C., and Meyerhof, W.E., *Phys. Rev.* A **44**, 7229 (1991).
6. Montenegro, E.C., and Meyerhof, W.E., *Phys. Rev.* A **46**, 5506 (1992).
7. Ricz, S., Sulik, B., Stolterfoht, N., and Kadar, J., *Phys. Rev.* A **48**, 1930 (1993).
8. Zouros, T.J.M., Lee, T.H., and Richard, P., *Phys. Rev. Lett.* **62**, 2261 (1989).
9. Zouros, T.J.M., Lee, T.H., Sanders, J.M., and Richard, P., *Nucl. Inst. Meth. Phys. Res.* B **79**, 166 (1993).
10. Montenegro, E.C., and de Pinho, A.G., *J. Phys. B: At. Mol. Phys.*, **15**, 1521 (1983).
11. Anholt, R., *Phys. Lett.* **114A**, 126 (1986).

Comment

After this talk was given, J. Macek remarked that the electron densities in the screening probability P_S, Eq. (3), and in the antiscreening probability P_A, Eq. (9), differ in an important aspect. In P_S, the density is the ground-state density, which is coherent. In P_A, the electron density is a transition density summed over all electronic states $n > 0$. These densities are incoherent, even though closure allows their sum to be expressed in terms of the ground-state density [line below Eq. (9)].

III. HIGHLY-CHARGED PROJECTILES AND TWO-CENTER DESCRIPTIONS

Classical-Quantum Correspondence for Ionization in Fast Ion-Atom Collisions

Joachim Burgdörfer and Carlos O. Reinhold

University of Tennessee, Knoxville TN 37996-1200, USA and Oak Ridge National Laboratory, Oak Ridge, TN 37831-6377, USA

Abstract. We analyze the interplay between classical and quantum dynamics in ionization of atoms by fast charged particles. The convergence to the classical limit is studied as a function of the momentum transferred to the electron during the collision, the impact parameter, the energy and angle of the emitted electron, and the initial state of the target. One goal is to assess the validity of exact classical (CTMC) methods and approximate classical models such as the Thomson model. Applications to data for electron ejection at large angles are presented. The connection between collisional ionization by charged particles and ionization by half-cycle pulses is discussed.

1. Introduction

The theoretical understanding of ionization caused by strong and non-periodic pulses is of fundamental interest. Such pulses can be delivered either by the time-dependent Coulomb field of fast ions passing by an atom or by ultrashort laser pulses with durations corresponding to a fraction of an optical cycle. For fast ion-atom collisions, well established approximations exist at high impact velocities (v_p) or small projectile charges (Z_p) for which the interaction of the target with the impinging ion can be treated as a small perturbation. For collision velocities comparable to the orbital velocity of target electrons and projectile charges of the order of or greater than the target nuclear charge (Z_t), more elaborate approaches are necessary which can treat the target and projectile nuclear fields on an equal footing and can account for the strong coupling among all reaction channels. It is under these circumstances that classical trajectory Monte Carlo (CTMC) methods[1-3] have become widely and successfully utilized. The principal assumption underlying these models is that during and after the collision all interacting particles evolve classically in time. Quantum mechanical features enter solely via the initial boundary conditions of the evolution, i.e. the atomic initial state.

A detailed understanding of the range of validity of these methods is of considerable interest for both fundamental and application-oriented reasons. For example, with the availability of state-selectively excited Rydberg states with sufficiently high target densities, the study of ion-Rydberg atom collisions has become routine [4,5]. In this regime, full quantum calculations are difficult to perform while the validity of classical methods for state-selective processes has not yet been established. Similarly, most non-perturbative calculations for energy loss of multiply charged particles in condensed matter employ classical approximations. For channeled particles, successive collisions with lattice atoms probe the "local", i.e. impact parameter dependent, energy transfer and ionization probability. The validity of classical methods for impact parameter specific processes remains an open question. Most recently, with advances in short-pulse laser technology the perturbation of atoms exposed to strong and ultrastrong laser fields has gained considerable attention[6]. As classical methods are, unlike quantum treatments, readily applied to atoms in strong electromagnetic fields, delineation of their validity as well as of their limitations is of considerable interest.

We focus in the following on the atomic response to a short but strong impulsive perturbation which is caused by the passing-by of a charged particle in fast ion-atom collisions or by an ultrashort half-cycle electromagnetic pulse. This regime is to be distinguished from low- to intermediate velocity atomic collisions or long laser pulses containing many optical cycles. We present an analysis of the classical character and limit of the quantum mechanics of ionization. We analyze the existence of a classical limit as a function of different parameters and identify regions in parameter space in which classical results mimic their quantum counterparts. We apply our analysis to two problems: the ejection of electrons into large angles in ion-atom collisions and the thresholds for ionization by ultrashort half-cycle electromagnetic pulses. Atomic units will be used throughout.

2. Classical Limit of the Liouville-Von Neumann Equation

Different equivalent formulations of classical dynamics can be recovered as the respective classical limit, formally to be taken as $\hbar \to 0$, of corresponding equivalent formulations of quantum dynamics. Such correspondences exist, for example, between Heisenberg's equation of motion and Hamilton's equation of motion or the Schrödinger equation and the Hamilton-Jacobi equation.

In principle, any of these approaches can be used to quantitatively explore the existence of and convergence to the classical limit. In practice, however, a judicious choice of the starting point on the quantum level is required since this limit is highly singular and non-uniform. One frequently explored avenue is the Feynman path integral approach[7] (Feynman propagator) whose semiclassical limit was given by Van Vleck[8] (Van Vleck propagator). It allows the decomposition of the S matrix and of the transition amplitude in terms of a classical transition amplitude multiplied by a phase factor containing the classical action [9],

$$S_{i \to f} = \sum_{\alpha} (p_{cl}^{\alpha}(i \to f))^{1/2} \exp\left[\frac{i}{\hbar} F_4(i, f, \alpha)\right], \qquad (1)$$

where F_4 is the reduced action along the classical path with path index α connecting the initial (i) and final (f) states and $p_{cl}^\alpha(i \to f)$ is the classical probability for the transition. Eq.(1) contains the so-called "primitive" semiclassical path interferences when more than one classically allowed path α connects the initial with the final state. For illustrative purposes we have displayed in Eq.(1) the $\hbar$ dependence of the interference phase to highlight the essential singularity of the classical limit as $\hbar \to 0$. Classical dynamics is therefore only recovered upon averaging $|S_{i \to f}|^2$ over infinitely rapidly oscillating phases by averaging (summing) over the initial (final) state variables which become continuous in the classical limit. The approach of the classical limit involves therefore non-commuting limiting operations of $\hbar \to 0$ and the long-time limit $t \to \pm\infty$, where from the latter the quantization conditions for the initial and final states are to be recovered. Despite the complexity of this task, considerable progress has been made along these lines [9].

We choose in the following a different approach which is based on the quantum dynamics of the density (or statistical) operator and its correspondence to classical phase space dynamics [10,11,12]. The density operator satisfies the Liouville-Von Neumann equation

$$i\frac{d}{dt}\hat{\rho} = [\hat{H}, \hat{\rho}]. \qquad (2)$$

In the following, we denote quantum variables by hats to distinguish them from the corresponding classical variables. The connection to the classical phase space flow is made through the Wigner function f_W, the Weyl transform of the coordinate representation of the density matrix, $\langle \vec{r} - \vec{s}/2|\hat{\rho}(t)|\vec{r} + \vec{s}/2\rangle$,

$$f_W(\vec{r}, \vec{p}, t) = \frac{1}{(2\pi)^3}\int d^3s\, e^{-i\vec{p}\vec{s}}\langle \vec{r} - \vec{s}/2|\hat{\rho}(t)|\vec{r} + \vec{s}/2\rangle. \qquad (3)$$

Application of the Liouville-Von Neumann equation to (3) leads to the quantum Liouville equation for the Wigner function [11],

$$\frac{\partial f_W}{\partial t} = \hat{L} f_W = -\vec{p} \cdot \nabla_{\vec{r}} f_W + \frac{2}{\hbar}\sin\left(\frac{\hbar}{2}\nabla_{\vec{r}} \cdot \nabla_{\vec{p}}\right) f_W U \qquad (4)$$

where the operator $\nabla_{\vec{r}}$ in the argument of the sine function is understood to act only on U while $\nabla_{\vec{p}}$ acts on f_W. Since we are interested in the $\hbar \to 0$ limit, we have explicitly displayed the dependence of Eq.(4) on $\hbar$.

The Wigner function is related to the diagonal elements of the density matrix, i.e. the densities in coordinate and momentum space by [10]

$$\hat{\rho}(\vec{r}, t) = \int d^3p\, f_W(\vec{r}, \vec{p}, t) \qquad (5)$$

$$\hat{\rho}(\vec{p}, t) = \int d^3r\, f_W(\vec{r}, \vec{p}, t). \qquad (6)$$

Furthermore, in the case that the density operator represents a pure state, the action can be recovered in the semiclassical limit from the momentum vector field

$$\langle \vec{p} \rangle_{\vec{r},t} = \nabla_{\vec{r}} S(\vec{r},t) = \frac{1}{\rho(\vec{r},t)} \int d^3p\, \vec{p}\, f_W(\vec{r},\vec{p},t), \tag{7}$$

where $S(\vec{r},t)$ is the phase of the wavefunction defined by $\psi(\vec{r},t) = |\psi(\vec{r},t)|e^{iS(\vec{r},t)}$. The phase $S(\vec{r},t)$ corresponds to the classical action provided the momentum vector field is single-valued. Since, typically, momentum vector fields are either multi-valued functions of $\vec{r}$ or do not even form a smooth Lagrangian manifold in non-separable systems, a more useful interpretation of Eq.(7) is that of a local expectation value of the momentum operator at $\vec{r}$.

The classical limit follows now formally from expanding the sine operator in Eq. 4 and keeping only the first term. The result is the classical Liouville equation

$$\frac{\partial f}{\partial t} = L f = -\vec{p} \cdot \nabla_{\vec{r}} f + \nabla_{\vec{r}} U \cdot \nabla_{\vec{p}} f \tag{8}$$

where L is the classical Liouville operator and $f(\vec{r},\vec{p},t)$ is the corresponding phase space distribution function of the electron. Note that in natural units this expansion can be recognized as an expansion in ascending powers of $\hbar$. Accordingly, Eq. (8) neglects terms of the order of $O(\hbar^2)$. Eq.(4) possesses, unlike Eq.(1), a convergent classical limit $\hbar \to 0$ and provides therefore a convenient starting point for the quantitative analysis of the classical-quantum correspondence. However, the complications with the singular behavior and the non-commutativity of the limits $\hbar \to 0$, $t \to \infty$ are preserved and reappear in the asymptotic initial conditions or final state representations of Eq.(4). The proper initial conditions for $t \to -\infty$ are stationary quantum density matrices which contain $\hbar$ to all orders through, in the semiclassical limit, quantization conditions of the form

$$\frac{1}{\hbar} \oint \vec{p} d\vec{r} = 2\pi(n+\beta) \tag{9}$$

where β is the Maslov index. One consequence of this difficulty is, that the Wigner function (Eq.(3)) does not, in general, provide appropriate initial conditions for the classical Liouville equation. For example, the Wigner distribution for an isolated hydrogen atom becomes dynamically unstable when its evolution is calculated using the classical Liouville equation in the absence of any perturbation. Thus, additional approximations are necessary to describe the initial phase-space distribution of the electron. The simplest and most frequently used initial phase-space distribution which is dynamically stable is the microcanonical ensemble proposed by Abrines and Percival [1]:

$$f_i(\vec{r},\vec{p}) = k\,\delta[E_i - p^2/2 + 1/r] \tag{10}$$

where $k = (-2E_i)^{5/2}/(2\pi)^3$ is a normalization constant. One of the most important properties of this distribution is that it exactly reproduces the momentum distribution of a statistical average of states of a given shell with principal quantum number n_i and binding energy $E_i = -1/(2n_i^2)$. Its position distribution, however, does not mimic the quantal probability distribution. The disagreement is most obvious in the classically forbidden region where Eq.(10) vanishes identically. A better representation of both the position and the momentum distributions can be achieved using a superposition of microcanonical ensembles with binding energies around E_i [13,14,15]. However this comes at the expense of an incorrect initial state energy distribution. The CTMC method [1-3] is simply a Monte Carlo technique to solve the classical Liouville equation (Eq.(8)) with compatible initial conditions such as Eq.(10). We refer to the CTMC method as an exact classical method since it solves the Liouville equation without involving further approximations to the dynamical evolution.

For short impulsive perturbations by fast collisions another property of the quantum and classical Liouville equations (Eqs.(4) and (8)) is of crucial importance: For short interaction times or large collision velocities ($v_p^{-1} \to 0$) classical and quantum mechanical variations of the position probability density and mean velocity field (i.e. Eqs. (5) and (7)) agree with each other up to second order in v_p^{-1}. The solution of the quantum Liouville equation after a short-time interaction (from $-T$ to T) with a fast moving projectile is given by [11],

$$f_q^{(1)}(r,p,T) = f_q(r,p,-T) + \frac{1}{v_p} \int_{-v_pT}^{+v_pT} dz'_p W_q(z'_p) f_q(r,p,-T) \quad (11)$$

with

$$W_q(z_p) f_q = 2\sin\left(\frac{1}{2}\nabla_r \cdot \nabla_p\right) f_q U_q(|r - b - z_p \hat{z}|). \quad (12)$$

where $z_p = v_p t$, $\vec{b}$ is the impact parameter, $f_q^{(1)}$ denotes a first-order solution in powers of v_p^{-1}, and we have used $(L_q - W_q) f_q(r,p,-T) = 0$. Consequently, the mean momentum transfer is given by

$$\Delta \vec{p}^{(1)}(r) \simeq \frac{1}{v_p} \lim_{v_p \to \infty} \int_{-v_pT}^{v_pT} dz_p \nabla_r U_p(|r - b - z_p \hat{z}|), \quad (13)$$

which agrees exactly with the corresponding classical result. This correspondence is completely analogous to the equivalence of the full Feynman propagator and the semiclassical Van Vleck propagator for infinitesimally small time intervals. As we anticipate from Eq.(13) and will illustrate in more detail below, the collisional momentum transfer $\Delta \vec{p}^{(1)}$ plays the role of a characteristic parameter in quantitatively elucidating the classical-quantum correspondence.

The following simple criteria can be formulated for which the convergence to the classical limit is expected. For the asymptotic initial and final states with (rationalized) De Broglie wavelength $\lambda_{i,f}$ and radii $\langle r \rangle_{i,f}$ we expect

$$\lambda_{i,f} \ll \langle r \rangle_{i,f} \quad (14a)$$

which reduces for hydrogenic states to the requirement of large quantum numbers

$$n_{i,f} \gg 1. \quad (14b)$$

The dynamical evolution during the collision is expected to proceed classically "on the average" if

$$\lambda_{\Delta p} \ll \langle r \rangle_{i,f} \tag{15a}$$

where $\lambda_{\Delta p} = 1/\Delta p$ is the wavelength of the impulsive perturbation probing the internal structure of the atom. For hydrogenic systems Eq.(15a) becomes

$$\Delta p\, n^2 \gg 1. \tag{15b}$$

If classical dynamics is sensitive to the internal structure of atoms on a length scale r_c small compared to the mean radius, $\langle r \rangle$ must be replaced by r_c in Eq.(15a). Furthermore, in elucidating the classical-quantum correspondence, the classical scaling invariance is very useful. Scaling invariances are based on the fact that the classical equations of motion are invariant under the scaling transformation $\vec{r}' = \vec{r}/n^2$, $\vec{p}' = \vec{p}n$, $t' = t/n^3$ [1,16] where n is in general an arbitrary scaling parameter. Our choice of the symbol indicates that we will use the principal classical action (principal quantum number) as the scaling parameter, i.e. we investigate the n dependence of the scaled dynamical observable. Upon proper scaling, classical dynamics is independent of n. Quantum corrections break the scaling invariance. Any remaining n dependence of quantum observables not accounted for by classical scaling is therefore a clear signature of non-classical contributions (i.e., the discreteness of n). In this context we introduce the scaled momentum transfer

$$\Delta p_0 = \Delta p / \langle p \rangle_i = n_i \Delta p \tag{16}$$

in units of the average orbital momentum of the initial state.

3. Slingshot Effect and Classical Impulse Approximation

Many features of the classical-quantum correspondence can be intuitively understood in terms of the slingshot effect, which is well-known from planetary science. One of the fastest and most efficient ways to transport a space probe (in atomic physics, the electron) to the outer fringes of the solar system and beyond (i.e., ionization) is to choose such an initial elliptic orbit around the sun (the nucleus) after take-off that features a (relatively) close encounter with another planet (the projectile) (Fig. 1). Within this three-body system of sun, planet and space probe, a two-body elastic collision between the space probe and the planet can lead to an energy gain of the two-body system (space probe-sun) such that the space probe acquires enough energy to leave the solar system (ionization). This turns out to be the essence of the description of ionization in fast collisions within classical dynamics.

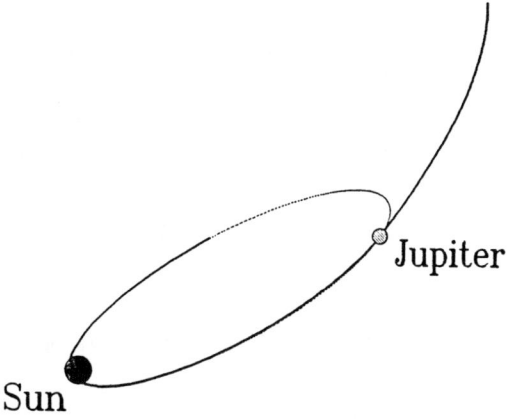

FIGURE 1. The planetary slingshot effect: Kepler ellipse of a space probe (i.e. electron) around the sun (nucleus) perturbed by an impulsive momentum transfer by Jupiter (incoming charged particle).

The slingshot transfers energy and angular momentum to the electron which are given by

$$\Delta E = \frac{1}{2}(\vec{p}(\vec{r}) + \Delta \vec{p})^2 - \frac{1}{2}\vec{p}(\vec{r})^2 \qquad (17)$$
$$= \Delta \vec{p} \cdot \vec{p}(\vec{r}) + \frac{1}{2}\Delta p^2$$

and
$$\Delta \vec{L} = \vec{r} \times \Delta \vec{p} \qquad (18)$$

where $\Delta \vec{p}$ is the momentum transferred to the electron by the passing by the projectile. Eqs.(17,18) correspond to the classical impulse approximation. Its validity hinges on the short interaction time T with the projectile on the time scale of the orbital period $T_n = 2\pi n^3$, i.e.

$$T/T_n = T_0 \ll 1 \qquad (19)$$

In the example of Fig. 1, the validity of Eq.(19) follows from the small mass ratio of Jupiter to sun, $M_j/M_s \approx 10^{-3}$. Only in the immediate vicinity of Jupiter during a fraction $\sim (M_j/M_s)^{1/2}$ of the orbital period the perturbing force by Jupiter dominates the gravitational attraction by the sun during which the energy ΔE is being transferred. For charged particle impact, the criterion (19) is satisfied if the collisions are fast (i.e., $v_p^{-1} \to 0$).

The important feature of the energy gain (or loss) by the slingshot (Eq.(17)) is the presence of a term dependent on the coupling of the local momentum on the orbit, $\vec{p}(\vec{r})$, at the point of the collision to the momentum transfer $\Delta \vec{p}$. This slingshot term is at the heart of both the success and the failure of classical approximations of the collisions. Estimating $|p(\vec{r})|$ by the average momentum $\langle p \rangle_i = n_i$, this term dominates in the energy transfer for small scaled momentum transfers

$$\Delta p \, n_i = \Delta p_0 \ll 1. \tag{20}$$

4. The Impulsively Perturbed Hydrogen Atom

We analyze first the impulsively perturbed ("kicked") hydrogen atom, i.e. a hydrogen atom subject to an instantaneous momentum transfer (or "kick") $\Delta \vec{p}$ which occurs at a time t_0. The Hamiltonian for this problem is[12]

$$H = \frac{p^2}{2} - \frac{1}{r} - \vec{r} \cdot \Delta \vec{p} \, \delta(t - t_0) \tag{21}$$

where $\vec{p}$ and $\vec{r}$ are the momentum and position vectors of the electron and where we have assumed an infinitely massive nucleus. Analogously to the gravitational slingshot discussed above, Eq.(21) can model realistic collision systems for fast projectiles for which the duration of the perturbing pulse is short compared to the characteristic times (inverse frequencies) of the atomic system. The beauty of the simple model described by Eq. 21 is the availability of exact solutions in both classical and quantum mechanics. The correspondence can therefore be investigated without resorting to any approximation. If we denote by $\psi(\vec{r}, t_0^-) = \phi_i(\vec{r})$ the wavefunction of the electron just before the kick, the exact wavefunction just after the kick will be

$$\psi(\vec{r}, t_0^+) = e^{i\Delta \vec{p} \cdot \vec{r}} \psi(\vec{r}, t_0^-), \tag{22}$$

i.e. the matrix elements of the time evolution operator in the atomic state representations reduce to inelastic transition form factors,

$$\langle \phi_f | U(\infty, -\infty) | \phi_i \rangle = \langle \phi_f | \exp(i \Delta p \cdot r) | \phi_i \rangle \tag{23}$$

Analytic expressions for this integral for arbitrary initial and final states can be found elsewhere [17]. The corresponding classical problem can be solved exactly as well. For ionization of a microcanonical ensemble of a shell n_i to a final state with energy E_f, the probability per unit energy interval is given by

$$\frac{dP_n^{Cl}}{dE_f} = \frac{8(-2E_i)^{5/2}(\Delta p)^5}{3\pi[(E_f - E_i + (\Delta p)^2/2)^2 - 2(\Delta p)^2 E_f]^3}. \tag{24}$$

This distribution (Eq.(24)) is compared in Fig. 2 with its quantum counterpart for a hydrogen atom initially in its ground state. Since the characteristic momentum of the electron in the ground state of a hydrogenic atom is $\langle p \rangle_i = 1$ the momenta of this figure can be considered as scaled momenta Δp_0. For $\Delta p = \Delta p_0 = 0.2$ the classical ionization probability is uniformly by far too small over the whole spectrum of energies of the emitted electron. On the other hand, for increasing values of Δp the classical ionization describes the dominant region of the spectrum increasingly accurately while discrepancies in the wings of the distribution persist.

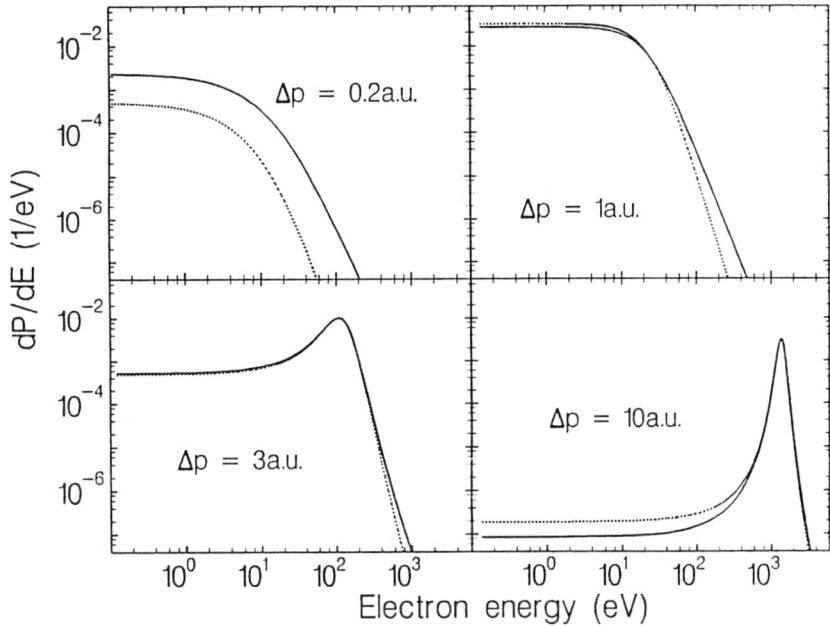

FIGURE 2. Quantum mechanical (full curve) and classical (broken curve) ionization probability density as a function of the energy of the ejected electrons for a hydrogen atom subject to an impulsive momentum transfer (Eq.(21)).

For $\Delta p_0 \geq 3$ the probability density of ionized electrons exhibits a noticeable peak at an energy $E_f \approx (\Delta p)^2/2 + E_i$ corresponding to the mean energy transferred to the electron. In a three-dimensional plot of the yield of electrons as a function of E_f and Δp, this peak would develop into a ridge which is the equivalent of the well known "Bethe ridge" [18-20]. This ridge signifies the energy-momentum dispersion relation of a free particle. The fact that classical dynamics is able to correctly describe the width of this "binary encounter" like peak is due to the fact that the slingshot term $\vec{p} \cdot \Delta \vec{p}$ (Eq.(17)) incorporates the information on the momentum distribution (Compton profile) of the initial state into the classical description. By the same token, however, the overestimate of the low-energy spectrum for $\Delta p_0 = 10$ is also a consequence of the very same slingshot term. This discrepancy, while quantitatively unimportant since the dominant region for ionization lies at much higher energies, can, nevertheless, give important insights into differences between classical and quantal ionization: We note first that for large Δp_0, the criterion for small wavelength of the perturbation (Eq.(15b)) is well satisfied and classical dynamics is expected to be valid "on the average." However, in order to produce a kinematically unfavored low- energy electron by a large pulse $\Delta \vec{p}$, a large orbital momentum $\vec{p}(\vec{r})$ antiparallel to $\Delta \vec{p}$ is required in the slingshot term (Eq.(17)). Since within classical mechanics the momentum distribution and position distribution are strongly correlated (via the Hamiltonian function) only segments of highly eccentric orbits at small distances from the nucleus, $r_c \ll \langle r \rangle_n$, can furnish the required large local momentum. The required distances r_c are, for large Δp compared to the mean momentum $\langle p \rangle$, smaller than

$$r_c \lesssim (\Delta p)^{-2}. \tag{25}$$

Such a tight localization is inconsistent with the uncertainty relation involving the momentum transfer (Eq.(15a)) in which we must replace $\langle r \rangle$ by r_c,

$$\Delta r \simeq r_c \gtrsim (\Delta p)^{-1} \tag{26}$$

The strong classical $r - p$ correlation is weakened in quantum mechanics due to delocalization of the electron over the size of the De Broglie wavelength associated with the momentum transfer. This leads to quantum suppression of ionization. The complete failure of classical dynamics for small scaled momentum transfer ($\Delta p_0 = 0.2$), on the other hand, is a consequence of the violation of the criterion (Eq.(15b)) involving the average size of the charge cloud and signifies the lack of dipole-allowed transitions in classical dynamics. In the limit of soft collisions which can be viewed as the absorption of a virtual photon by the atom, the momentum transfer is too small and the associated De Broglie wavelength too large to "resolve" the internal structure of the constituents of the hydrogen atom. Instead, the virtual photon interacts with the atom as a composite system. (Free electrons do not absorb or emit photons.) Quantum mechanically, the energy- momentum dispersion relation satisfied in the transition involves now the recoil of the nucleus rather than the momentum exchange between the external field (the "kick" or the projectile) and the electron. In other words, for $\lambda_{\Delta p} \gg \langle r \rangle_i$ almost all the energy of the virtual photon will be delivered to the electron while the corresponding momentum originates from the recoil of the nucleus. By contrast, classical dynamics requires energy and momentum transfer to the electron to be directly linked through the slingshot expression Eq.(17). Obviously, the fact that electron emission with energies large compared to $\Delta p^2/2$ occurs at all is due to the coupling term $\vec{p}(\vec{r}) \cdot \Delta \vec{p}$ which now strongly dominates.

The resulting breakdown of classical scaling invariance is shown in Fig. 3 for the ionization probability integrated over all final energies for different initial energy levels $n = 1, 3$ and 10 as a function of the scaled momentum transfer $\Delta p_0 = n\Delta p$. Note that if classical scaling invariances were to hold, results for all n-levels would lie on the same universal curve as classical results do. The larger n the larger the range of momentum transfers for which the quantum mechanical curve agrees with the scaling invariant classical result. Fig. 3 may be viewed as an illustration of the correspondence principle at work. In the limit $n \to \infty$ the quantum results converge to the same universal curve which, moreover, coincides with the classical result. The convergence is, however, non-uniform in Δp which is the cause of the breakdown of the classical approximation for finite n levels. Eq. (15b) gives the estimate $\Delta p_0 > n^{-1}$ for the region of convergence. In addition to this necessary criterion a more stringent criterion can be derived from the slingshot relation (Eq.(17)) requiring the classical free particle energy and momentum transfer to be sufficient to overcome the energy gap of adjacent quantum levels, $\Delta E = n^{-3}$ (as $n \to \infty$). This criterion which defines the region where the discreteness of the quantal excitation spectrum becomes unimportant leads to

$$\Delta p_0 \gtrsim (1/n)^{1/2} \tag{27}$$

in agreement with the results of Fig. 3.

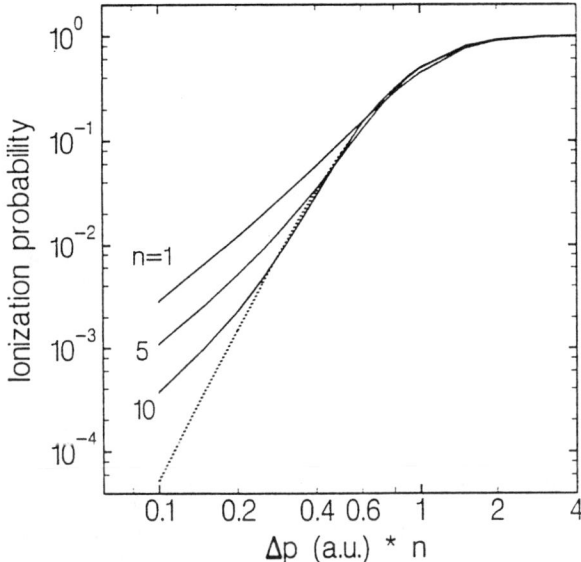

FIGURE 3. Quantum mechanical (full curves) and classical (broken curve) total ionization probability of a hydrogen atom in different excited levels with principal quantum number n as a function of the scaled momentum transfer $\Delta p_0 = n\Delta p$ (Eq.(21)). The classical results reduce to a single curve for all n levels.

5. Impact Parameter Dependence of Ionization

Fig. 4 displays the ionization probability for 5 MeVp + H(1s) collisions as a function of the impact parameter integrated over all energies and angles of the ejected electron. The quantum results in the first Born (B1) approxi-

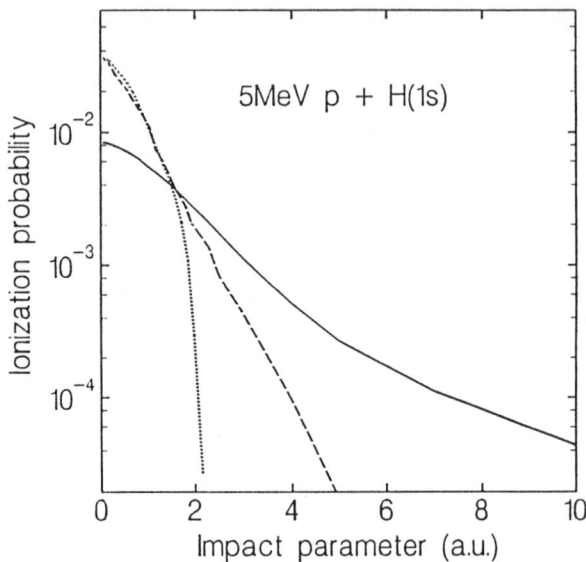

FIGURE 4. Total ionization probability as a function of the impact parameter in 5 MeV p + H(1s) collisions. The quantum mechanical B1 results (full curve) is compared with the CTMC results using an initial microcanonical distribution (dotted curve) and the initial distribution of Cohen [15] which reproduces the correct radial probability density of the atom (broken curve).

mation (see e.g. [21]), which is expected to be valid at this energy, is in sharp disagreement with the CTMC result over the whole range of impact parameters. The well-known failure at large impact parameters is the obvious consequence of the lack of dipole-allowed transitions in soft collisions. More interesting and less well-known is the equally drastic failure at small impact parameters. To verify whether this disagreement is due to the improper radial distribution associated with an initial microcanonical ensemble, we have also performed CTMC calculations with the initial ensemble proposed by Cohen[15] which reproduces exactly the radial distribution of H(1s). The resulting probabilities are seen to extend beyond the cutoff of the microcanonical probabilities at 2 au. However, the disagreement with the B1 calculations still persists for small impact parameters. The analysis of the classical ionization probability at zero impact parameter $P(b=0)$ for different substates in $n = 5$ (Fig. 5) reveals its relation to the slingshot effect and to the classical position-momentum correlation. The ionization probabilities are multiplied by the scaled impact energy in order to identify a E^{-1} dependence at high impact energies observed quantum mechanically for all substates. This dependence should not be confused with the dependence of the cross sections at high impact energies which results from the integration of the ionization probability over all impact parameters.

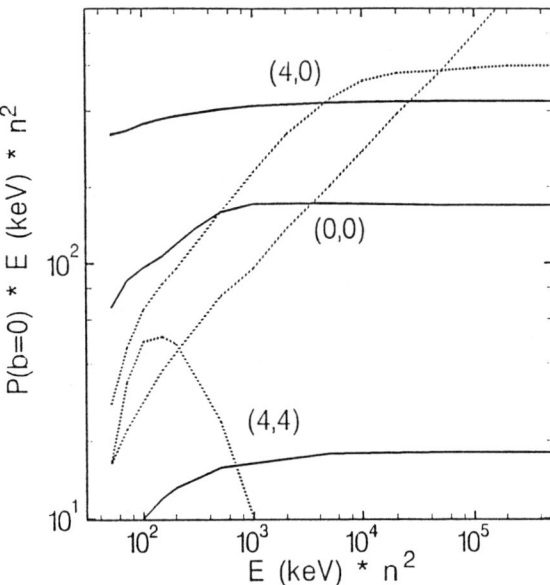

FIGURE 5. Total ionization probability times the scaled impact energy En^2 as a function of the scaled impact energy for p + H($n = 5, \ell, m$) collisions at an impact parameter $b \simeq 0$: first Born approximation (full curves); CTMC method (dotted curves). The numbers in the figure indicate the quantum numbers (ℓ, m) specifying the initial state of the electron within the level $n = 5$.

Depending on the shape and orientation of the classical orbits associated with the initial state of the electron, we find that ionization can be, within the CTMC approach, either classically enhanced or classically suppressed relative to the corresponding quantal $B1$ result. For the 5s state the ionization probability is drastically enhanced and decreases with increasing energy only as $(E)^{-1/2}$. This counterintuitive result is a direct result of the unphysical dominance of the $\vec{p} \cdot \Delta \vec{p}$ term of the energy transfer by the slingshot. s states corresponding to orbits with eccentricities close to 1 feature segments of the orbits close to the nucleus with large local $|p(\vec{r})| \to \infty$ as $r \to 0$. Zero-impact collisions remain therefore effective in ionization despite decreasing Δp with increasing collision energy E. This artifact is a direct consequence of the strong classical $r - p$ correlation which is absent in quantum mechanics (Eqs. (25,26)). The opposite trend can be observed for the state $\ell = n - 1 = 4$ and $m = 4$, i.e. for circular orbits with the orbital plane perpendicular to the beam axis. Classical ionization is strongly suppressed compared to quantum ionization. The reason is two-fold: the electron does not experience a close encounter with the projectile at zero impact parameter and therefore the collision approaches the distant-collision (dipole allowed) limit. Moreover, the residual classical momentum transfer $\Delta \vec{p}$ becomes at high speeds perpendicular to the local momentum of the circular orbit thereby switching off the dominant "slingshot" contribution. An interesting intermediate case arises for $\ell = 4$, $m = 0$, i.e. for circular orbits lying in the scattering plane. In this case, close encounters between the projectile and electron are still possible for zero-impact parameter collisions, however, $\Delta \vec{p}$ is orthogonal to $\vec{p}(\vec{r})$ at high speeds. The coupling term $\Delta \vec{p} \cdot \vec{p}$ in Eq.(17) therefore vanishes, resulting in the correct quantum mechanical energy dependence of $P_{n\ell}(0)$. The coefficient of proportionality is, however, different.

It is worth noting that, unlike the exact classical CTMC method approximate classical models such as the Thomson model do not contain the slingshot term $\vec{p} \cdot \Delta \vec{p}$. Remarkably, it is this additional, unjustified approximation which has led, coincidentally, to early successes in the description of ionization and of the stopping power [22] in agreement with quantum theory. Equally remarkable, the term $\vec{p} \cdot \Delta \vec{p}$ is also responsible for the failure of exact classical calculations for charge transfer from the ground state at high energies [23,24]. This shortcoming appears to be even more surprising in view of the fact that the well-known high energy limit of the electron capture cross section, $\sigma \sim v^{-11}$, results from double scattering (second-order) contribution which was discovered by Thomas [25] within the framework of classical mechanics. In his original work, Thomas neglected the initial momentum distribution of the captured electron (i.e., $\vec{p}(\vec{r}) = 0$) and hence, the slingshot term $\vec{p} \cdot \Delta \vec{p}$. In exact classical calculations it is precisely this term which gives rise to the first-order OBK-like [26] contribution to electron capture at high energies which relies on matching of the velocity between the projectile and the orbital speed. Because of the strong momentum-position correlation in the classical dynamics, electrons at small distances near the nucleus ($r \to 0$) provide strong momentum matching ($p(r) \to \infty$) for arbitrarily high speeds. Consequently, in exact classical calculations the OBK (or velocity matching) contribution overshadows the true classical Thomas double scattering contribution and gives an unrealistically large cross section at high energies. However, for electron capture from circular Rydberg states this contribution is reduced. This leads to an improved agreement with the experiment found in a recent CTMC calculation for circular Rydberg states [27].

6. Angular Distribution of Ejected Electrons

The results of the previous sections indicate that ionization becomes classically suppressed, which is for small momentum transfers. Thus, an improved description of the spectrum of ejected electrons in ion-atom collisions can be obtained by correcting the classical (Cl) spectrum for the quantum mechanical (QM) portion which is associated with classically suppressed ionization due to small momentum transfers [12]

$$\frac{d^2\sigma}{dE_f d\Omega} \simeq \frac{d^2\sigma^{Cl}}{dE_f d\Omega} + \frac{d^2\sigma^{QM}}{dE_f d\Omega}(q_{\min} < q < \Delta p_{cr}) \qquad (28)$$

where Δp_{cr} is the critical momentum transfer q below which ionization becomes classically suppressed, which is estimated by Eq.(27) and $q_{min} =$

$(E_f - E_i)/v_p$. The conceptual advantage of Eq.(28) is that it combines the "best of two worlds" since the quantum corrections come from a regime where perturbation theory is expected to be valid but classical dynamics fails while the non-perturbative regime can be treated exactly, though classically.

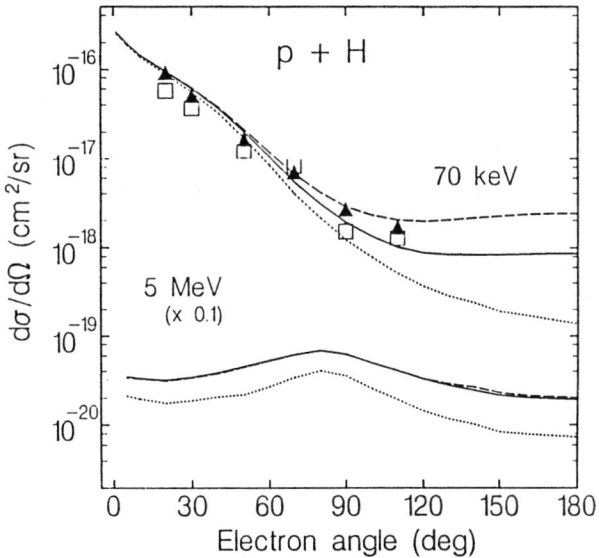

FIGURE 6. Single differential ionization cross section as a function of the angle of the ejected electron in 70 keV and 5 MeV p + H(1s) collisions. Experiments: atomic hydrogen from Rudd et al [28,29] for atomic hydrogen (open squares) and molecular hydrogen divided by a factor two (triangles). Theoretical results: CTMC (dotted curve), CTMC plus the small momentum transfer contribution of CDW-EIS (full curve) and B1 (broken curve) [12].

One application of Eq.(28) is shown in Fig. 6 for the singly differential cross section (SDCS) integrated over all electron energies. We compare the experimental data of Rudd et al.[28,29] with theoretical results as a function of the electron angle for 70 keV and 5 MeV p+H collisions. We display the standard CTMC results and the modified CTMC calculations corrected for the quantum contributions for small momentum transfer (Eq.(30)). At 70 keV all calculations agree with each other at forward angles whereas large discrepancies appear at backward angles. Standard CTMC results systematically underestimate the cross sections at large angles because of the lack of non-classical dipole emission channels. The important point to note is that at intermediate energy collisions (70 keV) dipole transitions do not constitute the dominant contribution to the total ionization cross section but do so only in the kinematically unfavored backward region. Therefore, classical dynamics is able to account for the dominant contribution emitted into the forward hemisphere, and hence, to the total ionization cross section but fails at backward angles. On the other hand, at high energies (5 MeV) the failure is uniform over all angles. This is to be expected in view of the fact that the mean momentum transfer at this energy $\langle \Delta p \rangle$ is small compared to $\langle p \rangle_i$, i.e., outside of the range of validity Eq.(27). This velocity regime closely mirrors the situation of Fig. 2a for small scaled momentum transfers $\Delta p_0 \ll 1$ where the classical dynamics fails over the whole spectral range.

7. Ionization of Rydberg Atoms by Half-Cycle Pulses

Very recently, Jones et al.[6] have measured the ionization of Na Rydberg atoms by subpicosecond "half-cycle" electromagnetic pulses. The atom is subject to a strong unidirectional electrical field confined to a very short time interval, whose duration T_p is shorter than or of the order of the classical orbital period T_n, n being the effective quantum level of the atom. This regime is of considerable conceptual interest as it provides a bridge between ionization by radiation fields and collisional ionization by charged particles at high energies.

A hydrogen atom subject to a pulsed electric field is described by the Hamiltonian

$$H = \frac{p^2}{2} - \frac{1}{r} - F(t)z \qquad (29)$$

where $F(t)$ denotes an external electric pulse which is directed towards the positive z-axis. Comparison between Eq. (29) and Eq. (21) shows the close connection to the "kicked" atom to which Eq.(29) converges in the impulsive limit of ultrashort pulses $T_p \ll T_n$. The corresponding "kick" or momentum transfer is given by $\Delta p = \int_{-\infty}^{\infty} F(t) dt$.

In Fig. 7 we display the data for the 10% ionization thresholds of Ref.[6] for different nd (m=0) states in terms of the scaled variables $T_0 = T_p/T_n$ and Δp_0 together with our classical and quantum calculations [30]. These calculations were performed for hydrogen using the fact that the quantum defect in d states of Na is very small ($\delta = 0.01$). Classical calculations were also performed for Na atoms using realistic core potentials. The quantum mechanical calculations have been performed for $n = 5$ and 10 and we display results for the regions of T_p for which convergence has been reached with the basis sizes used. All classical and quantum calculations for 10% ionization thresholds as well as the data lie on the same universal curve in the $(\Delta p_0, T_0)$ diagram, verifying the classical-quantal correspondence for ionization by short pulses. The limiting case $T_0 \gg 1$ of this curve which is equivalent to the $F \propto n^{-4}$ dependence for static electric fields. The coefficient depends, however, on the shape of the pulse and the substate populated. In the opposite limit of ultrashort pulses ($T_0 \ll 1$) the curve approaches the asymptote $\Delta p_0 \simeq 0.54$ giving rise to a dependence $\Delta p \simeq 0.54 n^{-1}$ a.u. . It is important to note that this asymptote is exact and is given by the evolution operator of the "kicked" atom (Eq.(23)), i.e. the inelastic form factor.

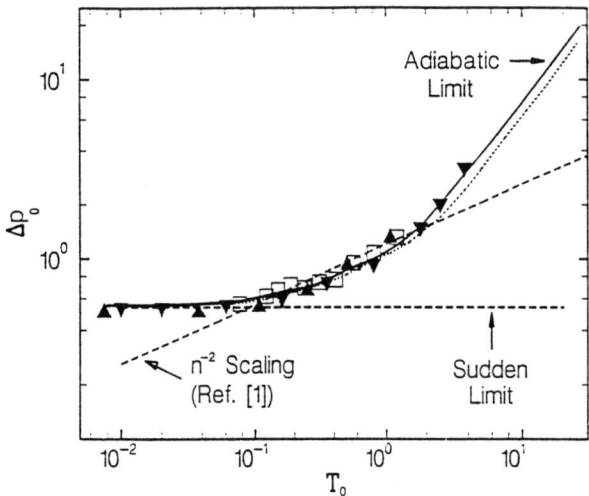

FIGURE 7. Scaled momentum transfer for 10% ionization threshold of H($n, \ell = 2, m = 0$) atoms as a function of the scaled pulse duration: classical results for inverse parabolic (solid line) and rectangular (dotted line) pulses, quantum mechanical result for ionization of 10d (solid triangles) and 5d (inverted solid triangles) states by rectangular pulses, and experimental data of Ref. 6 for Na(nd) atoms (equivalent to H($n, \ell = 2, m = 0$)) multiplied by a factor of 2.5 (open surfaces).

The almost perfect agreement between exact classical and quantum calculations appears surprising since the perturbation operator for a half-cycle pulse (Eq.(29)) is formally completely equivalent to the dipole coupling to an electric field of a distant collision, for which classical dynamics seems to fail. The reason is obviously the very different strength of the perturbation, i.e. Δp can be sufficiently strong to satisfy the validity criterion for classical dynamics. From Fig.3 it follows directly that 10% ionization (i.e. an ionization probability of $P_n = 0.1$) by ultrashort pulses requires a scaled momentum transfer of $\Delta p_0 = n\Delta p \simeq 0.5$. Except for very small n, the classical criterion (Eq.(27)) is well satisfied.

Another characteristic difference between short electromagnetic pulses and fast charged particle collisions originates from the time dependence of the polarization. In a fast charged-particle collision, the time-dependent electric fields consist simultaneously of a half-cycle pulse perpendicular to the beam axis $\hat{v}_p$, $F_\perp(t)$, and a full cycle pulse parallel to $\hat{v}_p$, $F_\parallel(t)$, out of phase relative to each other in the limit $v_p \to \infty$. Even for correspondingly strong perturbations (e.g., distant collisions by multiply charged projectiles), the longitudinal momentum transfer $\Delta p_\parallel$ can be small compared to $\Delta p_\perp$ because the two half-cycles of opposite polarity largely compensate each other. This may lead simultaneously to non-classical transitions driven by $\Delta p_\parallel$ and to dominant classical excitation driven by $\Delta p_\perp$. Future experimental and theoretical investigations of excitation and ionization by short laser pulses which can be produced with variable shape, duration, and polarity may provide an ideal tool to shed new light on the subtleties of the excitation dynamics.

8. Conclusions

Our study of classical-quantum correspondence in fast ionization by charged particles and half-cycle pulses identifies the momentum transfer during the impulsive interaction, Δp as the order parameter delineating the border between classical and quantum mechanical ionization. The critical momentum transfer Δp_{cr} between which classical mechanics fails is found to be of the order $\Delta p_{cr} \simeq n^{-3/2}$ for hydrogenic initial states corresponding to a scaled momentum $\Delta p_{0,cr} = \Delta p_{cr} \cdot n \simeq n^{-1/2}$. For smaller momentum transfers, the associated De Broglie wavelength is not sufficient to "resolve" the internal structure of the atom. Consequently, the atom interacts with the collisional perturbation as a composite system rather than in terms of a sequence of two-body collisions. The classical suppression of dipole-allowed transitions is one well known consequence of this limitation. Since a realistic ion-atom collision features a broad distribution of momentum transfers, the validity of classical methods depends on the observable under consideration. The convergence to the classical limit is non-uniform and is shown to depend on the relevant region in phase space. Classical dynamics is adequate in regions in phase space when $\Delta p_0 \gtrsim n^{-1/2}$ but fails in other regions when the classical momentum transfer is small. The breakdown of the CTMC calculation for ionization at backward angles, while providing an excellent description at forward angles (see Fig. 6), provides a beautiful illustration of this point. It is in this case easily possible to correct the CTMC method for the lack of the quantum contribution to ionization, by breaking the differential cross sections as a function of the momentum transfer down into the region above and below the critical value Δp_{cr} where quantum corrections set in. Such a method (Eq.(28)) has been shown to remedy the failure of CTMC at large emission angles.

Work supported in part by the National Science Foundation and by the U.S. Department of Energy, Office of Basic Energy Sciences, Division of Chemical Sciences, under Contract No. DE-AC05-84OR21400 with Martin Marietta Energy Systems, Inc.

References

1. Abrines, R., and Percival, I.C., Proc. Phys. Soc. London **88**, 861 (1966).
2. Percival, I.C., and Richards, D., Adv. At. Mol. Phys. **11**, 1 (1975).
3. Olson, R.E., and Salop, A. Phys. Rev. A.**16**, 531 (1977).
4. Hansen, S.B., Ehrenreich, T., Horsdal-Pedersen, E., MacAdam, K.B., and Dubé, L.J., Phys. Rev. Lett. **71**, 1522 (1993).
5. MacAdam, K.B., Nucl. Instr. Methods Phys. **B56/57**, 253 (1991); and private communication.
6. Jones, R.R., You, Y., and Bucksbaum, P., Phys. Rev. Lett. **70**, 1236 (1993).
7. Feynman, R., Rev. Mod. Phys. **20**, 367-387 (1948); Feynman, R.P., and Hibbs, A.R., *Quantum Mechanics and Path Integrals* (McGraw-Hill, NY, 1965).
8. Van Vleck, J., Proc. Natl. Acad. Sci. (USA) **14**, 178-188 (1928).
9. Miller, W.H., Adv. Chem. Phys. **25**, 69 (1974).
10. Carruthers, P. and Zachariasen, F., Rev. Mod. Phys. **55**, 245 (1983).
11. Reinhold, C.O. and Falcón, C.A., J. Phys. B **21**, 1829 (1988).
12. Reinhold, C.O., and Burgdörfer, J., J. Phys. B **26**, 3101 (1993).
13. Eichenauer, D., Grün, N. and Scheid, W., J. Phys. B **14**, 3929 (1981).
14. Hardie, D.J.W. and Olson, R.E., J. Phys. B **16**, 1983 (1983).
15. Cohen, J.S., J. Phys. B **18**, 1759 (1985).
16. Landau, L.D. and Lifshitz, E.M., *Mechanics* (Oxford: Pergamon, 1960).
17. Belkic, Dz., J. Phys. B **11**, 3529 (1978); Belkic, Dz., J. Phys. B: At. Mol. Phys. **14**, 1907 (1978); Belkic, Dz., Phys. Scr. **45**, 9 (1992).
18. Bethe, H., *Handbuch der Physik*, Vol. 24 (ed. H Geiger and K. Scheel (Berlin: Springer) p. 273 (1933).
19. Fano, U., Ann. Rev. Nucl. Sci. **13**, 1 (1963).
20. Inokuti, M., Rev. Mod. Phys. **43**, 297 (1971).
21. Kocbach, L., J. Phys. B **9**, 2269 (1976).

22. see e.g., Bohr, N. and Lindhard, J., Kgl. Danske Videnskab. Selskab. Mat.-Fys. Medd. **26**, No. 12 (1954); Jackson, J.D., *Classical Electrodynamics* (John Wiley and Sons, New York), Ch. 13 (1962).

23. Toshima, N., Phys. Rev. A **45**, R2663 (1992).

24. Schultz, D.R., Reinhold, C.O., and Olson, R.E., Phys. Rev. A **46**, 666 (1992).

25. Thomas, L.H., Proc. R. Soc. London A **114**, 561 (1927).

26. Oppenheimer, J.R., Phys. Rev. **31**, 349 (1928); Brinkman, H.C. and Kramers, H.A., Proc. Acad. Sci., Amsterdam, **33**, 973 (1930).

27. Wang, J. and Olson, R., Phys. Rev. Lett. **72**, 332 (1994).

28. Rudd, M.E., Phys. Rev. A **20**, 787 (1979).

29. Schultz, D.R., Olson, R.E., Reinhold, C.O., Gealy, M.W., Kerby, G.W., Hsu, Y., and Rudd, M.E., J. Phys. B **24**, L599 (1991).

30. Reinhold, C.O., Melles, M., and Burgdörfer, J., Phys. Rev. Lett. **70**, 4026 (1993); Reinhold, C.O., Melles, M., Shao, H., Burgdörfer, J., J. Phys. B. **26**, L659 (1993).

Close-Coupling Methods: A Critical Evaluation

C.D. Lin

*Department of Physics, Kansas State University,
Manhattan, Kansas 66506-2601 USA*

Abstract. The close-coupling method of treating ion-atom collisions is reviewed and evaluated in terms of the limitations. A new AO-MO matching method which can be easily generalized to collisions involving multi-electron systems is described. It is suggested that in the inner region the adiabatic MO approach is replaced by diabatic-by-sector MO basis functions.

INTRODUCTION

The simplest ion-atom collision consisting of an electron and two nuclei is a special class of reactions involving quantum mechanical three-body systems. For collisions in the energy range of a few keV's to a few hundred MeV's, the de Broglie wavelength of the heavy particle is small compared to the typical size or the interaction length of the atom, such that the motion of the heavy particle can be described classically. This simplification allows one to treat the motion of the electron as a one-body problem in a time-varying external field.

For collisions with energy greater than a few keV's, all the inelastic channels are open. Such a collision can result in the excitation, ionization and capture of the electron in different final states which are determined by various experimental means. The role of atomic collision theory is to predict the outcome of such a collision based on the solution of the time-dependent Schrödinger equation.

The classical three-body problem is not solvable analytically, nor is the quantum mechanical three-body problem. Thus the role of atomic collision theory is to find suitable approximations for different situations. The validity of the scattering theory is checked by comparing with precise experimental results.

The number of *ab initio* approaches for treating general ion-atom collisions is rather limited. These approaches can be divided into two major categories: the perturbative methods and the close-coupling methods. The former is useful for collisions at higher energies and the latter for collisions at low and intermediate energies. There are restricted theoretical methods for specific inelastic processes which are not addressed here.

Under special circumstances, the three-body collision system can be approximated as a two-body problem which can then be solved accurately. This occurs when the incident ion perturbs the target atom weakly and the first-order perturbation theory can be applied. For example, excitation or ionization at large impact parameters can be treated by the first-order plane wave Born approximation (PWBA). In this approximation, the projectile provides an equivalent photon field, and the problem is essentially a single center problem as in the photoabsorption of atoms. Account of the interaction beyond the first order theory is described as two-center effects.

Some of the two-center effects show prominently, such as electron capture to bound states or electron capture to the continuum where the effect cannot be accounted for unless specific two-center representation is used for the electronic wave functions. Other so-called two-center effects are less clearly defined in that there are no special observable structures. Any difference between the experimental result and that calculated by the first-order theory is viewed as the two-center effect.

In this article I will concentrate on the status of *ab initio* atomic collision theory for the simplest collision systems. Most experiments are carried out for target atoms with more than one electrons and thus additional approximations have to be made for the description of the atom. It is important to bear in mind the accuracy of a given collision theory for the simpler systems. Better agreement with experiments for complex systems should always be treated with caution when such agreement is not achieved for the simpler systems.

Despite of the good progress made in ion-atom collision theory in the last two decades, the major tool for collisions which are not perturbative in nature is still based on the close-coupling method. I will survey the various close-coupling methods and the recent developments, in particular, the development which sets ion-atom collision theory in the general framework of collision theories for other systems, such as electron-atom collisions, positron-atom collisions, photon-atom collisions and reactive scattering in chemistry. This latter aspect has not been actively pursued in the AMO fields so far.

THE CLOSE-COUPLING METHOD

A general approach of treating atomic collisions which has been found in many applications is the so-called close coupling method. In applying to ion-atom collisions, it is generally formulated in the time-dependent picture. Consider the simple one-electron system where the Hamiltonian for the electron in the field of two nuclei with charges Z_t and Z_p is

$$H = -\frac{1}{2}\nabla^2 - \frac{Z_t}{r_t} - \frac{Z_p}{r_p} \qquad (1)$$

where r_p and r_t are the distances of the electron measured from the projectile and from the target, respectively. The goal of the scattering theory is to solve the time-dependent Schrödinger equation

$$\left(H - i \frac{\partial}{\partial t}\right) \psi(\vec{r}, t) = 0 \qquad (2)$$

with known initial condition such that the amplitudes for ALL possible final states after the collision can be evaluated. For most atomic collisions in the keV to MeV energy region, all the inelastic processes are accessible and some channels are more important than others.

One way to solve Eq. (2) is to expand the time-dependent wave function in terms of a complete set of eigenfunctions. From the mathematical viewpoint, any complete set of basis functions is adequate. In actual applications two types of basis functions are commonly used--the molecular orbitals (MO) and the atomic orbitals (AO).

Expansion in Terms of Molecular Orbitals

In the conventional molecular orbital expansion, the time-dependent wave function is expanded as

$$\psi(\vec{r}, t) = \sum_j a_j(t) \chi_j(R(t); \vec{r}) \, e^{-i \int_{-\infty}^{t} U_j(R') dt'} \qquad (3)$$

where χ_j is the molecular orbital at an internuclear separation $R = R(t)$. The expansion (3) is called the perturbed stationary state (pss) approximation. It has been used as the basis for describing low-energy ion-atom and atom-atom collisions since the 40's. According to the pss approximation, dominant transitions occur at localized crossing points of the potential curves $U_j(R)$.

The pss approximation is known for years to have problems that expansion (3) does not satisfy the asymptotic boundary condition correctly for the scattering problems. To amend this deficiency, the basis functions in (3) is modified such that each static MO is replaced by a traveling MO where χ in (3) is replaced by

$$\tilde{\chi}_j(\vec{R}; \vec{r}) = \chi_j(\vec{R}; \vec{r}) \, e^{i f_j(\vec{v}, \vec{R}, \vec{r})} . \qquad (4)$$

The form of the phase factor $f_j(\vec{v}, \vec{R}, \vec{r})$ is not fully specified *a priori* except that it has to reduce to a plane wave in the asymptotic region. The lack of uniqueness in $f_j(\vec{v}, \vec{r})$ means that the result of a calculation depends on the specific form of translational factors used. There are few studies of this effect for specific collision systems. Despite of this limitation, calculations have been carried out for many collision systems, with or without electron translational factors and "good agreement" has been found for many systems, despite that such agreement is not always found for the simpler systems [1,2].

Expansion in Terms of Two-Center AO's

An alternative close-coupling approach is to expand the time-dependent wave function in terms of atomic orbitals on the two collision centers. Since the two centers are moving toward each other, these basis functions used are traveling atomic orbitals,

$$\psi(\vec{r},t) = \sum_j b_j(t) \phi_j(\vec{r},t) . \quad (5)$$

Substituting this expansion into (2), a set of coupled equations for the expansion coefficients are obtained,

$$i\sum_j S_{ij}(t) \dot{b}_j(t) = \sum_j G_{ij}(t) b_j(t) \quad (6)$$

where S_{ij} is the overlap matrix element and G_{ij} is the interaction matrix element.

The expansion (5) has the advantage that the basis functions are true atomic states which satisfy correct boundary conditions asymptotically. The undesirable features, however, are that the basis functions are not orthogonal and can pose numerical difficulty when the overlap is large.

The expansion in (3) and (5) in principle has to cover a complete set, including the ionization channels. In reality, for the close coupling method to be useful, it is desirable that the expansion in (3) or (5) can be easily truncated.

At low energy collisions, ionization is small and thus the continuum states can be excluded in the basis set in (3) or (5). This of course also eliminates the close-coupling approach from treating ionization processes at low energies. At intermediate energies where ionization becomes important, the expansion in (5) has to include these continuum channels.

A rigorous method for including these continuum states in the close-coupling formulation is still nonexistent. If the angular and energy distributions of the ejected electron are not measured, it has been established that pseudostates on both collision centers can be used to represent these continuum states approximately. Total ionization cross sections have been calculated using pseudostates.

In the two-center AO expansion method, the basis set is overcomplete since in principle only the set of basis functions from one center is adequate. The use of one-center AO expansion does not allow the account of charge transfer channels directly and the convergence of such a one-center expansion has never been established, in particular with respect to the orbital angular momentum. Use of large number of basis functions from each center is often believed to improve the accuracy of the calculation. However, evidences are mounting which indicate otherwise.

Evidences of the Limitation of the Two-Center AO Expansion

In Fig. 1 we show the n=2 excitation cross sections for p+H(1s) collisions carried out by Shakeshaft (3) almost twenty years ago. He used 35 orbitals on each collision center. In comparison with the experimental data of Park et al. (4) where only the relative cross sections are measured, one may wish to say that there is a general agreement between the calculation and the experiment. This ambiguous and noncritical statement fails to emphasize the structures shown in the theoretical results which are absent in the data. Such structures are not expected for collisions at higher energy, and may imply something more profound in the theoretical approach.

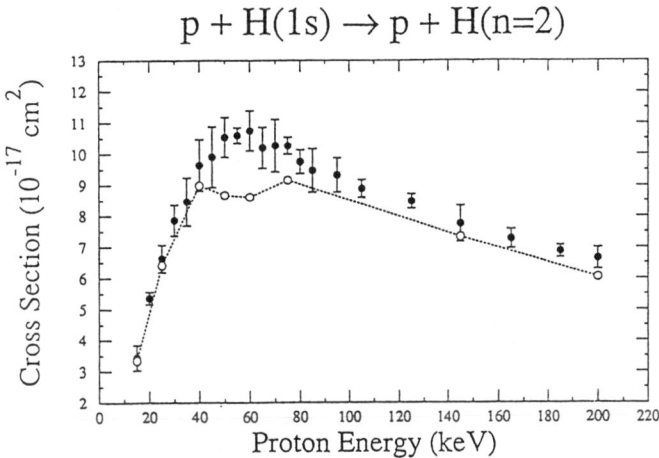

FIGURE 1. Excitation cross section to H(n=2) states for p+H(1s) collisions. The experimental data are from Park et al. (Ref. 4), and the calculations are from Shakeshaft (Ref. 3). Note the dip in the theoretical calculation in the 40-80 keV region.

This problem becomes even more serious as theorists try to "improve" the calculation to see if the remaining small discrepancy can be removed by using a larger basis set. In Fig. 2 we show the most recent calculation of Toshima (5) where he used about 150 basis functions on each collision center. The resulting structures in the total n=2 and n=3 excitation cross sections are most surprising. Whether the structure is related to the numerical instability or some unknown source due to the overcompleteness is not clear. Note it may be difficult to disentangle the two since overcompleteness implies basis functions which are nearly identical and thus the solution of the differential equations may become unstable. Similar conclusion has been found recently by Slim and Ermolaev (6).

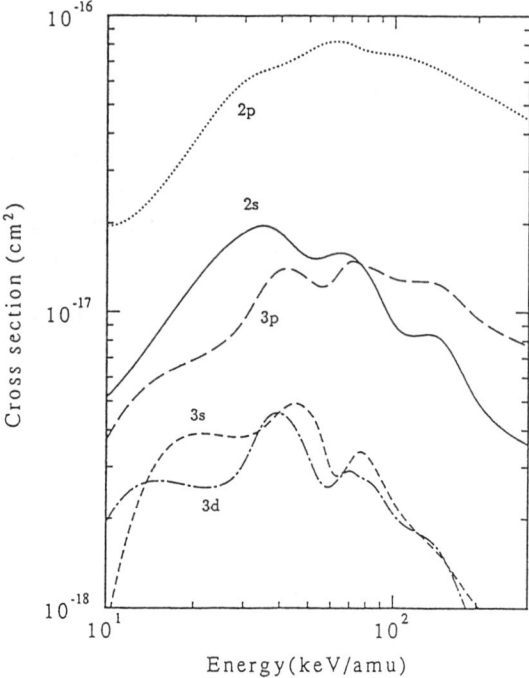

FIGURE 2. Excitation cross sections to H(nl) states for n=2,3 in p+ H(1s) collisions. The calculations are carried out using 150 atomic orbitals on each collision center. Bound states from n=1-5 are explicitly included and the remaining ones are pseudostates. Results from Toshima, private communication.

Another problem associated with a large AO basis set is that the truncation of the basis functions has to be done carefully. As an example, the ionization cross section for the collision of C^{6+} with H is shown in Fig. 3 [7]. In this case, the dominant process is electron capture to the n=4 states of C^{5+}, while electron capture to the n=1 and n=2 states of C^{5+} is not important. However, failure to include the latter states in the close coupling calculation gives large errors in the ionization cross section at low velocities, as demonstrated by the two curves in Fig. 3.

The previous sections address the limitation of the MO expansion and the AO expansion method. Despite of these shortcomings, the close-coupling method does serve to provide the best theoretical predictions on most of the phenomena in ion-atom collisions when the perturbation theory fails (Fritsch and Lin (8); Kimura and Lane (1)). However the "success" stories are often limited to the dominant channels only and partly because the accuracy in most collision

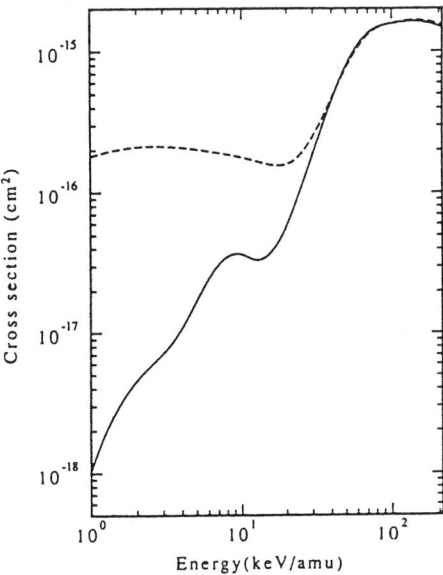

FIGURE 3. Ionization cross sections for $C^{6+}+H$. The difference between the full curve and the dashed curve is that the latter is calculated without the n=1 and n=2 states of C^{5+}. Note that these states are not populated at the end of the collision (from Ref. 7).

experiments is limited. The close-coupling method has not been well tested for the weaker collision channels. In Fig. 4 we show the excitation cross sections in $He^{++}+H$ collisions to the 2p state of H. The large basis-set calculations display some structures and the results are somewhat sensitive to the basis functions included (9-11). Incidentally, all the three close coupling calculations (9-11) show structures in the excitation cross sections near 10 keV, while the method based on the hidden crossing method (12) gives a smooth cross section. Unfortunately the experimental data of Hughes et al. (13) do not overlap with the data of Hoekstra and Beijers (cited in Ref. 9) and it is not clear whether the structure is indeed there. Note that the hidden crossing method has been used extensively to calculate ionization cross sections at low energies (14). It is essential to test its prediction for other well studied transitions such as the weak excitation processes discussed here.

Another "well-known" discrepancy for the simple $H^{+}+H$ collision is the emission cross sections from the Balmer-α line. The experimental results and the various theoretical calculations are shown in Fig. 5 (15,16). Note that the experimental data for the Balmer-α emission cross sections are 50% or more larger than ALL the close-coupling calculations. One would argue that it is likely that the experimental results are questionable since all the theoretical calculations converge rather well to each other. However it is important to realize that all the theoretical calculations use essentially the same close-coupling method using nearly identical basis functions.

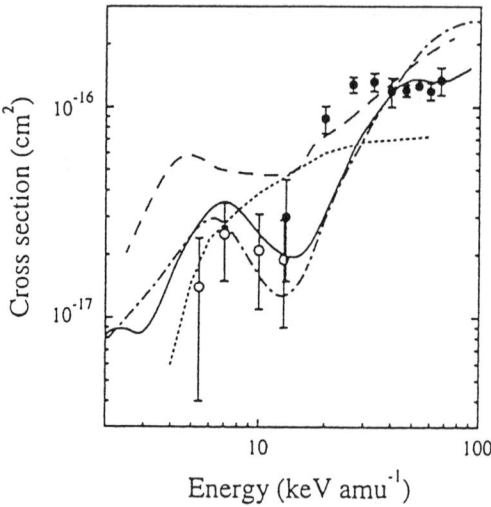

FIGURE 4. Cross sections for H(2p) formation in He^{2+}-H collisions. Full circles, experimental data from Hughes et al. (Ref. 13); open circles, data from Hoekstra and Beijers (cited in Ref. 9). Theoretical calculations: solid line, Fritsch et al. (Ref. 9); long-dashed lines, Bransden and Noble (Ref. 10); dash-dotted lines, Lundsgaard and Nielsen (Ref. 11); dotted lines, Krstic and Janev (Ref. 12).

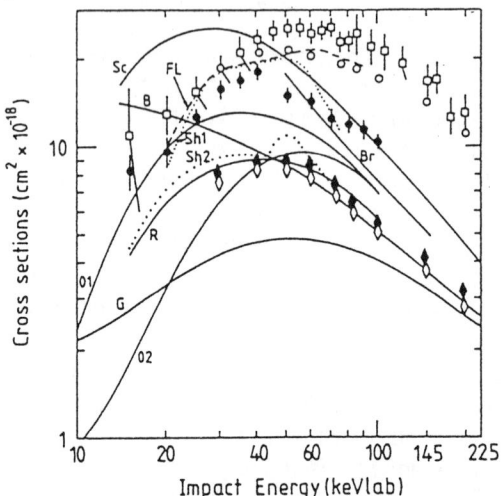

FIGURE 5. Cross sections for Balmer alpha emission from direct excitation to n=3 in p+H collisions. The open squares are data from Park et al. (Ref. 4) for the total n=3 excitation cross sections and the solid circles with error bars are the Balmer alpha emission cross sections from Donnelly et al. (Ref. 16). The rest are theoretical calculations, see detailed explanations in Fig. 1 of Ermolaev (Ref. 15). Note that the n=3 excitation cross sections are well reproduced by elaborate close coupling calculations, but the Balmer alpha emission cross sections are not in agreement with calculations.

The AO-MO Matching Method

Despite of the success of the close coupling methods based on expansion using atomic orbitals or molecular orbitals, the difficulties mentioned earlier remain. These difficulties tend to multiply when the methods are applied to collisions involving multi-electron collision systems.

A new approach which "avoids" the difficulties associated with the AO or MO expansions is the so-called AO-MO matching method. The idea is to expand the solution of Eq. (2) in two different sets of basis functions. In the inner region where the electron is shared by the two collision centers the electronic wave function is expanded in terms of molecular orbitals. In the outer region where the electron is moving with one or the other center the time-dependent electronic wave function is expanded in terms of traveling atomic orbitals on the two centers. At the boundaries (in time or in internuclear separation) the wave functions are matched.

This AO-MO matching method has been explored previously (17,18) for a number of collision systems, but its implementation is still rather tedious and not fully tested. The major drawback is in the inclusion of large number of MO basis functions in the inner region, especially in the calculation of nonadiabatic coupling terms including many MO's. It is known that these couplings vary with R rapidly in each avoided crossing region. Thus the MO's and the coupling terms have to be computed with very dense mesh points.

It is worthwhile to mention that the AO-MO matching method is analogous to the R-matrix method for time-independent problems except that the inner region is expanded in terms of MO's instead of treating as a "black box" as in the R-matrix method. The idea is that in the inner region where the electron is localized, it is possible to expand the wave function in a reasonable finite set of basis functions. Thus we do not include the electron translational factors in the inner region.

An important modification of the AO-MO matching method is currently under development. Instead of employing full adiabatic MO functions in the inner region, we use the so-called diabatic-by-sector method for this region. The inner region of R is divided into many small sectors. Within each sector, the basis functions are fixed. In general, these basis functions are chosen to be the molecular orbitals at the mid-point of the sector. Since the basis functions are fixed within each sector, the coupling terms are diabatic in nature and they are smooth functions of R within each sector. The propagation of wave function from one sector to another is obtained by matching the solution at the sector boundary.

In this diabatic AO-MO matching method, the number of basis functions calculated within each sector can be easily enlarged. Thus to solve the time-dependent Schrödinger equation (2), we first expand in terms of traveling atomic orbitals at a large negative time and integrate Eq. (2) with known initial condition as in the standard two-center AO expansion to an internuclear separation R_0 where the resulting electronic wave function is expanded in terms of the MO's of the first sector. The resulting coefficients serve as the initial conditions and the time-dependent Schrödinger equation is integrated within this first sector to reach R_1, where the wave function is expanded in terms of the MO's of the second sector. This procedure is repeated until the other end of the inner region is reached where the resulting wave function is expanded back in terms of traveling AO's. The resulting equation is integrated further out to a large distance to extract scattering amplitude for each channel.

The procedure indicated above has the advantage of all the "old" AO-MO matching method. It has the added strength that the calculation in the inner region becomes very efficient and thus the luxury of including more MO's. This would make it much easier to extend the present matching method to collisions involving more than one electrons.

It is appropriate to mention that this new AO-MO matching method can be viewed as a generalization of the hyperspherical close coupling method for the time-independent problem. The latter has been used fully in atom-diatom scattering (19), in photoionization of atoms (20) and in rearrangement collisions such as positron-hydrogen atom collisions (21,22) and muon transfer cross section calculations (23).

The Practical Limitations of the Close-Coupling Methods

The conventional MO expansion method has been used successfully to calculate cross sections for the dominant excitation and charge transfer channels. The accuracy depends on the size of the basis set used in the specific calculation. Similarly, the AO expansion method can be used successfully to obtain cross sections for the dominant channels near the velocity matching region. With the inclusion of pseudostates, the AO approach can be extended to the lower as well as the higher energy region. The validity of the two methods for the weaker channels is less clear. In this paper, we have documented the problem associated with the AO method for the weak channels. Similar tests for the MO approach have not been carefully examined so far. It is our opinion that the conventional AO or MO approaches have difficulties in treating the weak channels. The new AO-MO matching method would provide a much better possibility since it was designed to remove the major limitations of the AO and the MO methods directly. The error introduced at the matching is unavoidable but it is expected to be small if the experience from the hyperspherical close coupling method holds [19-22]. This is a subject that will be investigated in the near future.

There are still two major areas where the close-coupling method is not expected to be useful. The first one is the determination of double differential cross sections of the ejected electrons in an ionization event where the wave function in the asymptotic region cannot be described analytically. The two-center nature of ionization can be described using wave functions similar to those used in the continuum-distorted wave (CDW) approximation, but the calculations have only been done using perturbation theory. Thus it is difficult to apply CDW to ionizations of atoms by heavy multiply charged ions. Another difficulty of the close coupling method is when many excited states are populated in the collision. In a semiclassical calculation, the basis set has to include all the actual states populated. When a Rydberg level with principal quantum number n is populated, one needs to include roughly n^2 basis functions. In low energy collisions between a multiply charged ion of charge q with atomic hydrogen, the principal quantum number populated is roughly n (a more accurate value can be estimated using the classical overbarrier model, see Ref. 24). Since electron capture in general populates a few Rydberg levels even at low energies, the number of basis functions that need to be included grows rapidly. For ions with charge q greater than 10, the number of basis functions can easily reach a few hundreds. Thus there are few reliable close coupling calculations that have been carried out for ions with charge states greater than 10.

SUMMARY

In this paper we evaluate the status of ion-atom collision theories within the close-coupling approximation. For collisions where the initial electron cloud is strongly distorted and disrupted, there are very few tools available for theorists besides the close-coupling method. The direct numerical solution of the Schrödinger equation or the time dependent Hartree-Fock method have been shown to provide little new impetus to the field. Alternatively the classical method, in particular, the Classical Trajectory Monte-Carlo (CTMC) method, despite of its success in interpreting many experimental results, cannot be treated as an ultimate theory (it may be used as a guidance for formulating a better quantum theory). After all, the electron has been shown not described by classical physics since the beginning of this century.

In this paper we did not discuss perturbative approaches. In ion-atom collisions the perturbative approach is useful only at rather high energies. There is an energy range where the probabilities for the inelastic processes are small, but the first-order perturbation theory is not applicable. Perturbation theories beyond the first-order are seldom carried out accurately.

To include effects beyond what are included in the standard first order approximation, it is often much more convenient to perform a different partition of the Hamiltonian from which a new first-order calculation is carried out. With respect to the "standard" first-order theory this is equivalent to including some higher order effects. Such theories include the strong potential Born approximation, some form of impulse approximation and the different versions of the continuum distorted wave approximations. The usefulness of these approaches can only be checked against accurate experimental data since higher order calculations are often nearly impossible to carry out.

In conclusion, while progress has been made in atomic collision theories in the last two decades, there remain many difficulties in terms of carrying out full *ab initio* calculations for predicting the outcome of ion-atom collisions. In other words, experiments still have to be carried out and theoretical models are still indispensable.

ACKNOWLEDGMENT

It is a great pleasure to submit this paper in honor of Professor Gene Rudd. His dedication to high precision electron spectroscopy in ion-atom collisions provides the benchmark data for many simple collision systems to keep theorists "honest". I am also indebted to Professor N. Toshima for allowing me to show the results in Fig. 2 prior to its publication. This work is partially supported by the US Department of Energy, Office of Energy Research, Office of Basic Energy Sciences, Division of Chemical Sciences.

REFERENCES

1. Kimura, M. and Lane, N.F., *Adv. At. Mol. Opt. Phys.* 26, 79 (1989).
2. Errea, L.F., Harel, C., Jouin, H., Mendez, L., Pons, B., Riera, A., *J. Phys.* B27, 3603 (1994).
3. Shakeshaft, R., *Phys. Rev.* A18, 1930 (1978).
4. Park, J.T., Aldag, J.E., George, J.M. and Peacher, J.L., *Phys. Rev.* A14, 608 (1976).
5. Toshima, N., private communication (1994).
6. Slim, H.A., Ermolaev, A.M., *J. Phys.* B27, L203 (1994).
7. Toshima, N., *J. Phys.* B27, L49 (1994).
8. Fritsch, W. and Lin, C.D., *Phys. Rept.* 202, 1 (1991).
9. Fritsch, W., Shingal, R. and Lin, C.D., *Phys. Rev.* A44, 5686 (1991).
10. Bransden, B.H. and Noble, C.J., *J. Phys.* B14, 1849 (1981).
11. Lundsgaard K.F.V., and Nielsen, S.E. (to be published).
12. Krstic, P.S. and Janev, R.K., *Phys. Rev.* A47, 3894 (1993).
13. Hughes, M.P., Geddes, J. and Gilbody, H.B., *J. Phys.* B27, 1143 (1994).
14. Ovchinnikov, S., this volume.
15. Ermolaev, A.M., *J. Phys.* B24, L495 (1991).
16. Donnelly, A., Geddes J., and Gilbody, H., *J. Phys.* B24, 165 (1991).
17. Kimura, M. and Lin, C.D., *Phys. Rev.* A34, 176 (1986).
18. Kimura, M., *Phys. Rev.* A44, R5339 (1991).
19. Launay, J.M. and Le Dourneuf, M., *Chem. Phys. Lett.* 169, 473 (1989).
20. Tang, J.Z., Watanabe, S. and Matsuzawa, M., *Phys. Rev.* A46, 2437 (1992).
21. Zhou, Y. and Lin, C.D., *J. Phys.* B27, 5065 (1944).
20. Igarashi, A. and Toshima N., *Phys. Rev. A* (1994).
23. Fukuda, N., Ishihara, T. and Hara, S., *Phys. Rev.* A41, 145 (1990).
24. Ryufuku, H., Sasaki, K. and Watanabe, T., Phys. Rev. A21, 745 (1980).

The CDW-EIS Method

Roberto D. Rivarola*, Pablo D. Fainstein[†]
and Víctor H. Ponce[‡]

*Instituto de Física Rosario (CONICET and Universidad Nacional de Rosario), (2000) Rosario, Argentina, [†]Laboratoire de Chimie-Physique-Matière et Rayonnement, Université P. et M. Curie, Paris, France and [‡]Centro Atómico Bariloche and Instituto Balseiro, (8400) SC de Bariloche, Argentina.

Abstract. The electron ionization in ion-atom collisions is studied by using the Continuum Distorted Wave-Eikonal Initial State model. Some relevant aspects of this theory are reviewed. Two-center effects are analyzed.

Single electron ionization is one of the possible resulting reactions when bare nuclei impact on atomic targets. The electron is ejected in the simultaneous presence of the projectile and target fields. The relative movement of the nuclei can be described using the straight line version of the impact parameter approximation so that the internuclear potential can be excluded from our following analysis [1]. The long range coulombic behavior of the projectile and target fields must be taken into account in the initial and final wavefunctions which represent the entrance and final channels. This fact was realized by Cheshire [2], for the case of electron capture, who introduced distorted initial and final wavefunctions. Initial target and final projectile bound states were distorted by projectile and target continuum factors, respectively. So, in this approximation named Continuum Distorted Wave (CDW), two-center wavefunctions were used. A first order of the associated distorted wave series was employed with success to describe charge exchange reactions for different collision systems at high enough impact velocities [3]. Following this idea, Belkić [1] extended the CDW model to study single electron ionization. In such approximation the initial wavefunction is chosen as in the electron capture case, and the final wavefunction is proposed as a product of a plane wave and two continuum factors associated with the projectile and target fields and centered on the respective nuclei.

These distorted wavefunctions, described from the laboratory system verify the correct asymptotic behaviors

$$\lim_{t \to -\infty} \chi_i^+ = \phi_i(\boldsymbol{x}) \exp(-i\epsilon_i t) \exp(-i\frac{Z_P}{v} \ln(vs + \boldsymbol{v} \cdot \boldsymbol{s})) \tag{1}$$

for the entrance channel and

$$\lim_{t,x,s \to +\infty} \chi_f^- = (2\pi)^{-3/2} \exp(i\boldsymbol{k} \cdot \boldsymbol{x} - iE_k t)$$
$$\exp(i\frac{Z_T}{k} \ln(kx + \boldsymbol{k} \cdot \boldsymbol{x}) + i\frac{Z_P}{p} \ln(ps + \boldsymbol{p} \cdot \boldsymbol{s})) \tag{2}$$

for the exit channel.

In equation (1) and (2), ϕ_i and ϵ_i are the initial bound wavefunction and the corresponding electron orbital energy, Z_T and Z_P are the target and projectile nuclear charges, respectively, $\boldsymbol{x}$ and $\boldsymbol{s}$ are the electronic coordinates with respect to the target and projectile nuclei, respectively, $\boldsymbol{v}$ is the collision velocity, $\boldsymbol{k} = \boldsymbol{p} + \boldsymbol{v}$ is the momentum of the ejected electron with respect to a reference frame fixed on the target nucleus and $E_k = 1/2\, k^2$ its kinetic energy. It is clear from condition (2) that the emitted electron cannot be represented only by a plane wave, and that processes as *electron excitation to the continuum* and *electron capture to the continuum* cannot be separated.

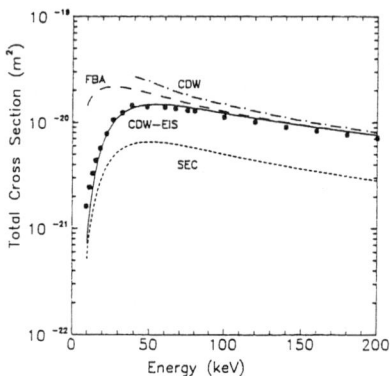

Figure 1: TCS for ionization of H(1s) by H$^+$ impact as a function of projectile energy. Theory: - -, FBA [8]; - · -, CDW [1]; ——, CDW-EIS [7]; - - -, SEC [9]. Experiments: •, [4] and [5].

The CDW approximation considers on an equal footing the presence of the projectile and target fields in the exit channel. However, like in electron capture, CDW total cross sections overestimate the experimental data at intermediate and low collision energies. We mean by intermediate the energetic domain for which the impact velocity is comparable to the initial electron orbital velocity. This overestimation is shown as an example in figure 1 where experimental [4, 5] and theoretical total cross sections (TCS) are presented for ionization of H(1s) by H$^+$ impact. The main reason for this disagreement is the failure in the normalization of the CDW distorted wavefunctions [6]. In the CDW approximation the initial exact wavefunction is replaced by the corresponding CDW distorted one at all collision times. Therefore, one fundamental condition for this replacement is the preservation of the unitarity of the norm for this distorted wavefunction at every time. In order to satisfy this condition, Crothers and McCann [7] introduced the CDW-EIS approximation.

The final distorted wavefunction was chosen as in the CDW model, but in the entrance channel the initial bound state was distorted by an eikonal phase associated with the projectile potential, as given by the asymptotic form presented in equation (1). However, it does not mean that the CDW-EIS approximation must be considered as a degradation of the CDW model. We are dealing with a three body problem where the electron must move in a combination of the projectile and target fields. So, the use of the projectile continuum factor to distort the initial wavefunction does not imply a better representation of the entrance channel than the use of the corresponding eikonal phase at intermediate internuclear distances. On the contrary, theoretical CDW-EIS results are in excellent agreement with experiments.

Calculations using the first order of the Born series (FBA) [8], where the ionization is represented as an excitation from a non-distorted bound state to a non-distorted continuum state of the target, are also shown. CDW-EIS and FBA are in close agreement at high enough energies but FBA overestimates the measurements at intermediate impact velocities. Symmetric Eikonal calculations (SEC) from Fainstein et al [9] that result from replacing the target continuum factor by the corresponding eikonal phase in the CDW-EIS approximation are also included.

In figure 2, TCS corresponding to the CDW-EIS approximation [10] and experiments [11, 12] are in very good agreement for ionization of He$^+$ by H$^+$ impact.

Calculations of total cross sections have been extended for ionization of multielectronic targets [9], reducing the problem to a one active electron model. The initial bound state is represented by a Roothan-Hartree-Fock wavefunction [13] and the interaction between the ejected electron and the residual target in the exit channel is approximated by a coulombic potential of effective nuclear charge $\xi_T = (-2n^2\epsilon_i)^{1/2}$, following the prescription given by Belkić et al [3] for electron capture. TCS calculated with the CDW-EIS model are in close agreement with experimental data [14, 15] for impact of

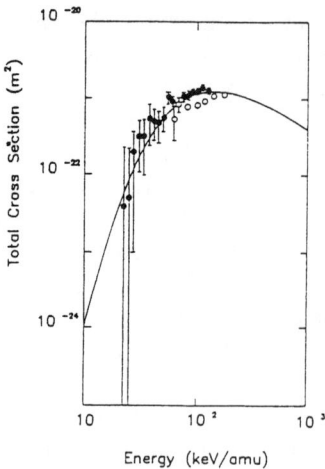

Figure 2: TCS for ionization of $He^+(1s)$ by H^+ impact as a function of projectile energy. Theory: ——, CDW-EIS [10]. Experiments: •, [11]; ∘, [12].

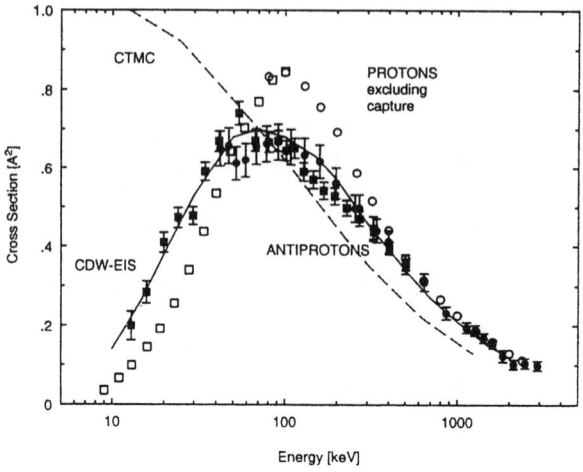

Figure 3: TCS for ionization of He by antiproton impact. Experiments: ■ [15] and • [14]. Theory: ——, CDW-EIS [9]; - - -, CTMC [16]. Data for ionization by proton impact are also shown, □ and ∘, [17, 18].

antiprotons on He targets (see figure 3). Classical trajectory Monte Carlo (CTMC) calculations [16] are also shown in figure 3 and for a further comparison experimental TCS for the case of proton impact are also included [17, 18]. The proton data excludes the electron capture channel contributions. The comparison between the measurements for proton and antiproton impact evidences an effect associated with the projectile nuclear charge. At high impact energies TCS for both projectiles coincide because the initial electronic distributions cannot adapt to the projectile potential. At intermediate velocities the electron cloud reacts as a whole to the attractive or repulsive projectile potential and is polarized. For protons the electron is attracted and for antiprotons is repelled so that the probability of ionization for a given impact parameter is higher for a positive charge projectile than for a negative one. This effect is named the *polarization* effect. At lower energies the active electron adapts to the projectile potential and moves in a dynamical quasi-molecular state. At close internuclear distances, where the emission takes place, the electron evolves in the combined fields of the target and projectile potentials, so that its binding energy increases (decreases) for $Z_P > 0$ ($Z_P < 0$) and consequently the emission probability decreases (increases). This effect is termed the *binding* effect. Basbas et al [19, 20] have studied these effects for the cases of impact of positive charge projectiles on heavier targets, and they were interpreted like deviations from FBA predictions.

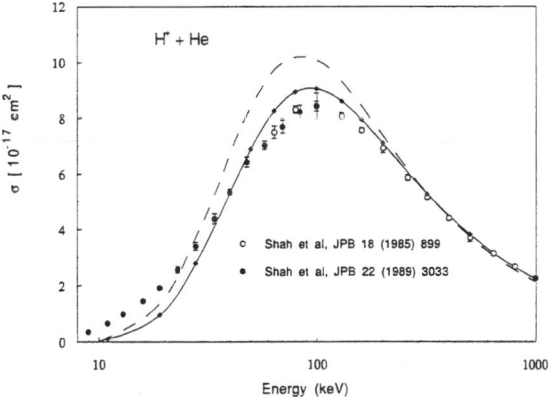

Figure 4: TCS for ionization of He by proton impact. Theory (see the text): - - -, CDW-EIS [9]; ——, CDW-EIS [21]. Experiments: • and ○, [17, 18].

In figure 4, experimental TCS for the H$^+$+He system are compared with two different CDW-EIS calculations. The difference between both calculations resides in the choice of the bound and continuum target wavefunctions.

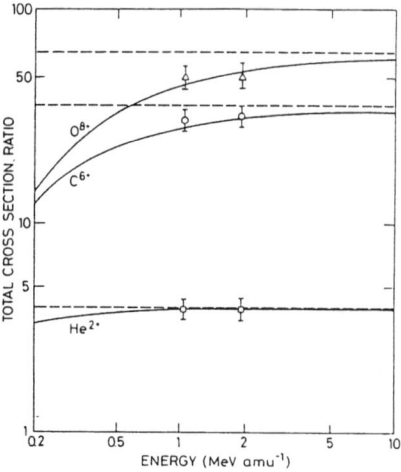

Figure 5: Ratio between the total ionization cross sections for He^{2+} ($\square$), C^{6+} ($\circ$) and O^{8+} ($\triangle$) impact on He and the total ionization cross section for p^+ impact. - - -, FBA; ——, CDW-EIS [9,24]. Experiments are from reference [25].

When these wavefunctions are chosen as for the antiproton case an overestimation of the measurements is obtained at intermediate energies. As it has been shown by Fainstein et al [21] this disagreement is reduced if the active electron-residual target interaction is represented by the same Hartree-Fock-Slater potential in the entrance and exit channel as it was done in the FBA by Madison [22] and Manson et al [23]. Non-distorted bound and continuum target wavefunctions result thus orthogonal. Fainstein et al [21] have also shown that calculations of TCS for antiproton impact obtained in the CDW-EIS approximation, using Hartree-Fock-Slater target wavefunctions, do not modify the results presented in figure 3.

In the FBA the TCS for ionization in collisions between multiply charged ions and atomic targets can be obtained by using the known Z_P^2-scaling law. In order to analyze deviations from this scaling law, we represent in figure 5 the ratio between the TCS for ionization of He by impact of a bare projectile of nuclear charge Z_P with the TCS for ionization of the same target by impact of protons with the same velocity. CDW-EIS predictions [9, 24] and experimental data [25] are compared for impact of He^{2+}, C^{6+} and O^{8+} ions. The agreement between theory and measurements is very good. Deviations from the Z_P^2-law become more pronounced as the nuclear charge Z_P increases. The CDW-EIS model reproduces the experimental TCS ratios because it takes into account the distortion introduced by the long-range Coulomb potential of the projectile. One is tempted to identify this

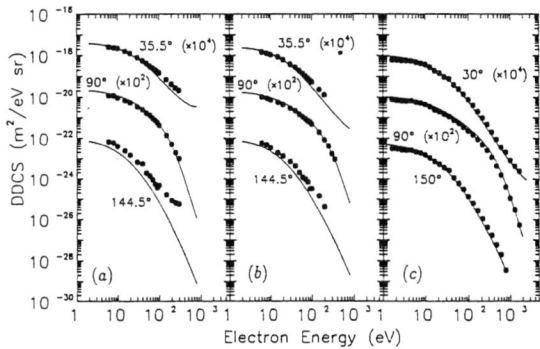

Figure 6: DDCS for ionization of He by (a) 1 MeV amu^{-1}, (b) 1.84 MeV amu^{-1} and (c) 5 MeV amu^{-1} C^{6+} impact for fixed electron emission angles and as a function of electron energy. Theory: —–, CDW-EIS [26,27]. Experiments: (a) and (b), [25,28] and (c), [29].

distortion effect with the binding one introduced by Basbas et al [19, 20]. In their works K-shell vacancy production was studied for asymmetric systems with $Z_P << Z_T$ so that in such cases it is valid to develop a perturbative series in powers of Z_P. Moreover the capture channels can be neglected. But for the cases here analyzed, for which $Z_P \geq Z_T$, Z_P cannot be used as a *small* parameter for a perturbative expansion.

Doubly differential cross sections (DDCS) give a detailed information on the ionization processes and constitute a severe test for the validity of the theoretical approximations. DDCS are usually given as a function of the energy E_k and of the solid scattering angle subtended by the ejected electron. In figure 6, DDCS for ionization of He by impact of C^{6+} ions are presented at three different collision energies: 1 MeV/amu, 1.84 MeV/amu and 5 MeV/amu. DDCS are plotted as a function of E_k for fixed values of the electron polar scattering angle θ. The agreement between CDW-EIS predictions [26, 27] and experiments [25, 28, 29], which is very good in all cases, improves as the collision energy increases.

In order to study deviations from the Z_P^2-law given by FBA, the ratios between DDCS for impact of 5 MeV/amu C^{6+}, O^{8+} and Ne^{10+} on He and DDCS for impact of protons with the same velocity on He, all divided by Z_P^2, have been studied experimental [29] and theoretically [26]. The same ratios were also studied by Stolterfoht et al [30] for impact of 25 MeV/amu Mo^{40+} on He and by Pedersen et al [25, 28] for impact of He^{2+}, C^{6+} and

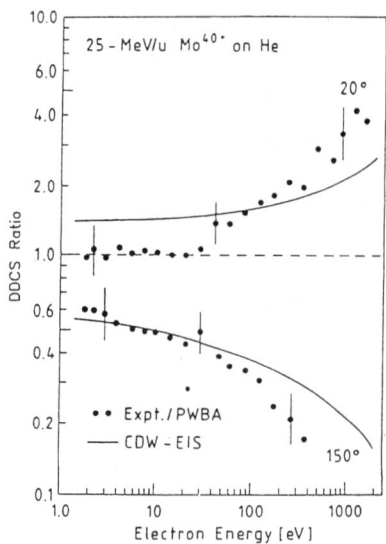

Figure 7: Ratio between DDCS for ionization of He by 25 MeV amu^{-1} Mo^{40+} and H$^+$ impact with the same velocity, divided by Z_P^2, as a function of electron energy for $\theta = 20°$ and $150°$. Theory: ——, CDW-EIS; - - - -, FBA. Experiments: •. Experimental and theoretical data extracted from reference [30].

O^{8+} on He at 1.00 and 1.84 MeV/amu collision energies. In figure 7 the results for Mo^{40+} are presented as a function of E_k and for two fixed angles θ, one of 20 degrees in the forward direction and another of 150 degrees in the backward direction. FBA calculations which correspond to a *single center* representation give a constant value 1.0, independently of θ and E_k. CDW-EIS ratios show larger (lower) values than 1.0 for 20 (150) degrees scattering angle. For $\theta < 90°$ ($\theta > 90°$) the ratio increases (decreases) as a function of E_k due to the attraction provoked by the Coulomb potential of the highly charged ion, which distorts the final electronic distribution. The same behavior is observed for the ratio between experimental DDCS for Mo^{40+} and theoretical FBA-DDCS for proton impact, as shown in figure 7. We must note the fact that at the high energies considered, the FBA is expected to give a good representation of the reaction for the case of proton projectiles. This observed effect has been termed *two-center electron emission* (TCEE) [30]. It is possible to show that if the eikonal distortion is introduced in the initial state as in CDW-EIS, but the final state is chosen as in FBA, the DDCS ratio depends on the projectile charge and collision velocity but not on E_k and θ [31]. Moreover, if the initial and final wavefunctions are

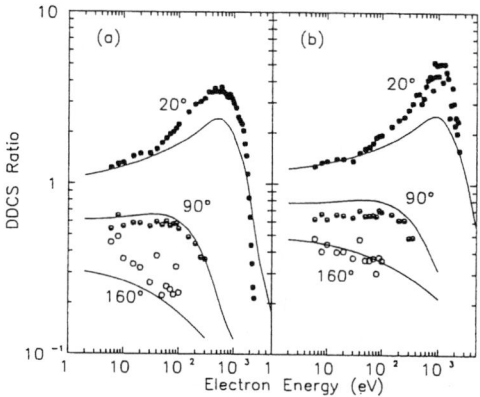

Figure 8: Ratio between DDCS for ionization of He by (a) 1 MeV amu^{-1} and (b) 1.84 MeV amu^{-1} C^{6+} and H$^+$ impact with the same velocity, divided by Z_P^2 as a function of electron energy for $\theta = 20°$, $90°$ and $160°$. Theory: ——, CDW-EIS. Experiments: •, ■, ∘, [25,28].

chosen as in FBA and CDW, respectively, the resulting DDCS ratio does not reproduce the observed dependence.

In figure 8, we present the DDCS ratio for impact of 1 MeV/amu and 1.84 MeV/amu ions of C^{6+} on He at $\theta = 20, 90$ and 160 degrees. Up to a certain energy E_k the behavior is similar to the Mo^{40+} case. But, for 20 degrees as E_k increases, the ratio presents a maximum with a strong decreasing slope as E_k continues increasing. The position of the maximum depends on the emission angle [28], shifting to lower E_k when θ increases from 0 degrees. The maximum is associated with the electron capture to continuum peak (ECC) that appears at $\theta \simeq 0$ degrees at final electron velocities close to the projectile velocity. In the experiments from Stolterfoht et al [30] this effect has not been observed because the measured energies E_k were not high enough to exhibit the maximum. The DDCS ratio decreases at higher energies reaching the value 1 at energies E_k corresponding to the binary encounter peak. The ratio continues decreasing under the value 1 given by the FBA approximation. This effect could be attributed to the following fact. When the final electron velocity is much larger than the collision one, the ejected electron feels an effective nuclear charge (which results from the combination of the projectile and target nuclear charges) larger than Z_T (the

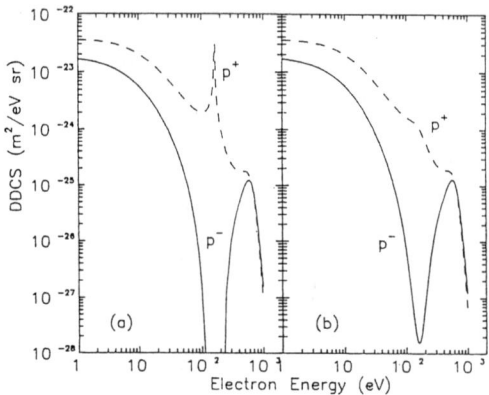

Figure 9: DDCS for ionization of He by 300 keV antiproton and proton impact as a function of electron energy for (a) $\theta = 0°$ and (b) $\theta = 10°$. Theory: ———, CDW-EIS for antiproton impact; - - - -, CDW-EIS for proton impact.

one assumed to be felt by the ejected electron in the FBA).

Interesting physical information is obtained if DDCS are compared for protons and antiprotons colliding with He targets, where two attractive or one repulsive and one attractive potentials are combined. In figure 9, DDCS are presented for these projectiles with a 300 keV collision energy at scattering angles $\theta = 0$ and 10 degrees [32]. For the proton impact case the ECC peak is observed at $\theta = 0$ degrees while at 10 degrees only a hump remains. This behavior is well known from the experiments of Rudd et al[33] and the calculations of Salin [34]. At larger angles the structure is not observed. For the antiproton case a dip appears at $\theta = 0$ degrees, and at $\theta = 10$ degrees a pronounced valley still remains. It is clear that in the proton case there is a preference of the electrons to move with a velocity close to the projectile one, but in the antiproton case the electrons are repelled at those velocities. As the DDCS depends on the density of the states of the attractive or repulsive Coulomb projectile potential, and these densities are different for both cases, DDCS show a very different behavior for proton or antiproton impact. In figure 10, single differential cross sections (SDCS) are shown as a function of θ for 200 keV protons and antiprotons colliding with He targets. The CDW-EIS calculations [35] are in close agreement with experimental data [36] for proton projectiles. In the forward (backward) direction the SDCS for antiproton impact is lower (larger) than the SDCS for proton impact due to the repulsive potential of the negatively charged particle. CTMC calculations [38] support this behavior. The ratio of SDCS for impact of multicharged

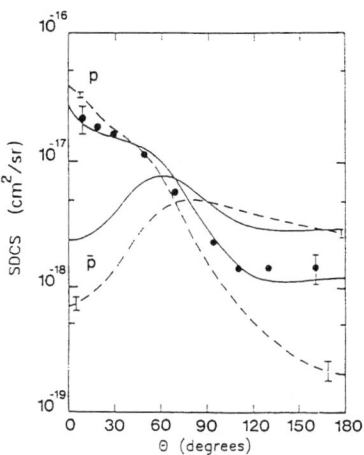

Figure 10: SDCS for ionization of He by 200 keV antiproton and proton impact as a function of electron emission angle. Theory: ——, CDW-EIS [35]; - - - -; CTMC [38]. Experiments: •, [36].

bare ions on He atoms and SDCS for impact of protons at the same velocities and targets, all divided by Z_P^2, has been also studied [24, 25]. CDW-EIS and experimental results show that at forward (backward) angles this parameter increases (decreases) as Z_P increases.

In the case of multicharged particles on lighter targets it has been also shown that the strong potential produced by the projectile affects the height and position of the binary encounter peak (BEP) in DDCS [37]. In a classical analysis, the BEP, which appears at $\theta < 90$ degreees, arises from the binary collision between the impinging ion and the active electron. From the classical laws of energy and momentum conservation it is possible to show that the BEP is produced at electron momenta close to $k = 2v\cos\theta$. If the binding energy ϵ_i is considered, the BEP is shifted to lower energies by a factor proportional to ϵ_i. Within this binary collision picture the DDCS is the Rutherford cross section averaged over the initial state momentum distribution [39]. In this binary encounter approximation (BEA), the DDCS is proportional to Z_P^2, which is the Z_P^2-scaling law also given by FBA. Experimental results show that this scaling law is valid for relatively small Z_P ($Z_P < 10$) [25, 28, 40], but deviations are observed for larger values of Z_P. This behavior is confirmed in figure 11 where the cross section maxima σ^+ are represented as a function of the projectile charge state for impact of 3.6

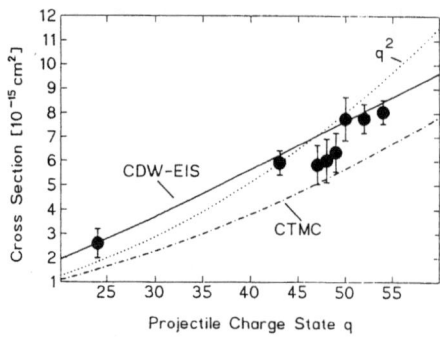

Figure 11: He single ionization cross sections σ^+ for 3.6 MeV amu^{-1} Au^{q+} impact as a function of the projectile ionic charge (full circles); dotted curve, q^2 dependence; full curve, CDW-EIS calculations [42]; chain curve, CTMC calculations [43].

MeV/amu-Au^{q+} on He. Experiments [41] are in good agreement with CDW-EIS prediction [42]. CTMC calculations [43] deviate also from the q^2 power law indicated by a dotted curve in the figure.

The absolute value of the BEP shift from the classical prediction, for impact of 1 MeV/amu bare projectiles on He, as a function of Z_P and at $\theta = 0$ degrees, is shown in figure 12. The CDW-EIS calculations show an increasing of the shift as Z_P increases. This behavior is supported by the simple tunneling (TM) [37] and Bohr and Linhard (BL) [25, 44] models. The TM and BL representation of the ionization reaction give a BEP shift proportional to $Z_P^{1/2}$. The experimental point for $Z_P = 9$ [40] is in agreementwith the CDW-EIS predictions. The dependence of the BEP shift with the collision energy and with electron scattering angle has been studied by different authors [25, 28, 37, 45].

In conclusion, some relevant aspects of the CDW-EIS approximation have been reviewed. The electron ionization channel resulting from ion-atom collisions is a two-center reaction and must be represented by a theory that includes the simultaneous action of the projectile and target fields on the active electron. This is the case for the CDW-EIS model.

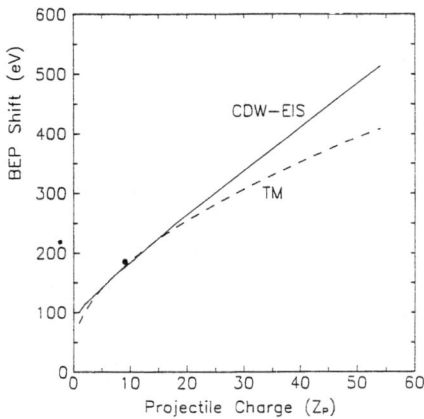

Figure 12: BEP shift from the classical prediction, for impact of 1 MeV/amu bare projectiles on He, as a function of Z_P at a scattering angle $\theta = 0$ degrees. Theory: ——, CDW-EIS [37]; - - -, TM [37]. Experimental data from reference [40].

References

[1] Belkić, Dž., *J. Phys. B: At. Mol. Phys.* **11**, 3529 (1978).

[2] Cheshire, I. M., *Proc. Phys. Soc.* **84** 89 (1964).

[3] Belkić, Dž, Gayet, R. and Salin, A., *Phys. Rep.* **56** 279 (1979).

[4] Shah, M. B., and Gilbody, H. B., *J. Phys. B: At. Mol. Phys.* **14** 2831 (1981).

[5] Shah, M. B., Elliot, D. S., McCallion, P., and Gilbody, H. B., *J. Phys. B: At. Mol. Phys.* **20** 2481 (1987).

[6] Crothers, D. S. F., *J. Phys. B: At. Mol. Phys.* **15** 2061 (1982).

[7] Crothers, D. S. F., and McCann, J. F., *J. Phys. B: At. Mol. Phys.* **16** 3229 (1983).

[8] Bates, D. R., and Griffing, G., *Proc. Phys. Soc. A* **66** 961 (1953).

[9] Fainstein, P. D., Ponce, V. H., and Rivarola, R. D., *Phys. Rev. A* **36** 3639 (1987).

[10] Martínez, A. E., Deco, G. R., Rivarola, R. D., and Fainstein, P. D., *Nucl. Instr. Meth. B* **43** 24 (1989).

[11] Rinn, K., Melchert, F., Rink, K., and Salzborn, E., *J. Phys. B: At. Mol. Phys.* **19** 3717 (1986).

[12] Watts, N. F., Dunn, K. F., and Gilbody, H. B., *J. Phys. B: At. Mol. Phys.* **19** L355 (1986).

[13] Clementi, E., and Roetti, C., *At. Data Nucl. Data Tables* **14** 177 (1974).

[14] Andersen, L. H., Hvelplund, P., Knudsen, H., Moller, S. P., Pedersen, J. O. P., Tang-Petersen, S., Uggerhoj, E., Elsener, K., and Morenzoni, E., *Phys. Rev. A* **41** 6536 (1990).

[15] Hvelplund, P., Knudsen, H., Mikkelsen, U., Morenzoni, E., Moller, S. P., Uggerhoj, E., and Worm, T., *J. Phys. B: At. Mol. Phys.* **27** 925 (1994).

[16] Schultz, D. R., *Phys. Rev. A* **40** 2330 (1989).

[17] Shah, M. B., and Gilbody, H. B., *J. Phys. B: At. Mol. Phys.* **18** 899 (1985).

[18] Shah, M. B., and Gilbody, H. B., *J. Phys. B: At. Mol. Phys.* **22** 3037 (1989).

[19] Basbas, G. E., Brandt, W., and Laubert, R., *Phys. Rev. A* **7** 983 (1973).

[20] Basbas, G. E., Brandt, W., and Laubert, R., *Phys. Rev. A* **17** 1655 (1978).

[21] Fainstein, P. D., Gulyás, L., and Salin, A., unpublished (1994).

[22] Madison, D. H., *Phys. Rev. A* **8** 2449 (1973).

[23] Manson, S. T., Toburen, L. H., Madison, D. H., and Stolterfoht, N., *Phys. Rev. A* **12** 60 (1975).

[24] Fainstein, P. D., Ponce, V. H., and Rivarola, R. D., *J. Physique Coll.* **50** C1 183 (1989).

[25] Pedersen, J. O. P., Hvelplund, P., Petersen, A. G., and Fainstein, P. D., *J. Phys. B: At. Mol. Phys.* **24** 4001 (1991).

[26] Fainstein, P. D., Ponce, V. H., and Rivarola, R. D., *J. Phys. B: At. Mol. Phys.* **21** 287 (1988).

[27] Fainstein, P. D., Ponce, V. H., and Rivarola, R. D., *J. Phys. B: At. Mol. Phys.* **24** 3091 (1991).

[28] Pedersen, J. O. P., Hvelplund, P., Petersen, A. G., and Fainstein, P. D., *J. Phys. B: At. Mol. Phys.* **23** L597 (1990).

[29] Platten, H., Schiwietz, G., Schneider, T., Schneider, D., Zeitz, W., Musiol, K., Zouros, T., Kowallik, R., and Stolterfoht, N., in *Proc. 15th. Int. Conf. on the Phys. of Electronic and Atomic Collisions (Brighton)*, 1987.

[30] Stolterfoht, N., Schneider, D., Tanis, J., Altevogt, H., Salin, A., Fainstein, P. D., Rivarola, R. D., Grandin, J. P., Scheurer, J. M., Andriamonje, S., Bertault, D., and Chemin, J. F., *Europhys. Lett.* **4** 899 (1987).

[31] Reading, J. F., and Fitchard, E., *Phys. Rev. A* **10** 168 (1974).

[32] Fainstein, P. D., Ponce, V. H., and Rivarola, R. D., *J. Phys. B: At. Mol. Phys.* **21** 2989 (1988).

[33] Rudd, M. E., Sautter, C. A., and Bailey, C. L., *Phys. Rev.* **151** 20 (1966).

[34] Salin, A., *J. Phys. B: At. Mol. Phys.* **5** 979 (1972).

[35] Fainstein, P. D., Ponce, V. H., and Rivarola, R. D., *J. Phys. B: At. Mol. Phys.* **22** L559 (1989).

[36] Rudd, M. E., Toburen, L. H., and Stolterfoht, N., *At. Data Nucl. Data Tables* **18** 413 (1976).

[37] Fainstein, P. D., Ponce, V. H., and Rivarola, R. D., *Phys. Rev. A* **45** 6417 (1992).

[38] Olson, R. E., and Gay, T. J., *Phys. Rev. Lett.* **61** 302 (1988).

[39] Gryzinski, M., *Phys. Rev.* **115** 374 (1959).

[40] Lee, D. H., Richard, P., Zouros, T. J. M., Sanders, J. M., Shinpaugh, J. L., and Hidmi, H., *Phys. Rev. A* **41** 4816 (1990).

[41] Berg, H., Ullrich, J., Bernstein, E., Unverzag, M., Spielberg, L., Euler, J., Schardt, D., Jagutzki, O., Schmidt-Böcking, H., Mann, R., Mokler, P. H., Hagmann, S., and Fainstein, P. D., *J. Phys. B: At. Mol. Phys.* **25** 3655 (1992).

[42] Fainstein, P. D., results presented in Datz, S., Hippler, R., Andersen, L. H., Dittner, P. F., Knudsen, H., Krause, H. F., Miller, P. D., Pepmiller, P. L., Rosseel, C. T., Schuch, R., Stolterfoht, N., Yamazaki, Y., and Vane, C. R., *Phys. Rev. A* **41** 3559 (1990).

[43] McKenzie, M. L., and Olson, R. E., *Phys. Rev. A* **35** 2863 (1987).

[44] Bohr, N., and Linhard, J., *K. Dansk. Vidensk. Selsk. Mat. Fys. Meddr.* **28** N° 7 (1954).

[45] González, A. D., Dahl, P., Hvelplund, P., and Fainstein, P. D., *J. Phys. B: At. Mol. Phys.* **26** L135 (1993).

One- and Two-Center Electron Emission in Energetic Ion-Atom Collisions

N. Stolterfoht

*Hahn-Meitner-Institut Berlin, Bereich Festkörperphysik Glienickerstr. 100,
D-14109 Berlin, Germany*

Abstract: Fundamental mechanisms for electron production in ion-atom collisions are reviewed. Electron emission is discussed in terms of categories involving zero, one and two atomic centers. The centers are defined as heavy nuclear charges which interact strongly with the ejected electron. The binary encounter process, attributed to zero or one center, is dominated by two-body kinematics. The soft collision process, which corresponds to dipole transitions with small momentum transfer, is associated with one center. Various two-center phenomena are discussed. Electron capture to continuum, giving rise to a cusp-shaped peak near $0°$, is viewed as a focusing effect in the Coulomb field of the receding projectile. Two-center electron emission, involving the effect of both target and projectile charge on the outgoing electrons, is shown to be visible at forward as well as at backward emission angles.

INTRODUCTION

The process of electron ejection in ion-atom collision is of fundamental importance for many fields of basic research and applications. Accordingly, ion-induced electron emission received a great deal of attention during the past decades. It has first been studied systematically within the pioneering experiments by Kuyatt, Jorgensen, Rudd and collaborators (1-3) in the early 60's. In that work, ejected electrons were measured in nearly complete ranges of electron energies and emission angles. By these studies the community was introduced into the field of electron emission in ion-atom collisions. It became evident that the measurements of low-energy electrons at energies of a few eV are extremely difficult and that a large number of the tedious problems were previously solved in a meritorious manner (1-3).

This article was prepared for the conference on two-center effects held in honor of Eugene Rudd. I would say that the early experimental data, being as old as 30 years, can still compete in quality with the results measured to date. The experimental work provided deep insight into the dynamics of ion-atom collisions and the structure of individual atoms. Various electron production mechanisms were discovered in those early experiments. This fruitful work motivated several groups to study the details of electron production mechanisms in ion-atom collisions. After 30 years of intense studies of ion-induced electron emission it is certainly worthwhile to look back at some basic features of the field.

The present article is devoted to the fundamental mechanisms responsible for the emission of electrons in ion-atom collisions. As several mechanisms can produce electrons in heavy ion-atom collisions, it is useful to find general categories under which they can be summarized. In this work, the electron production mechanisms are discussed in terms of collision centers consisting of heavy nuclei whose fields interact *strongly* with the ejected electron. As ion-atom collisions involve (at least) two nuclei, we distinguish primarily the cases of one and two centers. However, also the case of zero center, where both heavy particles interact *weakly* with the electron, will be considered. The work performed within the last three decades is so extensive that no attempt is made to give a complete summary. Rather, I shall point out a few general features which are essential for our understanding of ion-induced electron emission.

ELECTRON EMISSION FROM CENTERS

In recent years considerable attention has been devoted to two-center effects in ion-atom collisions accompanied by electron emission (4-8). In view of these recent studies it appears useful to discuss first the conceptual aspect of a center in more detail. Before treating two-center phenomena, the cases of zero and one center will be discussed.

Formally, a center is associated with a heavy nucleus whose interaction with the electron requires a description beyond first-order perturbation theory. The principles of the center concept are schematically displayed in Fig. 1 which shows the removal of an electron in a perturbation caused by the incident ion. Examples for which the outgoing electron is unaffected, deflected in the field of the target

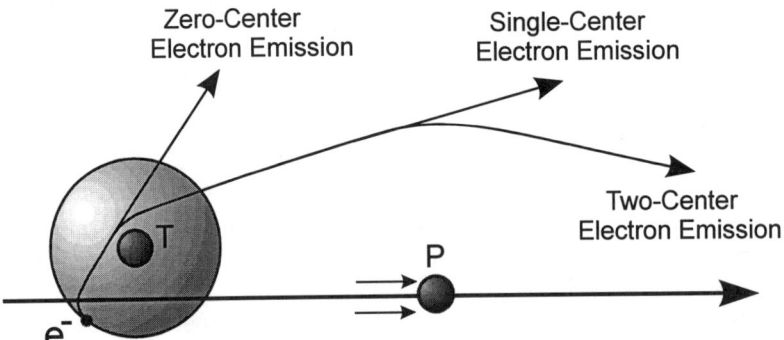

Figure 1: Mechanisms for electron emission from different centers provided by the target (T) and the projectile (P)

nuclear charge, and deflected by both target and projectile charges involve zero, one, and two centers, respectively.

To apply this definition unambiguously it is useful to set up a convention for the treatment of the nucleus, to which the electron is bound before the collision. For instance, the precollisional interaction may be disregarded completely. Likewise, the effect on the bound electron before the collision may be considered as a single (weak) interaction. This latter choice is adopted here. Certainly, both choices are arbitrary. They are considered to allow for the introduction of the zero-center case, where both projectile and target nuclei affect the electron in a perturbation. In particular, when the electron is bound to the target before the collision, the zero-center case implies a negligible effect of the target nucleus during the collision. It is noted that neglecting the effect of the nuclear charge, to which the electron is initially bound, is well known as the free-electron or impulse approximation.

In the following, the cases involving different centers are discussed in more detail. Two-body interactions, where both zero- and one-center cases exist, are

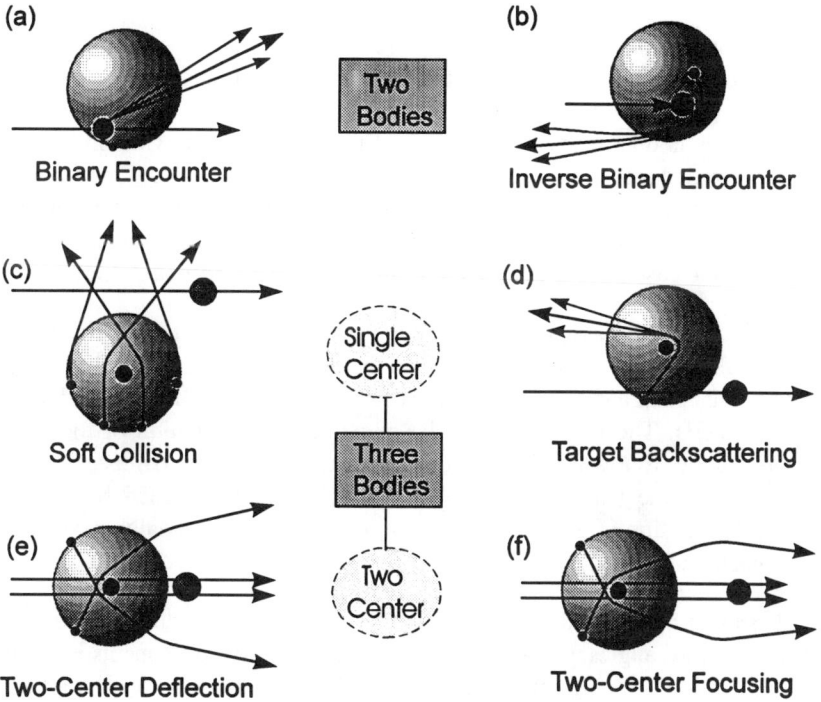

Figure 2. Scenario of two- and three-body mechanisms for electron emission where zero, one, and two centers are important.

considered first. Then, three-body interactions are treated including the one- and two-center case. A summary of the electron emission mechanisms is shown in Fig. 2. For simplicity, only ionization of the target is discussed here. Analogously, the electron production mechanisms occur for projectile ionization (9-12). It is noted that the following discussion applies for energetic projectiles whose velocity is larger than that of the associated bound electron.

TWO BODIES: BINARY ENCOUNTER EMISSION

When the interaction of the outgoing electron with the target nucleus is neglected, the ionization process is determined by a two-body mechanism that is referred to as binary-encounter (BE) emission (Fig. 2a). This process, where the target nucleus is only responsible for the initial velocity distribution of the electrons, is described by the binary-encounter theory (13). Binary-collisions are usually associated with a single center. It should be realized, however, that for a weakly interacting projectile, the two-body problem reduces to a zero-center case (see Fig. 1). Formally, this zero-center case is defined by the Born approximation involving a *plane wave* for the final state of the continuum electron (14).

The interaction of a bare projectile with an electron may be treated within the theoretical framework already established in the fundamental studies by Rutherford (15) in 1911. It is noted that a bare projectile does not seem to provide a center. This is due to the remarkable fact that the two-body Coulomb problem can be solved exactly by first-order perturbation theory (14). Thus, a violent binary collision, involving a *bare* projectile center, yield the same result as the zero-center case. On the contrary, for a *dressed* projectile it is necessary to distinguish the cases where the projectile does or does not form a center.

For sufficiently fast projectiles, the binary collision event gives rise to a pronounced BE peak that can be observed in the double differential ionization cross sections. This peak has clearly been identified in the early work by Rudd and collaborators (2,3). The location of the BE maximum is a function of the electron emission angle following from two-body kinematics. The BE peak shape is determined by the Compton profile which, in turn, is governed by the initial velocity distribution of the ejected electron. For an electron initially at rest the BE peak vanishes at angles larger than $90°$.

It should be realized, however, that the BE process produces electron ejection at backward angles, in particular by the high-velocity components involved in the initial velocity distribution. In this case the projectile approaches as a relatively slow particle, providing a Coulomb field, which may significantly change the momentum direction of a rapidly orbiting electron (Fig. 2b). In fact, this scattering event is associated with a binary-encounter process, where the role of the projectile and the target electron is reversed. The inverse binary-encounter process may produce electrons that are leaving the target atom with a velocity as large as their initial velocity. This process is similar to the *sling-shot* mechanisms pointed out by Burgdörfer elsewhere in this proceedings (16).

When the incident ion carries electrons, the projectile nuclear charge is partially screened. The non-Coulombic field of a dressed projectile gives rise to unexpected effects when it interacts strongly with the target electron, or, with other words, when the projectile becomes a center. This projectile single-center case is displayed in Figs. 3a and 3b. As the electrons undergo violent interactions with the projectile They are likely to be scattered at 180° giving rise to BE emission at 0° in the laboratory frame. It was realized only recently by Richard et al. (17) that a dressed projectile can scatter electrons by 180° more effectively than bare projectiles. At small distances, the screened Coulomb potential is steeper than that produced by a bare ion so that the force, equal to the derivative of the potential, becomes relatively large. Hence, the BE peak at the observation angle of 0° is expected to be more pronounced for a dressed projectile than for a bare one. This has clearly been found in various experimental and theoretical studies performed recently for BE electron emission at 0° (17-21).

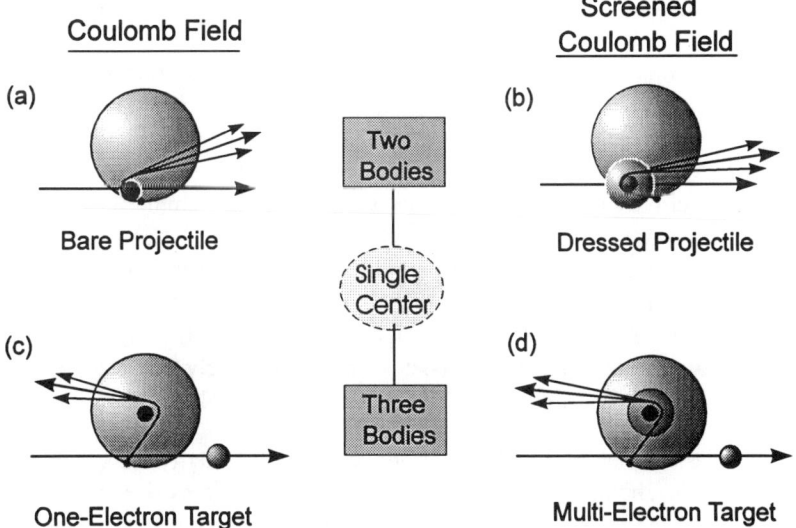

Figure 3. Comparison of electron scattering near 180° in the field of a bare (a) and a dressed (b) projectile nucleus. In (c) and (d) the corresponding backscattering is shown for the target nucleus. Note that large angle scattering is more probable for the dressed heavy particle than for the bare ones.

Finally, it is noted that the binary encounter maximum disappears in the single differential cross section integrated over the electron emission angle. This is due to the fact that the binary-encounter maximum is continuously varying in energy as the electron emission angle varies. Hence, the single differential cross section is determined by the top part of the binary-encounter maximum. Consequently, the single differential cross section is dominated by single-center effects.

THREE BODIES: SINGLE-CENTER PHENOMENA

Single-center phenomena may also be associated with the target atom. They become important when the outgoing electron is significantly influenced by the field of the target nucleus (Fig. 1). The influence of the target nucleus increases with decreasing electron energy. At the low-energy limit, the interaction of the target nucleus with the active electron may be stronger than the interaction between the active electron and the projectile. This occurs for glancing collisions giving rise to the soft-collision (SC) maximum (3,11). The weak projectile interaction producing the soft-collision electrons allows for the application of perturbation theory. Therefore, soft collisions are described by the first Born approximation where the initial (bound) state and a final (continuum state) of the electron is centered at the target nucleus (22,23). The use of target-centered wave functions is in accordance with the case of a single-center formed by the target nucleus.

Soft collisions refer to a mechanism where the momentum transfer by the projectile reaches its minimum. This small-momentum transfer mechanism has been first pointed out by Bethe (22) in 1932. He provided evidence that this mechanism produces low-energy electrons in dipole-type transitions that are well known in the field of photoionization. However, it should be emphasized that pure dipole transitions are produced only by very fast projectiles. Generally, monopole transitions, characteristic for charged particle at lower incident velocities, play a significant role. The interference between the monopole and dipole term gives rise to an asymmetry in the angular distributions around 90°. This asymmetry, also enhanced by two-center effects, have been seen in rather early studies (2,3,24,25). Examples for angular distributions (25,26) are given in Fig. 4.

In a classical picture, the SC electrons of near zero energy are ejected close to 90° (Fig. 2c). The initial electron velocity distribution gives rise to a broadening of the electron emission at 90°. However, this broadening effect is generally insufficient to explain the wide angular distribution around 90° which is typical for electrons at the low-energy limit. The broad angular distribution may be understood since the deflection of the low-energy electrons in the field of the (screened) target nucleus is taken into account. As indicated in Fig. 2c, in the nuclear target field they are likely to lose their "memory" of the initial 90° direction and, hence, the outgoing electrons are distributed into a wide angular range. It should be pointed out, however, that classical pictures are likely to fail in this case, since the quantum-mechanical aspect plays a dominant role for soft collisions.

Fig 2d shows a specific case, where a high-energy electron is initially emitted in the forward direction and, then, is backscattered in the field of the target nucleus. This process contributes to the backward emission of high-energy electrons, in addition to the inverse binary encounter mechanism discussed above. As expected, the probability for the target backscattering decreases with increasing electron energy.

The target backscattering bears some resemblance to the production of the binary encounter electron at 0° that also implies backscattering of the electron by about 180°. In Fig. 3 the two types of processes are compared. The backscattering becomes more effective as the Coulomb field of the target increases. Hence in a multi-electron target atom (e.g., Ar), backscattering is larger than in a single electron target (H). The backscattering picture explains the finding by Rudd and collaborators (2,3) that scaled hydrogenic wave functions used in conjunction with the Born approximation yielded theoretical cross sections for electron emission which underestimated strongly the corresponding experimental results at backward angles. Later, Hartree-Slater wave functions were used in the calculations by

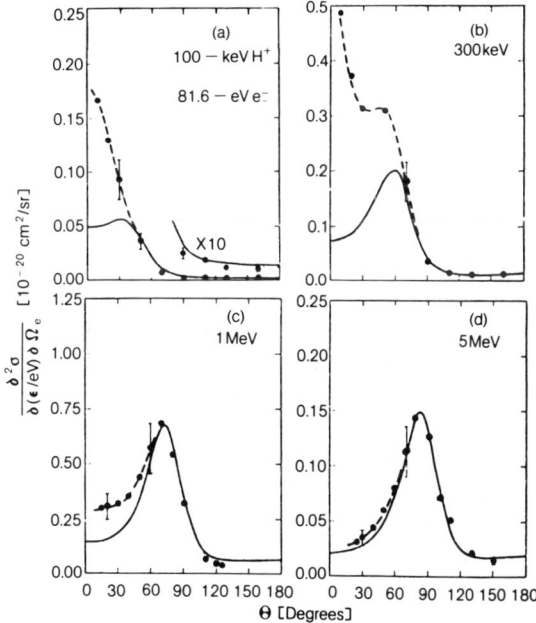

Figure 4. Angular distributions of 81.6 eV electrons ejected in H$^+$ + He collisions for different impact energies. Experiment: Points and theory: Solid line. From Ref.(25).

Madison et al. (24) and Manson et al. (25) and excellent agreement was found between theory and experiment at angles larger than about 90°. This can clearly be seen in Fig. 4 which shows calculations by Manson et al. in comparison with the experimental data from different laboratories at Lincoln, Richland, and Berlin (25).

As outlined above, the Born approximation describes adequately the production of low-energy electrons by dipole-type transitions. However, it is important to note that the Born approximation also correctly describes the production of the binary-encounter peak by bare projectiles. This may be surprising since binary encounter electrons are produced in violent collisions for which perturbation theory generally fails to be valid. To explain this seeming contradiction, it is recalled that the two-body Coulomb problem yields the same result when calculated either in first- or higher-order perturbation theory (14).

Fig. 4 provides information for the validity of the Born approximation including Hartree-Slater wave functions in both initial and final states. As mentioned above, excellent agreement between theory and experiment is obtained at backward angles, however, significant deviations are observed at forward angles. These discrepancies are attributed to two-center effects. It is important to note that for the relative high projectile energies of 5 MeV, the theory agrees very well with experiment at all angles (Fig. 4d). This gives confidence that the Born approximation can be used as a standard to verify two-center effects as is done in the following Section.

THREE BODIES: TWO-CENTER PHENOMENA

Two-center effects become important when the outgoing electron is increasingly influenced by both the target and the projectile nucleus. In this case the electron may undergo simultaneous or successive deflections in the two-center field of the heavy particles (Fig. 1). Two-center phenomena are observed in regions of the electron spectra where they are not shadowed by dominant one-center effects. As pointed out in the previous Section, two-center effects are identified from measured cross sections by comparing with the corresponding model results from the Born approximation representing the contribution from a single-center. Alternatively, two-center effects may be identified exclusively from experiment by comparing results for high incident charge states with equal velocity data for proton impact. This method, however, works only for energies not lower than a few MeV/u, since low-energy protons give also rise to two-center effects (Fig. 4).

Two-center effects have been observed by various groups. Enhanced electron emission at 10° has already been seen by Rudd and collaborators (2), using proton impact at energies of several hundred keV. This observation motivated Crooks and Rudd (27) to study electron emission at an angle of 0°. Hence, they discovered a pronounced cusp-shaped peak which they attributed to the process of electron capture to the continuum (ECC). Two-center effects are expected also in saddle-point electrons "stranded" along the internuclear axis where the combined Coulomb potential of the collision partners has a minimum (4,5,28). Moreover, two-center electron emission (TCEE) has been observed at both forward and backward angles in fast collisions involving highly charged projectiles (7,29-31). Finally, two-center effects are found to influence even soft- and binary-collision electrons [7,32,33]. In the following the ECC and TCEE mechanisms will be discussed in more detail.

Electron capture to the continuum

The electron-transfer-to-continuum mechanism involves electrons traveling at 0° with the velocity equal to the projectile velocity. This mechanism produces a cusp-like peak at the electron energy corresponding to the matching of the projectile and electron velocities. The electron cusp was predicted by Salin (34) and Macek (35) who showed that the ECC mechanism requires a description beyond the Born approximation based on target-centered wave functions. After its first observation by Crooks and Rudd (27), many groups have devoted their attention to the ECC process. The details of those studies lie outside the scope of this article. For more information the reader is referred to the review by Breinig et al. (36). Here, only a few fundamental aspects of the ECC process shall be pointed out.

The ECC implies two center phenomena, as long-range forces of both the projectile and target atoms play an important role in the formation of the electron cusp. A plausible scheme to explain the production of the cusp electrons is the classical double-scattering mechanism by Thomas (37,38). From two-body kinematics it follows that an electron with a velocity, equal to the projectile velocity, is produced in a binary encounter collision at an angle of 60° (Fig. 2f). This electron would separate rapidly from the projectile so that the process of electron capture becomes impossible. However, if it is backscattered by 60° in the field of the target nucleus, the electron may travel on a trajectory parallel to the projectile direction. In this case, the projectile interacts strongly with the electron (Fig. 2f).

The classical scattering mechanism suggests that the electrons are focused in the forward direction. It should be realized that this focusing takes place on a time scale large in comparison with that of the collision (39). If the focusing is directed to a point at infinity, the electron intensity exhibits a singularity at 0°. Hence, the cusp profile characteristic for the ECC process may be produced. It is noted that the focusing phenomenon is the classical counterpart to the process of electron capture to the continuum.

The picture of electron focusing is supported by classical-trajectory Monte-Carlo (CTMC) calculations (40,41). To demonstrate the focusing aspect in the cusp peak formation, Reinhold and Olson (40) terminated their calculations *before* the collision partners have separated completely. Fig. 5 shows CTMC results for the cusp-peak formation in 100-keV H^+ + He collisions where the integration of the Hamiltonian is terminated at different finite internuclear distances. It is seen that internuclear distances of more than several thousand a.u. are needed to achieve the characteristic cusp peak observed for the ECC process. In spite of limitations of the classical picture, the present consideration shows that the electron-projectile interaction at large internuclear distances plays an essential role for the formation of the cusp electrons.

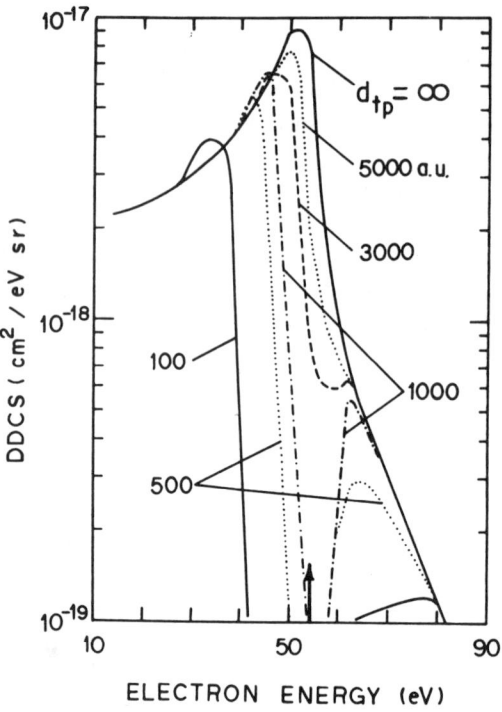

Figure 5: Cusp electron production in 100-keV H^+ on He collisions calculated using the CTMC method. The data are evaluated with a Hamiltonian whose integration is stopped at different internuclear distances d_{tp}, as indicated From Ref.(40).

In view of the cusp peak formation, interesting questions remain open. Besides the Thomas mechanism, based on a classical double-scattering event, the capture to the continuum may proceed via a single-step event incorporated in the first-order theory by Brinkmann and Kramers (42). This mechanism involves a weak interaction by the target and, hence, it suggests that ECC corresponds to a single-center phenomenon associated with the projectile. On the other hand, the cusp peak is generally observed to be asymmetric which is commonly attributed to the influence of the receding target atom (36). This shows that the influence of the target atom cannot be neglected in the post-collision region. Hence, in any case, cusp electrons are influenced by the two-center field of the collision partners.

Two-center electron emission

Two-center electron emission (TCEE) may be understood as a generalization of the ECC process. Rather than focusing, TCEE involves a deflection of the outgoing electron in the receding projectile field (Fig. 2e). Electron deflection is particular important for electrons ejected at forward angles. However, TCEE implies more general cases. When an electron is emitted at backward angles, it also feels the projectile field (in addition to that of the target atom). Hence, two-center effects may be present not only for electrons ejected near $0°$.

Evidently, two-center effects are expected to become visible in regions of the electron spectra where the single-center phenomena diminish (7). One such region occurs in the domain between the soft collision and the binary encounter maxima. To separate TCEE from the other mechanisms it is useful to utilize high-energy projectiles, because the peaks due to soft and binary-encounter collisions increase in separation as the projectile energy increases. Thus, the energy region where the TCEE occurs is expanded. Also, it should be recalled that TCEE is observable only in doubly-differential electron spectra, since single-differential electron spectra are governed everywhere by single-center phenomena.

Measurements of two-center electron emission in high-energy collisions have recently been performed in various laboratories (7,29-31). The experimental results exhibit characteristic features that can be seen in Fig. 6. It shows electron emission cross sections, measured for 25-MeV/u Mo^{40+}+He collisions (7), in relation to the corresponding theoretical predictions from the Born approximation. As before, the Born approximation is used as a standard to verify two-center effects. In accordance with ECC, the experimental results are found to be enhanced at the forward angle of $30°$. In addition, the experimental data are seen to be significantly reduced at the backward angle of $150°$.

It may be anticipated that the comparison of the experiment with the Born approximation is not sufficient to verify unambiguously a two-center effect by a fast moving projectile. However, the observed enhancement at forward angle and reduction at backward angles is reproduced by theoretical results from the *continuum-distorted wave* approximation including *eikonal initial states* (CDW-EIS). This theory is well suited to model two-center effects by inclusion of a final wave function centered both at the projectile and the target atom (8,43). The good agreement between experiment and the CDW-EIS theory reveals a decisive manifestation of two-center effects in the present collision system (Fig. 6).

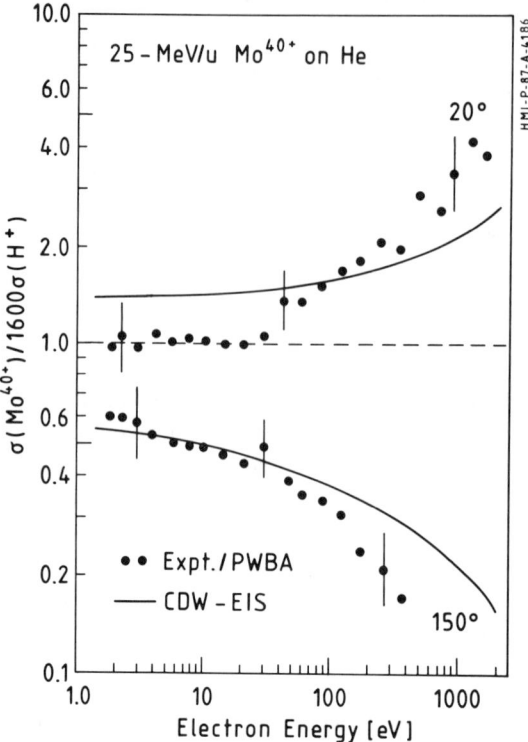

Figure 6: Ratio of double differential cross sections for electron emission by 25-MeV/u Mo^{40+} and H$^+$ impact on He observed at forward (20°) and backward (150°) angles. The points refer to the ratio of experimental cross sections and corresponding results obtained using the Born approximation (PWBA). The solid line refers to calculations using the CDW-EIS method, see e.g., (44). From Ref. (7).

The enhancement of the TCEE cross sections at forward angles and their reduction at backward angles can be visualized as an attraction of the outgoing electrons by the charged projectile (7,31). The enhancement at forward angles may be understandable from the intuitive picture cultivated for many years to explain the cusp electrons at $0°$. However, the relatively strong reduction of the cross section at backward angles may be unexpected. Evidently the outgoing electron "feels" the projectile even if it moves rapidly in a direction opposite to that of the projectile. This is likely to be caused by the fact that a high projectile charge is used in the experiments. However, the surprising observation from the theoretical data is that the reduction at backward angles is *larger* than the enhancement at forward angles (Fig. 6). This finding is not understood at present. Further experimental and theoretical work appears to be needed to elucidate two-center effects for the electron emission at backward angles.

The picture of post-collision attraction of the electron by the projectile should be used with some caution. One should not overlook the role of the Coulomb field of the target. Two-center effects are produced by the simultaneous or successive interaction of the ejected electron with the projectile and the target atom. As pointed out in the previous discussions, electrons outside the regions of the binary encounter maxima are produced by interactions involving a significant target nuclear field. This field is active before or during the post-collision interaction by the projectile. For instance, as noted before, the target field is usually considered as the cause for the asymmetry of the cusp peak. It is felt, however, that further work is needed to clarify the role of the collision partner in two-center electron emission. It appears worthwhile to investigate the question of whether the essential interactions of the collision partners happen simultaneously or successively in the electron emission process.

CONCLUDING REMARKS

In this article, a few aspects of electron emission in ion-atom collisions were treated. The discussion applies for energetic projectiles whose velocity is larger than the bound electron velocity. Looking back at the field, it becomes evident that an enormous amount of work has been performed during the last three decades. I feel that this work has primarily been motivated by the excitement emerging from the research of electron-production mechanisms. The attempt to understand three-body interactions involves great challenges for the associated scientists. In addition, ion-induced electron emission is of fundamental importance for many applied fields, including the technology of ion production, plasma and fusion research, astronomy, and various branches of solid state physics. It is safe to state that the research about ion-induced electron emission has provided a meritorious service to scientists from other fields.

During the past 30 years, the great research effort has caused a remarkable progress in the field of ion-induced electron emission. Various fundamental mechanisms have been understood. Hence, it may be legitimate to look back with the question in mind of what the future will look like. A general answer to this question cannot be given. In the past few years, surprisingly new aspects have been discovered for mechanisms that were believed to be fully understood. An example is the enhanced large-angle scattering of electrons by dressed projectiles. Also it should be kept in mind that the investigation of the electron-electron interaction phenomena, resulting in electron emission, is still in its infancy. Moreover, important branches of the field are far from being explored, such as electron emission in slow ion-atom collisions and the dynamics of electron production in ion-solid interactions.

Acknowledgment

When I constructed an electron spectroscopy apparatus in the late 60's, the publications by Gene Rudd and collaborators were practically the only ones available in the field of ion-induced electron emission. From these pioneering publications I received the basic information necessary to perform the tedious work of measuring electrons from ion-atom collisions. Therefore, I consider Gene Rudd as my teacher although I did not meet him before I graduated. Later, I was fortunate to become one of his collaborators. I am convinced that the care and honesty, Gene Rudd devoted to his research work, have set standards for the following generation. I shall never forget the important contributions Gene Rudd provided to my work.

REFERENCES

1. Kuyatt, C.E. and Jorgensen, T. Jr, *Phys. Rev.* **130,** 1444 (1963)
2. Rudd, M.E. and Jorgensen, T. Jr., *Phys. Rev.* **131,** 666 (1963),
3. Rudd, M.E., Sautter, C.A. and Bailey, C.L., *Phys. Rev.* **151,** 20 (1966)
4. Olsen, R.E., *Phys. Rev. A* **33,** 4397 (1986).
5. Gay, T.J., Gealy, M.W., and Rudd, M.E., *J. Phys. B* **23,** L823-L828 (1990).
6. Meckbach, W., Focke, P.J., Goni, A.R., Suárez, S., Macek, J., and Menendez, M., *Phys. Rev. Lett.* **57,** 1587 (1986).
7. Stolterfoht, N. and Schneider, D., Tanis, J., Altevogt, H., Salin, A., Fainstein, P.D., Rivarola, R., Grandin, J.P., Scheurer, J.N., Andriamonje, S., Bertault, D. and Chemin, J.F., *Europhys. Lett.* **4,** 899 (1987).
8. Fainstein, P.D., Ponce, V.H. and Rivarola, R., *J. Phys. B* **24,** 3091-3119 (1991)
9. Wilson, W.E. and Toburen, L.H., *Phys. Rev. A* **7,** 1535 (1973)
10. Burch, D., Wieman, H. and Ingalls, W.B., *Phys. Rev. Lett.* **30,** 823 (1973)
11. Stolterfoht, N. and Schneider, D., Burch, D., Wiemann, H. and Risley, J.S., *Phys. Rev. Lett.* **33,** 59 (1974)

12. Drepper, F. and Briggs, J.S., *J. Phys. B* **9**, 2063 (1976)
13. Gryzinski, M., *Phys. Rev.* **138**, 305, 322 and 336 (1965)
14. Madison, D.H., and Merzbacher, E., in *Atomic Inner-Shell Processes*, ed. by B. Crasemann, Academic Press, New York, 1975, p. 1-72
15. Rutherford, E., *Phil. Mag.* **21**, 669 (1911)
16. Burgdörfer J., *this issue*
17. Richard, P., Lee, D. H., Zouros, T. J. M., Sanders, J. M. and Shinpaugh, J. L., *J. Phys* B **23**, L213 (1990)
18. Olson,R.E., Reinhold, C.O., and Schultz, D.R., *J. Phys. B* **23**, L455 (1991)
19. Quinteros, T. and Reading, J.F., *Nucl. Instr. Methods B* **53**, 363 (1991)
20. Gonzales, A.D., Dahl, P., Hvelplund, P., and Taulbjerg, K., *J. Phys. B.* **25**, L573 (1992)
21. Bhalla C.P. and Shinghal R., *J. Phys. B* **24**, 3187 (1991)
22. Bethe, H., *Ann. Phys. (Leipzig)* **5**, 325 (1930).
23. Bates, D.R. and Griffing, G.W., *Proc. Roy. Soc. (London)* **66A**, 961, (1953)
24. Madison, D.H., *Phys. Rev. A* **8**, 2449-2455 (1973)
25. Manson, S.T., Toburen, L.H., Madison, D.H. and Stolterfoht, N., *Phys. Rev. A* **12**, 60 (1975)
26. Stolterfoht, N., in *Topics of Current Physics*, Vol. **5**, Springer Verlag, Berlin, 1978, p. 155-199
27. Crooks, G.B. and Rudd, M.E., *Phys. Rev. Lett.* **25**, 1599, (1970).
28. DuBois, J.R., *Phys. Rev. A* **48**, 1123 (1993)
29. Platten, H., Schiwietz, G., Schneider, T., Schneider, D., Zeitz, W., Musiol, K., Zouros, T.J.M., Kowallik, R. and Stolterfoht, N., *XVth International Conference on the Physics of Electronic and Atomic Collisions*, Abstracts, Brighton, 1987, p. 437)
30. Schneider, D., DeWitt, D., Schlachter, A.S., Olson, R.E., Graham, W.G., Mowat, J.R., DuBois, R.D., Loyd, D.H., Montemayor, V. and Schiwietz, G., Phys. Rev. A **40**, 2971 (1989)
31. Pedersen, J.O., Hvelplund, P., Petersen, A.G. and Fainstein, P.D., *J. Phys. B.* **23**, L597 (1990)
32. Pedersen, J.O., Hvelplund, P., Petersen, A.G., and Fainstein, P.D., *J.Phys. B* **24**, 4001 (1991)
33. Hidmi, H.I., Richard, P., Sanders, J. M., Schöne, H., Giese, J.P., Lee, D.H., Zouros, T.J.M, and Varghese, S.L., *Phys. Rev. A* **48**, 4421 (1993)
34. Salin, A., *J. Phys. B* **5**, 979 (1972).
35. Macek, J., *Phys. Rev. A* **1**, 235 (1970).
36. Breinig, M., Elston, S.B., Huldt, S., Liljeby, L., Vane, C.R., Berry, S.D., Glass, G.A., Schauer, M., Sellin I.A., Alton, G.D., Datz, S., Overbury, S., Laubert, L., and Suter, M., *Phys. Rev. A* **25** 3015 (1982).
37. Thomas, L.H., *Proc. Roy Soc.* **114**, 501 (1927).
38. Briggs, J.S., and Macek, J.H., *Adv. At. Mol. and Opt. Phys.* **28**, 1-74 (1991)
39. Swenson, J.K., Havener, C.C., Stolterfoht, N., Sommer, K., and Meyer, F.W., *Phys. Rev. Lett.* **63**, 35 (1989).
40. Reinhold, C. and Olson R.E., *Phys. Rev. A* **39**, 3861-3870 (1989)
41. Montemayor, V.J. and Schiwietz, G., *J. Phys. B* **22**, 2555-2565
42. Brinkman H.C. and Kramers, *Proc. Acad. Sci. Amsterdam* **33**, 973 (1930)
43. Crothers D.S.F. and McCann J.F., *J. Phys. B* **16**, 3229 (1983)
44. Fainstein P.D. and Rivarola R.D., *J. Phys. B* **20**, 1285 (1987)

IV. M. E. RUDD'S CONTRIBUTIONS TO ATOMIC PHYSICS

Proton-Atom Collisions: Contributions of M. E. Rudd

L. H. Toburen[*]

National Research Council/National Academy of Sciences
2101 Constitution Avenue, N.W.
Washington D.C. 20418

Abstract. Beginning with his initial studies of the angular dependence of the spectra of electrons emitted in ion-atom collisions, the first measurements to provide a detailed and comprehensive description of the collisional ionization process, M. Eugene Rudd contributed to an impressive list of "firsts" in the study of collision physics. In 1963, Gene published the first observation of a two-center phenomena in collision physics, although it was several years before the features he observed in the spectra of ejected electrons were clearly interpreted as contributions from electron-capture-to-the-continuum, a two-center phenomena. He contributed firsts in studies of doubly differential cross sections, inner-shell- and auto-ionization, interactions involving dressed projectiles, and interactions of ions and photons with surfaces. He also refined the experimental techniques to provide data of improved reliability in many areas where others had made pioneering studies including measurements of doubly-differential cross sections for incident electrons and total-ionization and charge-transfer cross sections for ion impact.

INTRODUCTION

Those who have had the opportunity to work with Gene Rudd have long recognized the degree of dedication and ingenuity that he brings to his many areas of interest in science. Whether he is reading the latest book on new scientific discoveries while at a remote camp ground in the wilderness of the Pacific Northwest, or discussing the potential implications of the latest experimental findings over a sandwich at the lunch table, the excitement of physics is always in the air when Gene is present. Gene has the reputation for performing experiments with the utmost precision and accuracy, and for seeking a thorough understanding of the physical phenomena leading to the spectral features he has observed. I have enjoyed our numerous discussions regarding experimental techniques and the underlying physics of ion-atom collisions, as well as the opportunity to share the excitement of several collaborative projects.

Gene began his career in the study of low-energy electrons before many of the technical advances that were later to enable this field to flourish. For those of us who were initiating our research programs in the early 1960's BC (before channeltrons®), the detection of low-energy electrons was an art that few mastered.

I remember trying to detect low-energy electrons at that time, with varying degrees of success, using thin-window gas-flow proportional counters equipped with high voltage to preaccelerate electrons through the thin detector windows and encountering the gas leaks and spurious electron counts common to such devices. These are not techniques that would lend themselves to accurate spectral measurements in the confined quarters of vacuum systems normally encountered in the measurement of angular resolved electron spectra. Gene Rudd, undeterred by the common conception at that time that secondary-electron emission devices of the discrete-dynodes type were of unknown and varying efficiencies, was able to master the use of such detectors to accurately and reproducibly measure electron spectra for electron energies as low as a few electron volts (1). Although he was also one of the first to use the "new" continuous channel electron multipliers (2), many of his important contributions to the field of doubly-differential electron emission cross sections were made using discrete-dynode electron multipliers to detect the electrons. They offered advantages of high counting rates and large detector apertures that were unavailable with the early channel electron multipliers.

The skill and innovation of Gene Rudd as an experimentalist cannot be over-emphasized. However, the theoretical insights and understanding he exhibited in seeking an interpretation of spectral features in the data he obtained must also be noted. Gene was among the first to exploit the Born approximation showing that, although there had been good agreement between cross sections calculated using the plane-wave Born theory and *total* ionization cross-section measurements, the details of the cross sections obtained from this theory for *differential* ionization cross sections were often in wide disagreement with measurement. Such discrepancies led to the discovery of electron-capture-to-the-continuum (ECC) as an important contributor to the spectra of electrons ejected at small angles with respect to, and with velocities comparable to, the passing ion. He also utilized binary encounter theory as a convenient and straight forward classical method of predicting single differential cross sections (3). Throughout the years Gene conducted extensive collaboration with theoretical specialists to arrive at insightful descriptions of the "physics" of the collision process.

When reviewing his publications, one cannot fail to be impressed at finding that three of Gene's first six papers were published in the Physical Review Letters, and this occurred within only a few years of his leaving graduate school. The following sections present descriptions, necessarily brief, of some of the contributions Gene Rudd has made in the field of proton-atom collisions. I realize that this is not an exhaustive list of his accomplishments, and I may inadvertently leave out what some investigators may feel are the most important. But even with these shortcomings I'm sure we all can agree that even this is an impressive list of accomplishments. As indicated, this discussion will focus on the contributions by Gene Rudd, more comprehensive reviews of the general field of research in proton-atom/molecules collisions can be found in works by numerous authors, including those of Rudd (4,5), Stolterfoht (6), and Toburen (7).

INNER-SHELL- AND AUTO-IONIZATION

For those who know Professor Rudd's work primarily through his contributions to the field of continuum electron spectroscopy and the measurement of doubly-differential ionization cross sections, his accomplishments in high-resolution spectroscopy of Auger and auto-ionization electrons might come as a surprise. In fact, Gene's second career publication, a Physical Review Letter, described the autoionizing lines of helium excited by protons and H_2^+ ions (8). The data from this work, illustrated in Fig. 1, clearly show the influence of electron exchange on the selection rules for excitation when exciting helium atoms by molecular ions. The $2s2p(^3P)$ line, at about 33.8 eV, that is missing in excitation by H^+ is clearly evident when excitation is by H_2^+ ions. Although he was not the first to study autoionization using excitation by molecular ions, Rudd was the first to have sufficient spectral energy resolution to clearly identify the states involved. He subsequently published the first observation of the $(2p^2)^1S$ line at 37.6 eV and several lines in the $(2sns)^1S$ series that range from 33.24 to 40.13 eV, and the $(2snp)^3P$ series with lines from 33.76 to 40.13 eV (9).

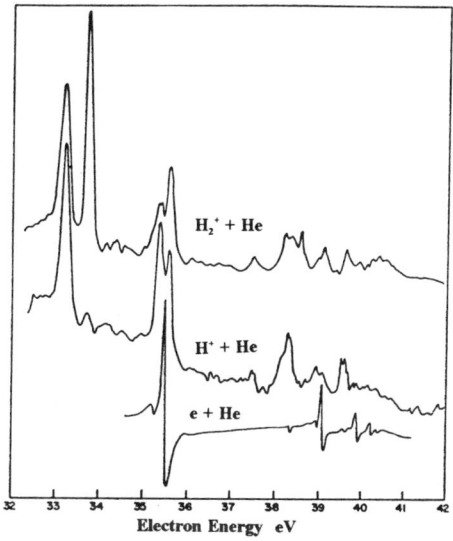

Figure 1. Autoionization spectrum of helium under H^+, H_2^+, and electron impact; ion energies are 50 keV, electron energy 4 keV. Data reproduced from Rudd and Macek (4).

Professor Rudd's studies of autoionization have included investigation of the energetics and line shapes of electrons emitted from autoionizing states, as well as of the influence of the energy and type of incident ion and the emission angle of the ejected electron on the autoionization electron spectra. The data in Fig. 2 show the autoionization spectra in the region of the $(2s^2)^1S$ state of helium as a function of the exciting proton energy (10). These data illustrate the strong influence of the exciting proton energy on the asymmetry of the line shape. This asymmetry arises from the interference between an autoionizing state and the background continua as had been pointed out by Fano (11) earlier. Gene also explored autoionization in other atoms and by other ions, (see, for example, references 12 and 13). These studies resulted in the observation of many previously unseen autoionization lines and the first observation of electron spectra Doppler-shifted in energy owing to the velocity of the atom from which the electrons were emitted.

Gene published numerous papers with students and colleagues that described

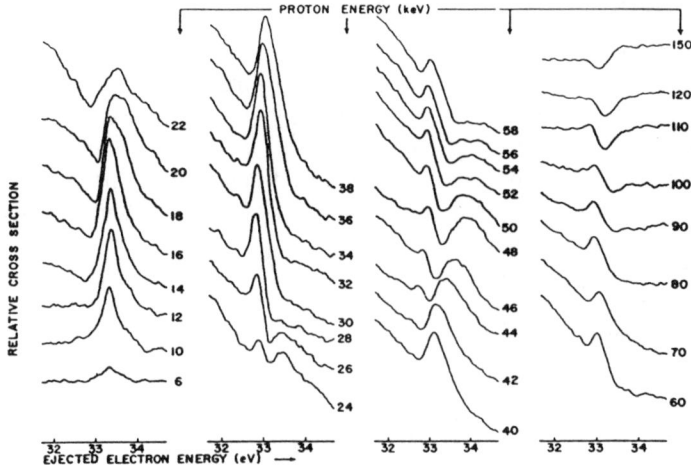

Figure 2. Electron spectrum from the $2s^2\,{}^1S$ state of helium excited by protons; data from Schowengerdt and Rudd (10).

Auger-electron spectra from excitation of inner-shells of atoms by protons and heavier ions (see, for example, references 14-18). For low-Z elements the accuracy of inner-shell ionization cross sections determined from measured Auger-electron yields can be greatly enhanced over those from x-ray measurements because the Auger yield may be nearly unity whereas the fluorescence yield is small and uncertain. In addition to the study of transition energies and line

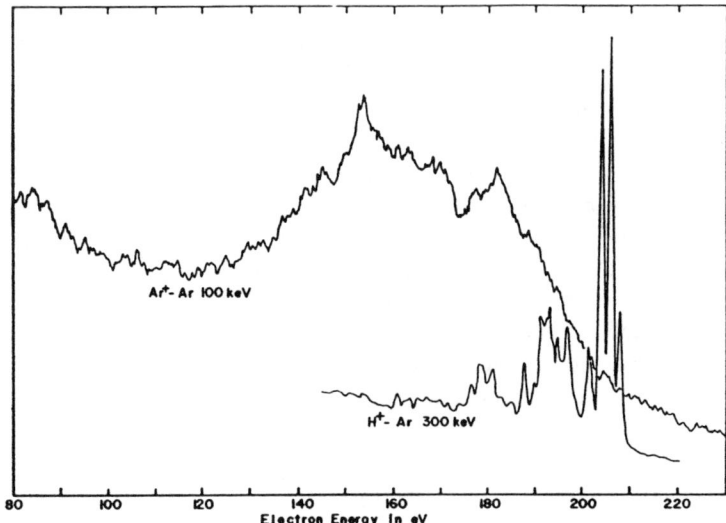

Figure 3. L-Auger spectra for argon excited by protons and argon ions; data of Rudd et al. (13).

shapes, including the effects of Doppler broadening, these measurements leading to the determination of total inner-shell cross sections were instrumental in testing the early stages in the development of the promotion model of inner-shell vacancy production (19).

Gene was also one of the first to explore the effects of multiple inner-shell vacancies on the Auger electron spectra and the Doppler shifts of Auger line energies emitted from moving ions. By use of Doppler line shifts and the comparison of spectra from heavy ions to those excited by protons and electrons he was able to interpret much of the complex spectra originating from heavy ion impact. The comparison of L-Auger spectra from argon produced for excitation by protons and argon ions is shown in Fig. 3. The spectrum for argon ion impact clearly demonstrates the broadening of the Auger spectrum owing to the shifts in the Auger electron energy that occurs when they are emitted from multiply ionized atoms. Throughout his career Gene Rudd has played an important role in the study of autoionization and Auger-electron spectroscopy

DOUBLY DIFFERENTIAL CROSS SECTIONS

The first measurements of the energy and angular distributions of continuum electrons ejected from atoms and molecules by proton impact were actually published by Kuyatt and Jorgensen (20); prior to that work there had only been isolated studies of electron spectra measured at a few selected angles. However, it was Gene Rudd who would continue the studies of continuum electrons initiated when he and Kuyatt were students of Professor Jorgensen, and who would provide the broad range of high-quality cross sections to which the rest of the field was to be compared. Gene's first publication (1), only a few months after appearance of the Kuyatt and Jorgensen paper, presented the data that was to produce the first divergent view from the accepted understanding of the collisional ionization mechanism. He had first tried to discount the features, "humps" as he referred to them, that were observed on the low-energy portion of the spectra of electrons emitted at 10°, shown in Fig. 4, as experimental artifacts. But try as he might, no alterations to the instrumentation, or to the

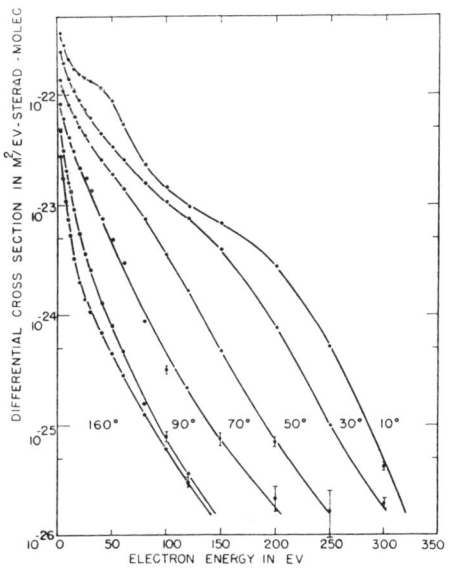

Figure 4. Electron emission in 100 keV H^+ - H_2 collisions; data of Rudd and Jorgensen (1).

experimental techniques, would remove them. A careful analysis of the energy and angular dependence of these doubly-differential cross sections indicated that the humps contributed to an enhancement of the spectra primarily for small emission angles and for ejected electron energies that corresponded to electron velocities comparable to the incident proton. These angular distributions are illustrated in Fig 5; data such as these led to the discovery that a new ionization mechanism, that of electron-capture-to-the-continuum was contributing to the production of free electrons in ion-atom collisions. These results provided strong evidence of the utility of doubly differential electron emission cross sections to explore the details of ionization mechanisms in ion-atom collisions.

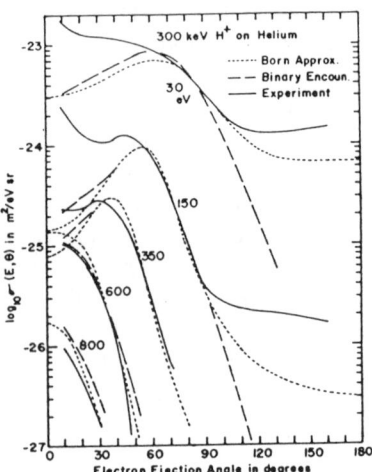

Figure 5. Comparison of measured and calculated cross sections; data from Rudd and Macek (4).

From the kinematics of the contributions of electron-capture-to-the-continuum to the spectra of ejected electrons, it was apparent that the physics underlying this process was best examined by looking at the spectra of electrons emitted at 0°, where their contribution was greatest. Professor Rudd constructed a device for such measurements and compared the spectra he measured at 0° to the current theory; these data are reproduced in Fig. 6. Such data clearly illustrate the inadequacy of the plane-wave Born approximation for description of electron emission at, or near, 0°. Although it would be years before the process of electron-capture-to-the-continuum would be fully understood (and led to a new field of research referred to by some as "cuspology"), Rudd's studies, early in his career, had put the field on a solid footing and clearly demonstrated the utility of doubly-differential cross sections (DDCS)

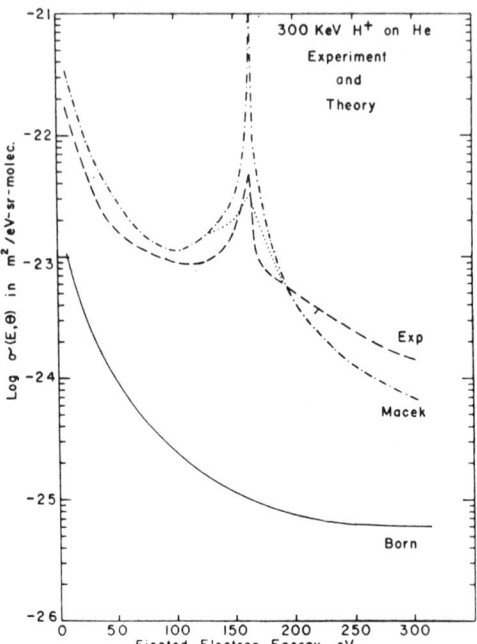

Figure 6. Electrons ejected at 0° in 300 keV H$^+$-He collisions. Data of Crooks and Rudd (2), theory of Macek (21); dot-dash is 0°, dotted curve 1.4°.

186

as an analytic tool for understanding the collision process.

Rudd continued to expand the scope of his studies of DDCS, first by extending the measurements to emission of electrons ejected from different atomic and molecular targets and then to expanding the energy range of the protons. Electron emission cross sections were measured for targets of H_2 and He (1, 22); and N_2, O_2, Ne, and Ar (23) in the energy range 50-300 keV. These data were instrumental in developing an understanding of the systematics of electron emission cross sections as a function of the atomic and molecular binding energy. His data for these systems showed that the cross sections exhibited very nearly the same energy and angular distributions for different molecular targets (except hydrogen), this is illustrated in Fig. 7 for ionization of molecular oxygen and nitrogen by 300 keV protons.

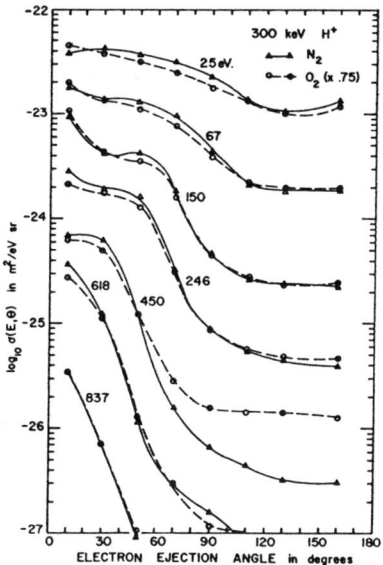

Figure 7. Angular distribution of electrons ejected from O_2 and N_2; data of Rudd (23).

The only notable difference in the shapes of the angular distributions shown in Fig. 7 is for emission of 450 eV electrons at large angles. Here the oxygen cross sections are considerable larger than those for nitrogen. This occurs because at 450 eV the electrons are ejected with energies near the peak of the K-Auger electron emission spectrum in oxygen. The isotropic emission of the Auger electrons enhances the cross sections for oxygen markedly, over the nitrogen cross sections, in the backward direction. None of the ejected electron energies shown in Fig. 7 are near the nitrogen K-Auger energy of about 365 eV, thus we do not see its influence on the emission cross sections.

For proton energies less than a few hundred keV there is no reliable theoretical technique for calculating double-differential electron emission cross sections, so Gene extended his measurements to lower proton energies to provide data to better understand the ionization process for such low-energy ions and to encourage further theoretical development. One might note, that if studies of low-energy electrons are considered difficult, measurements of electron emission for ionization by very-low-energy protons is doubly so; low-energy protons transfer only small amounts of energy to electrons (thus only low-energy electrons are ejected) and the emission cross sections are quite small. The maximum energy of an electron that one can detect from ionization of helium by a 5 keV proton is only about 75 eV. Again, however, Gene Rudd was successful in designing and conducting these very difficult experiments and was able to measure the first double-differential cross sections for protons with energies as low as 5 keV (24, 25). An example of the results of these low energy ion measurements is shown in Fig. 8 where the single differential cross sections (SDCS), integrated with respect to emission angle, are shown for ionization of N_2 by protons of 5 to 70 keV. For proton energies less than about 50 keV the SDCS tend to a nearly constant value at ejected electron energies near zero and then exhibit an approximately exponential decrease with increasing ejected electron energy. Spectra for different atomic and molecular targets were found to have similar shapes and Rudd was able to fit the resulting cross sections with an expression of the general type $\sigma(E) \propto \exp[-\alpha E/(IT)^{1/2}]$, where E is the ejected electron energy, I is the ionization potential of the ejected target electron, $T = E_p/1836$ with E_p being the proton energy, and α is a dimensionless constant near unity. This expression is consistent with the molecular promotion model of ionization and has led to successful application of that model in describing ionization by low-energy protons (24).

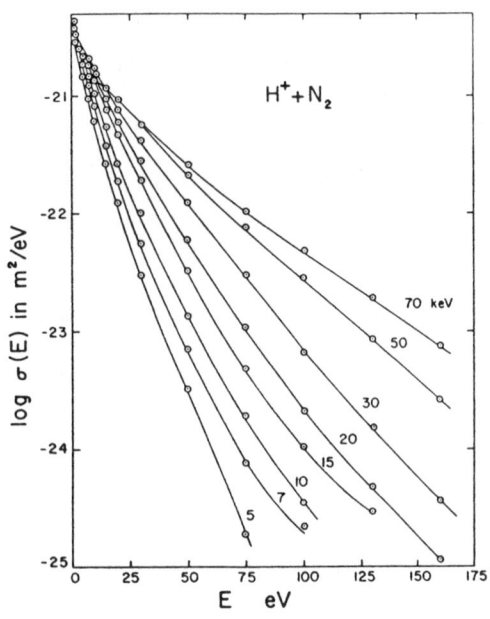

Figure 8. Energy distributions of electrons ejected from N_2 by low-energy protons; data of Rudd (24).

Gene Rudd has never been one to take the easy way out when it comes to answering basic questions in collision physics. But who, other then Gene, would dare undertake the task of measuring DDCS from *atomic* hydrogen. Certainly no one before him had considered it possible to test the theory for ion-atom collisions with measured DDCS for this, the simplest 3-body collision; the tough questions of atomic target density, molecular dissociation efficiency, and the effects of stray electric and magnetic fields would scare most others away from such a task. Needless to say, Gene designed and built the experimental system and successfully measured electron spectra from H^+ - H collisions. As one might also expect, he found the unexpected, and published another Physical Review Letter. What he found was that there was a large enhancement in the emission cross section for ejection of electrons into large angles with respect to the proton beam when measurements were conducted for an atomic hydrogen target over those for a molecular hydrogen target. A comparison of the DDCS for 70 keV protons on H and H_2 is shown in Fig. 9; since the molecular hydrogen is dissociated by a radio-frequency (RF) field, the curve designated as RF ON implies that the data were measured for an atomic hydrogen target, whereas the

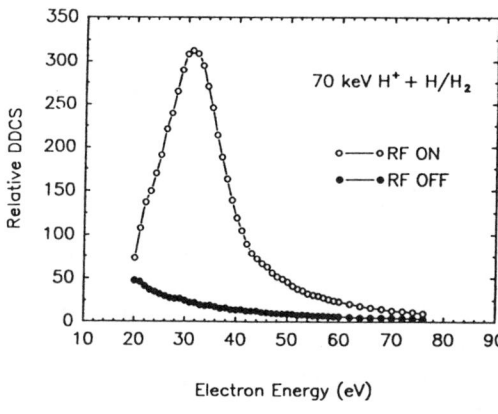

Figure 9. Spectrum of electrons ejected at 160° by H^+ from H (RF ON) and H_2 (RF OFF); data of Rudd et al. (26).

RF OFF indicates that the data were obtained from a molecular hydrogen target. The spectra of ejected electrons taken at different proton energies from, 30 to 100 keV, showed no difference in the energy of the peak in the distribution observed at 160° for atomic hydrogen targets. However, the magnitude of this enhancement is maximum for protons of 70-100 keV and decreases to a small value at a proton energy of 30 keV (higher energy protons were not explored).

The study of ionization by low-energy protons from targets other than atomic hydrogen have also resulted in some features that have yet to be understood. For example, Rudd and Madison (25) studied the proton energy dependence of DDCS for ionization of helium by proton energies from 5 to 2000 keV. When looking at the dependence of the cross sections for emission of low-energy electrons they found distinct dips in the cross sections near the condition in which the ejected electron and the proton have similar velocities. This is illustrated in Fig. 10, where arrows mark the position of the equal velocity condition. The dip in the cross sections is found for predominantly low-energy electrons, for large emission angles, and for protons in the energy range of 100- to 300-keV. For higher and lower

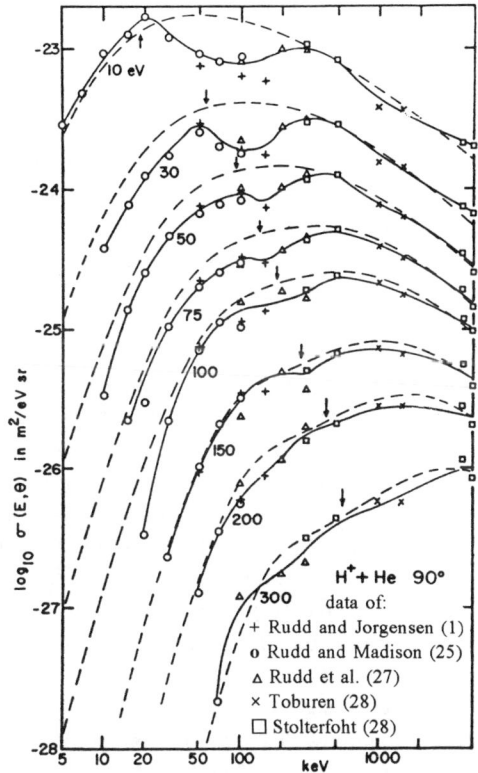

Figure 10. DDCS for electrons of various energies ejected at 90°; arrows are equal velocity H^+ and e^-; dash lines are Hartree-Fock Born calculations.

proton energies the agreement between experimental results and the calculations based on the plane-wave Born approximation is quite good. Only in the areas of these dips is the divergence from theory remarkable.

OTHER CONTRIBUTIONS

As noted at the outset, there is inadequate space within the limitations of this paper to adequately describe the breadth of the contributions that Gene Rudd has made to the field of proton-atom/molecule collision, and certainly too little to discuss his contributions to the study of electron and heavy-ion collisions. For example, nothing has been said about his measurements of target recoil ions produced by fast proton collisions with atomic and molecular targets. Gene's measurements of the energy and angle of recoil ions (29, 30) were some of the first such measurements to provide the rich information available from doubly-differential cross sections. One should also not forget that, in addition to studies of doubly-differential cross sections for the emission of electrons by proton and heavy particle impact, Rudd and his students also made extensive studies of doubly-differential cross sections for ionization of atomic and molecular targets by electrons. These measurements expanded the range of detection of both secondary and scattered electrons (31-33).

One certainly should acknowledge Gene Rudd's contributions to the study of total-ionization cross sections. He made the first systematic study of total-ionization cross sections that spanned the range from low energy collisions to the high energy region; from 5 to 4000 keV (34, 35). He also made measurements of charge-transfer and total-ionization cross sections for helium ions for energies from 5 to 450 keV/u (36, 37). In addition, he also made the first measurements of cross sections for production of vacuum ultraviolet radiation emitted in collisions of protons with a number of atomic and molecular targets (38)

In addition, Gene has contributed extensively to the development of theoretically based models of differential and total cross sections (39-42), but that will be discussed later in other reports at this meeting. These models are particularly important for assimilation of the vast quantities of data that one can generate when exploring the broad range of parameters available to the study of doubly-differential cross sections and for developing an understanding for the systematics governing their properties. From a practical point-of-view these models enable one to gain information on the systematics of ionization cross sections and to interpolate/extrapolate data where experiments have yet to be performed. Insights gained from these models support our understanding of such diverse phenomena as energy degradation in the atmosphere and the biological effects of ionizing radiation. These models have been particularly useful for incorporating the ion and electron impact cross sections into computer programs designed to describe the properties of energy deposition and transport for the slowing down of fast charged particles traversing biologically relevant media (43).

Although it is not the central topic of this paper, one can hardly conclude a discussion of Gene Rudd's contributions to ion-atom/molecule collisions without some recognition of his contribution to the field of heavy ion collisions. Briefly, we mentioned his use of molecular ions and the heavy ions Ar^+ and Ne^+ in his studies of autoionization and Auger electron emission in earlier sections of this report. But one should also recognize that he was the first to study the doubly-differential cross sections for neutral particle impact (44). This work has particular significance to me because of an experience our laboratory in 1991. We were analyzing some cross sections we had recently measured for a series of neutral and charged particles and thought we were seeing some new phenomena in the ejected electron spectra. We saw what looked like contributions to the electron ejection cross sections from interactions between the projectile electrons and the target electrons which led to an increase in the very-low-energy portion of the measured electron spectra. Effects of projectile electrons had been documented for the ejection of fast electrons in collisions involving very heavy ions, but we felt that we were seeing the systematic effects of projectile electrons for the first time in the ejection of very-low-energy electrons. As we formulated our interpretation of these data I recalled seeing a paper describing something similar for neutral hydrogen impact that had been published earlier by Gene Rudd, so I went to my office to search through the reprints to see if he had seen the same effects. Indeed, as we should have expected, Gene had seen the same effect we were observing, and he had published it 10 year earlier (44).

REFERENCES

*The views and opinions expressed are solely those of the author and do not reflect those of the National research council.

1. Rudd, M. E., and Jorgensen, Jr., T., *Phys. Rev.* **131**, 666-675 (1963).
2. Crooks, G. B., and M. E. Rudd, *Phys. Rev. Lett.* **25**, 1599-1601 (1970).
3. Rudd, M. E., and Gregoire, D., "Energy distribution of electrons from ionization of helium and hydrogen by proton collisions: comparison of classical theories and experiment," in *Physics of the One- and Two-electron Atoms*, Bopp, F., Ed., Amsterdam, North Holland, 1969, pp. 795-800.
4. Rudd, M. E., and Macek, J. H., *Case Studies in Atomic Physics* **3**, 47-136 (1972).
5. Rudd, M. E., *Radiat. Res.* **64**, 153-180 (1975).
6. Stolterfoht, N., "Excitation in energetic ion-atom collisions accompanied by electron emission," in *Topics in Current Physics V.: Structure and Collisions of Ions and Atoms*, Sellin, I. A. Ed., Berlin, Springer-Verlag, 1978, pp. 155-199.
7. Toburen, L. H., "Atomic and molecular physics in the gas phase," in *Physical and Chemical Mechanisms in Molecular Radiation Biology*, Glass, W. A, and Varma M., Eds. New York, Plenum Press, 1991, pp. 51-94.
8. Rudd, M. E., *Phys. Rev. Lett.* **13**, 503-505 (1964).
9. Rudd, M. E., *Phys. Rev. Lett.* **15**, 580-581 (1965).
10. Schowengerdt, F. D., and Rudd, M. E., *Phys. Rev. Lett.* **28**, 127-130 (1972).
11. Fano, U., *Phys. Rev.*, **124**, 1866-1872 (1961).

12. Schowengerdt, F. D., Smart, S. R., and Rudd, M. E., *Phys. Rev. A* **7**, 560-566 (1973)
13. Rudd, M. E., Jorgensen, Jr., T., and Volz, D. J., *Phys. Rev. Lett.* **16**, 929-930 (1966); and *Phys. Rev.* **151**, 28-31 (1966).
14. Volz, D. J., and Rudd, M. E., *Phys. Rev. A* **2**, 1395-1403 (1970).
15. Rudd, M. E., Fastrup, B., Dahl, P., and Schowengerdt, F. D., *Phys. Rev. A* **8**, 220-225 (1973).
16. Rudd, M. E., *Phys. Rev. A* **10**, 518-521 (1974).
17. Dahl, P., Rødbro, M., Fastrup, B., and Rudd, M. E., *J Phys. B* **9**, 1567-1579 (1976).
18. Dahl, P., Rødbro, M., Hermann, G., Fastrup, B., and Rudd, M. E., *J Phys. B* **9**, 1581-1599 (1976).
19. Cacak, R. K., Kessel, Q. C., and Rudd, M. E., *Phys. Rev. A* **2**, 1327-1331 (1970).
20. Kuyatt, C. E., and Jorgensen, Jr., T., *Phys. Rev.* **130**, 1444-1455 (1963).
21. Macek, J., *Phys. Rev. A* **1**, 235-237 (1970).
22. Rudd, M. E., Toburen, L. H., and Stolterfoht, N., *At. Data Nucl. Data Tables* **18**, 413-432 (1976).
23. Crooks, J. B., and Rudd, M. E., *Phys. Rev. A* **3**, 1628-1634 (1971).
24. Rudd, M. E., *Phys. Rev. A* **20**, 787-796 (1979).
25. Rudd, M. E., and Madison, D. H., *Phys. Rev.* **14**, 128-136 (1976).
26. Rudd, M. E., Gealy, M. W., Kerby, III, G. W., and Hsu, Ying-Yuan, *Phys. Rev. Lett.* **68**, 1504-1506 (1992).
27. Rudd, M. E., Sautter, C. A., and Bailey, C. L., *Phys. Rev.* **151**, 20-33 (1966).
28. Manson, S. T., Toburen, L. H., Madison, D. H., and Rudd, M. E., *Phys. Rev. A* **12**, 60-69 (1975).
29. Itoh, A., and Rudd, M. E., *Phys. Rev. A* **35**, 66-69 (1987).
30. Itoh, A., and Rudd, M. E., *Phys. Rev. A* **35**, 19937-1938 (1987).
31. Rudd, M. E., and DuBois, R. D., *Phys. Rev. A* **16**, 26-32 (1977).
32. DuBois, R. D., and Rudd, M. E., *Phys. Rev. A* **17**, 843-848 (1978).
33. Bolorizadeh, M. A., and Rudd, M. E., *Phys. Rev. A* **33**, 882-887 (1986).
34. Rudd, M. E., DuBois, R. D., Toburen, L. H., Ratcliffe, C. A., and Goffe, T. V., *Phys. Rev. A* **28**, 3244-3257 (1983).
35. Rudd, M. E., Goffe, T. V., DuBois, R. D., and Toburen, L. H., *Phys. Rev. A* **31**, 492-494 (1985).
36. Rudd, M. E., and Goffe, T. V., *Phys. Rev. A* **32**, 2138-2133 (1985).
37. Rudd, M. E., Itoh, A., and Goffe, T. V., *Phys. Rev. A* **32**, 2499-2500 (1985).
38. Rudd, M. E., *J. Quant. Spectrosc. Radiat. Transfer* **31**, 387-395 (1984).
39. Rudd, M. E., *Radiat. Res.* **109**, 1-11 (1987).
40. Rudd, M. E., *Phys. Rev. A* **38**, 6129-6137 (1988).
41. Rudd, M. E., *Phys. Rev. A* **44**, 1644-1652 (1991).
42. Rudd, M. E., Kim, Y.-K., Madison, D. H., and Gay, T. J., *Rev. Mod. Phys.* **64**, 441-490 (1992).
43. Wilson, W.E., Radiat. Res. **140**, 375-381 (1994).
44. Rudd, M. E., Risley, J. S., Fryar, J., and Rolfes, R. G., *Phys. Rev. A* **21**, 506-514 (1980).

Electron Transfer to Continuum States

J. H. Macek

Department of Physics and Astronomy, University of Tennessee, Knoxville, TN 37996-1501
and
Oak Ridge National Laboratory, Post Office Box 2009 Oak Ridge, TN 37831

Gene Rudd's analysis of doubly differential cross sections for the ionization of He atoms by proton impact suggested that electrons were being carried along by the proton for a short period of time after being ejected from the target region. Normally, this would represent an electron capture event in which an excited state of atomic hydrogen is formed. Because the electron ends up ionized it was recognized that these states of the proton must be continuum states. This insight was confirmed by observations of the continuum electron capture (CEC) cusp when the electron velocity equals the proton velocity in the final state. The impact of this idea upon the theory of ionization at high energies is reviewed.

I. INTRODUCTION

It is a great honor for me to speak at this symposium commemorating Gene Rudd's accomplishments in physics. These accomplishments are many and are amply recounted by other speakers at this symposium. My talk concentrates on the discovery of what is now known as the continuum capture cusp. Probably we should call it the Rudd peak in honor of its discoverer. In any event, the occasion recalls the great fun Gene, his students, and I had trying to understand ionization in ion-atom collisions. It is a pleasure to recount the discovery of the continuum capture mechanism, not the least because Gene was my patient mentor during my first ventures into this field. I am forever indebted to him for bringing me into his research program to see doubly differential cross sections through his eyes. Atomic units are used in this manuscript although Gene always advocated SI units.

II. DISCOVERY OF THE CONTINUUM CAPTURE CUSP

Gene's work first came to the attention of the general physics community when he observed (1) Auger electrons produced in $Ar^+ + Ar$ collisions, thereby

© 1996 American Institute of Physics

confirming the Fano-Lichten (2) model that accounted for structure in the energy loss of Ar^+ ions observed by Kessel and Everhart (3). The research that Gene did on this process is well known. That process represents a mechanism for ionization quite different than the one of interest here. There is a connection, however, in that one of Gene's early discoveries was that Auger states of both the target ion and projectile ion were excited (4). These were distinguished experimentally by the Doppler shift of the projectile electrons. While this is a simple observation, it was quite important at the time in that it confirmed another aspect of the Fano-Lichten model, namely, electron promotion occured in a region of space where electrons from the projectile and from the target participated equally. The Physical Review Letter on projectile Auger electrons excited much interest in Professor Fano's group at the National Bureau of Standards. It represented elegant, convincing, evidence for the model, so much so that Professor Fano remarked at the time, "I wish I had thought of that!" I suspect that the Doppler effect set Gene to thinking along the lines that eventually led to the theory of continuum electron capture.

During one of our many discussions about doubly differential cross sections in my first summer at the University of Nebraska, Gene showed me his analysis of the data that he had published together with Sautter and Baily (5). He had developed a First Born Approximation program to compute the theoretical doubly differential cross section (DDCS) in order to interpret the measurements. The computed and measured electron distributions generally agreed fairly well except for those electrons ejected at small and large angles. There was an excess electrons at small angles for almost all electron velocities, but the relative excess was most pronounced at electron velocities equal to the projectile velocity. This led Oldham (6) to suggest that the electron-proton interaction was playing a role not well represented in the first order calculations. To analyse the role of the projectile, Gene plotted the position of a peak in the electron velocity distribution vs the variable $v\cos\theta$ where v is the projectile velocity, and θ is the angle of ejection of the electron relative to the beam direction. Such a plot is shown in Fig.(1) where you can see that the peak position is linearly related to $\cos\theta$. This is not the plot that Gene actually showed me, since atomic units are used whereas Gene always employed SI units.

This figure was the key to the puzzle; the extra electrons appeared to come from the projectile just like Doppler shifted Auger electrons. Gene therefore posed the question, "Are there states of the hydrogen atom that would produce this Doppler shifted feature" We discussed this idea at length and exhasted several possiblities; high Rydberg states subsequently stripped by secondary collisions would not appear at angles as large as 30° and a substantial neutral component of the primary beam was unlikely. But Gene's analysis suggested that the basic assumption of the First Born Approximation, namely that the electrons in the final state are target eigenstates, was incorrect. Perhaps one should try using projectile eigenstates in the First Born Approximation. Gene encouraged me to work out this idea, and it turned out

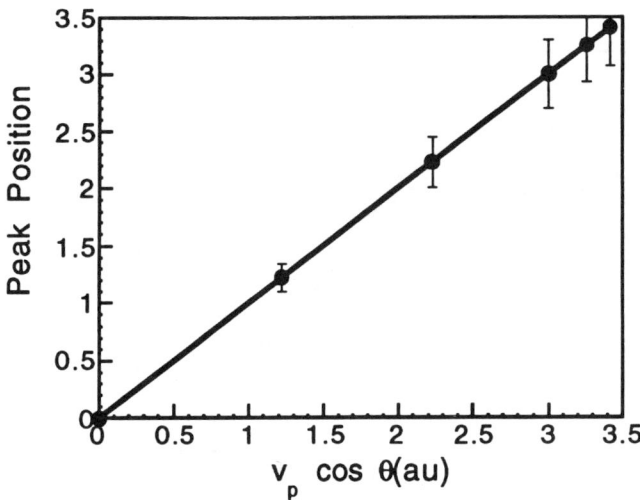

FIG. 1. Plot of the peak position of excess electrons vs $v\cos\theta$.

to give the right shape to the electron angular distribution for electrons whose velocity matched the primary beam velocity of the protons.

Fig.(2) shows our comparison of the data of Ref. (5) with computations based on this picture. Because the angular distribution agreed well in shape, we were encouraged that we were on the right track. My first paper using this idea was rejected since the use of projectile eigenstates was, in the words of the referee, "purely ad hoc." Fortunately, Francisco Pratts had taught me about Faddeev's equations during my stay at the National Bureau of Standards so it was easy to use these equations to justify projectile eigenstates and the computations were eventually published (7). Meanwhile Gene noted that the calculations must predict an infinite cross section when the electron velocity exactly matches the proton velocity in the final state. He immediately saw that this provided a way to probe the continuum capture mechansim experimentally.

The problem of experimentally measuring electrons ejected at zero degrees was posed to a very capable graduate student, Jeff Crooks. He was given a two week timetable to check out the zero degree electron velocity distribution experimentally. He just completed it in the two week time frame, and his results (8) are shown in Fig.(3). They confirmed the infinite cross section at **k** = **v**, where **k** is the electron velocity in the laboratory frame, and firmly established the "continuum electron capture" mechanism.

After this seminal discovery, Gene's research focused on more global questions related to ionization by charged particle impact. However, the con-

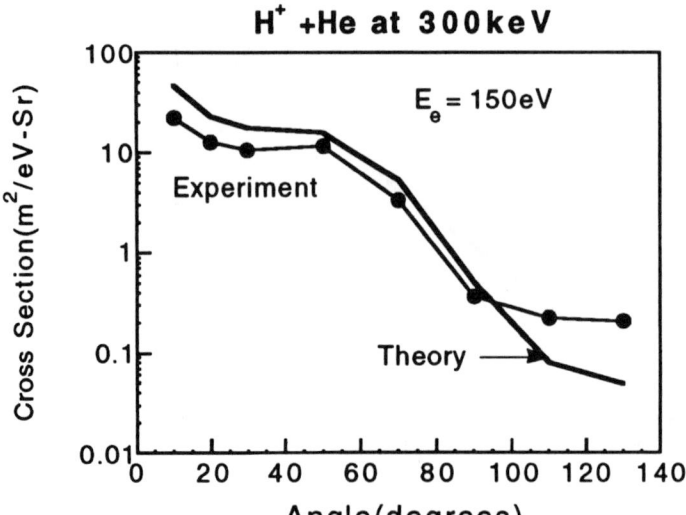

FIG. 2. Plot of measured and calculated DDCS for the ionization of He by 300 keV proton impact. The measurements are from Ref. (5) and the calculations are from Ref. (7).

tinuum capture cusp was part of that global problem. We both sought a description of the cusp that did not depend upon detailed calculations involving all sorts of approximations. To this end Gene recalled a key formula from his Doppler shift studies, namely the transformation from the lab to the projectile frame. This transformation is implicit in the replacement of target electron wave functions by projectile electron wave functions, but does not imply a particular approximation for the cross section.

The tranformation of the DDCS from the lab(unprimed) to the projectile(primed) frame is

$$\frac{d^3\sigma}{dE_{\mathbf{k}} d\Omega_{\mathbf{k}}} = \frac{k}{k'} \frac{d^3\sigma'}{dE_{\mathbf{k'}} d\Omega_{\mathbf{k'}}} \quad (1)$$

where the electron wave vector $\mathbf{k'}$ in the projectile frame relates to the wave vector $\mathbf{k}$ in the target or lab frame according to

$$\mathbf{k'} = \mathbf{k} - \mathbf{v} \quad (2)$$

If for some reason the projectile frame DDCS is non-zero when $k' = 0$ it immediately follows from Eq.(1) that the lab frame cross section is infinite when $\mathbf{k} = \mathbf{v}$. Now an essential feature of charged particles is that they bind electrons in states of any principal quantum number n' and such states are always populated to some extent by electron capture reactions. As n' becomes

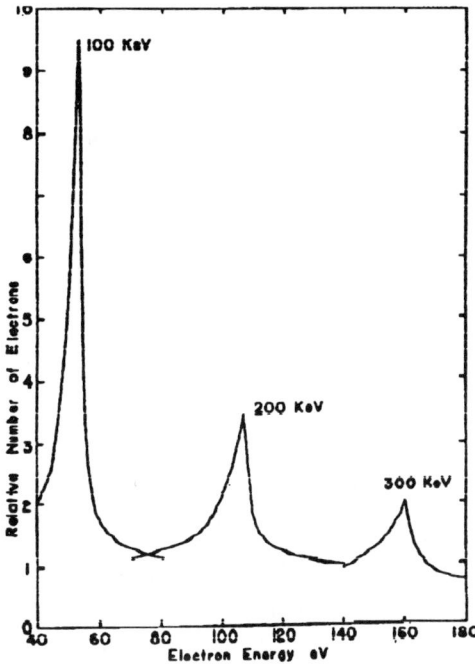

FIG. 3. Energy distribution of electrons ejected from He at $0°$ by proton impact showing the continuum capture cusp. From (8).

infinite this capture process extrapolates smoothly across the threshold, which separates states of high n' from continuum states of vanishingly small $E_{\mathbf{k'}}$. It follows from continuity that the cross section for ejecting electrons moving with zero velocity in the *projectile* frame is non-zero. Eq.(1) then predicts a $1/|\mathbf{k}-\mathbf{v}|$ singularity in the lab frame cross section, as verified experimentally in Ref. (8). A similar cusp is predicted for projectiles which carry electrons that are stripped off in collisions. The cusp was actually observed by Lucas and co-workers (9) for electrons ejected when protons passed through thin carbon foils. The cusp observed in this case could have been caused by a two step process where the proton captures an electron which is stripped off in a subsequent collision. In retrospect the two step process is probably unlikely, but there is no way to distinguish the two possiblities for foil targets. In any event, this model-independent prediction of the cusp seemed to satisfy Gene that the process was well understood, at least when considered as an isolated feature (10). But this cusp posed great conceptual problems for theory, so in the rest of this talk I will describe the status of efforts to build a comprehensive theory of ionization which naturally incorporates the continuum cusp.

III. FURTHER DEVELOPMENTS

A. The Bethe ridge and the elastic scattering model

As background to a comprehensive theory it is useful to look at the Galilean invarient cross section, defined as

$$\frac{d^3\sigma}{dk^3} = \frac{1}{k} \frac{d^3\sigma}{dE_\mathbf{k} d\Omega_\mathbf{k}}, \qquad (3)$$

plotted vs the projection $k_\parallel$ of $\mathbf{k}$ on the $\mathbf{v}$ direction and the magnitude of k perpendicular to this direction, namely $k_\perp = [k^2 - k_\parallel^2]^{1/2}$, as given by the First Born Approximation (Fig.4).

Notice that the only prominant feature is the semicircular ridge centered at $\mathbf{k} = \mathbf{v}$ with radius v. This is just the binary encounter or Bethe ridge charactarized by the absence of target ion recoil in the final state, i.e. all of the momentum lost by the projectile is transferred to the electron. To a first approximation there are no other features.

The Bethe ridge feature is aptly described by a simple model called the elastic scattering model (ESM). In this model the target is regarded as a source of electrons moving with a velocity $-\mathbf{v}$ in the projectile frame which scatter quasi-elastically from the projectile into some other direction with $k' \approx v$. This picture qualitatively describes the surface in Fig.4 and is quantitatively more accurate than the Born approximation when $k_\parallel > v/2$. The elastic scattering model misses the cusp which tends to fill in the valley in the vicinity of $k' = 0$ or $\mathbf{k} = \mathbf{v}$.

B. A free-free transition model and the continuum capture cusp

To reconcile the intuitively attractive ESM with the apparently contradictory cusp, consider an alternative interpretation of the quasi-elastic scattering. Fig.5 illustrates schematically an electron of wave vector $\mathbf{s} - m\mathbf{v}/\hbar$ initially bound in a target eigenstate. The electron's momentum distribution about the value $-\mathbf{v}$ is represented by $\mathbf{s}$, which is to be integrated over. The electron wave function in the field of the projectile is the continuum function $\psi^+_{\mathbf{s}-\mathbf{v}}$. In the final state the electron wave function is $\psi_{\mathbf{k}'}$. The ESM considers that this transition is a quasielastic scattering. If we use that picture no cusp structure is predicted.

Fig.(5) suggests a more general interpretation in terms of free-free transitions between projectile eigenstates

$$E_{\mathbf{s}-\mathbf{v}} \to E_{\mathbf{k}'} \qquad (4)$$

under the action of some transition operator. This operator emerges from Briggs' (11) expansion of capture amplitudes in powers of the projectile charge

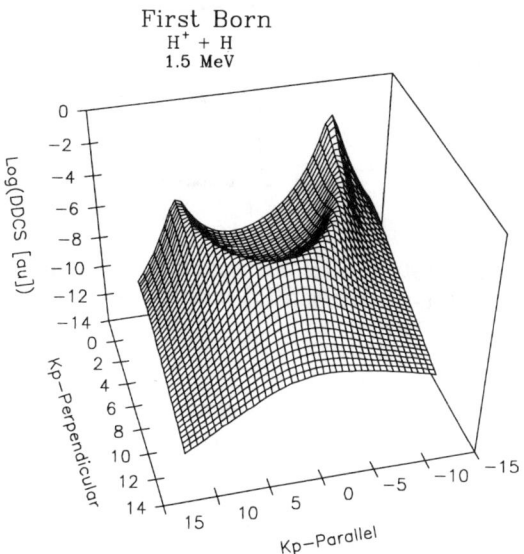

FIG. 4. Plot of the First Born Approximation Galilean invariant cross section vs components of electron momentum parallel and perpendicular to the incident proton velocity for 1.5 MeV proton impact on atomic hydrogen.

Z_T. Using this expansion one finds that the appropriate transition operator is

$$\tilde{V}_T(\mathbf{s} + \mathbf{J}) \exp[i(\mathbf{s} + \mathbf{J}) \cdot \mathbf{r}] \quad (5)$$

where **J** is the recoil momentum of the ionized target in the final state and $\tilde{V}_T(\mathbf{p})$ is the Fourier transform of the electron-target nucleus interaction. The corresponding cross section in this approximation, known as the Distorted-wave Strong Potential Born (DSPB) approximation, is

$$\frac{d^3\sigma}{dE_\mathbf{k} d\Omega_\mathbf{k}} = (2\pi)^4 \mu K_i \int d\Omega_{\mathbf{K}_f} \left| \int d^3s < \psi^-_{\mathbf{k}'} | \exp[-i(\mathbf{J} + \mathbf{s}) \cdot \mathbf{r}] | \psi^+_{\mathbf{s}-\mathbf{v}} > \right.$$
$$\left. \times \tilde{V}_T(\mathbf{J} + \mathbf{s}) \tilde{\varphi}_i(\mathbf{s}) \gamma_k(\mathbf{s}) \right|^2, \quad (6)$$

where $\gamma_k(\mathbf{s})$ is a quantity of nearly unit magnitude for $s \approx 0$ but is rapidly varying for large values of s. To a first approximation this quantity is omitted in calculations of the cross section.

Fig.(6) shows the Galilean invariant cross section calculated using Eq.(6). Note that the continuum capture cusp appears as well as the more pronounced Bethe ridge. The accuracy of Eq.(6) is better illustrated in Fig.(7), which compares the computed and measured (12) DDCS at zero degrees for the ionization of He by 1.5 MeV proton impact. The calulations use Eq.(6) with

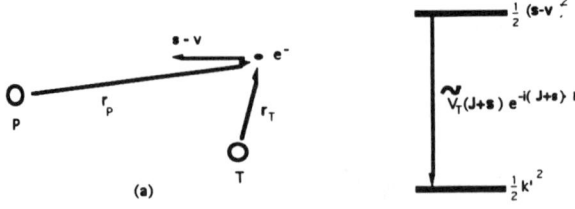

FIG. 5. Schematic illustration of an initially bound electron (a) scattering quasi-elastically from a projectile and (b) making a free-free transition to a final continuum state of the projectile.

an approximate He initial state. Since unobserved quantum numbers are summed over approximately using unitarity, the calculations refer to inclusive cross sections. The agreement is quite good over the electron energy range from 100 eV up to energies well above the position of the binary encounter peak at . Contrary to the ESM, both the cusp and the binary encounter peak are accurately reproduced.

That the cross section of Eq.(6) is infinite at the cusp follows from the use of the projectile eigenstates $\psi_{\mathbf{k}'}^-$ which incorporate a normalization factor proportional to $1/\sqrt{k'}$ for small k'. It is known that this normalization factor must not be present near the binary encounter peak, otherwise the cross section is much too large. This is the conceptual difficulty alluded to earlier–it seems that the cusp, which requires projectile eigenstates, is incompatible with the Bethe ridge, which appears to require target eigenstates. This apparent contradiction is resolved by the free-free transition picture.

Macek and Taulbjerg (13) have shown that Eq.(6) actually gives the elastic scattering model in the region where the momentum of the recoil ion $\mathbf{J}$ tends to zero. Specifically, they show that when $\mathbf{J} \approx 0$ one has $k' \approx v$ and

$$\frac{d^3\sigma}{dE_\mathbf{k} d\Omega_\mathbf{k}} = \frac{(2\pi)^4}{v^2} 2k \int_0^\infty \int_0^{2\pi} |T^{elas}(\mathbf{k}', -\mathbf{J} - \mathbf{v})|^2 |\tilde{\varphi}_i(-\mathbf{J})|^2 d\phi_\mathbf{J} J_\perp dJ_\perp \quad (7)$$

where $\mathbf{J}$ is resolved into components parallel $\mathbf{J}_\parallel$ and perpendicular $\mathbf{J}_\perp$ to the incident velocity vector $\mathbf{v}$;

$$\mathbf{J} = \mathbf{J}_\parallel + \mathbf{J}_\perp \quad (8)$$

and T^{elas} is the T-matrix for elastic scattering of an electron by the projectile. The magnitude of the parallel component is given by

$$J_\parallel = \frac{1}{v}\left(\frac{1}{2}k'^2 - \frac{1}{2}v^2 + |\varepsilon_i|\right) \quad (9)$$

while all values of $\mathbf{J}_\perp$ are integrated over. The free-free transition picture yields the elastic scattering model when $J \approx 0$; thus accurate descriptions of

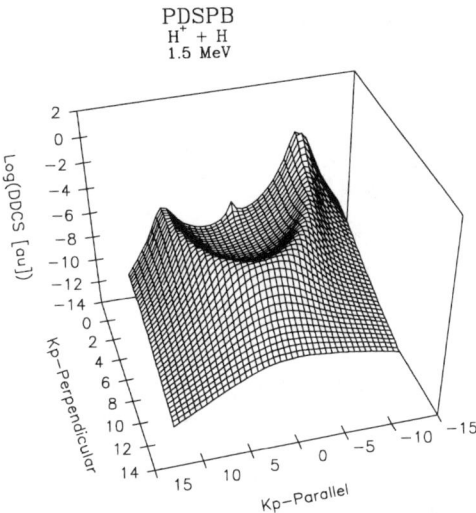

FIG. 6. Plot of the DSPB Galilean invariant cross section vs components of electron momentum parallel and perpendicular to the incident proton velocity for 1.5 MeV proton impact on atomic hydrogen.

both the continuum capture cusp and the binary encounter peak emerge from Eq.(6). It must be emphasized that the standard Born approximation is still needed to describe electrons that have low velocity in the target frame. Even so, the theory of ionization by fast charged particles is in a fairly satisfactory state since we have available two good approximations which give reasonably accurate cross sections in their domains of applicablility. The computations needed to evaluate Eq.(6) are rather complicated, but Madsen and Taulbjerg (14) have been able to obtain a simpler expression valid in the vicinity of the binary encounter peak. They show that Eq.(7) actually incorporates a multiplicative factor $\zeta_{k'}(-\mathbf{J} - \mathbf{v})$ which equals unity near the binary encounter peak where $k' \approx v$ but equals the singular normalization factor $2\pi Z_P/k'$ near $k' = 0$ which is responsible for the cusp. In essence they show that the continuum capture mechansim even modifies the binary encounter peak slightly through the factor $\zeta_{k'}$. This demonstrates rather directly that ionization by charged particle impact cannot be fully understood without incorporating Gene's insight that electrons are carried along by the positively charged projectile in the ionization process.

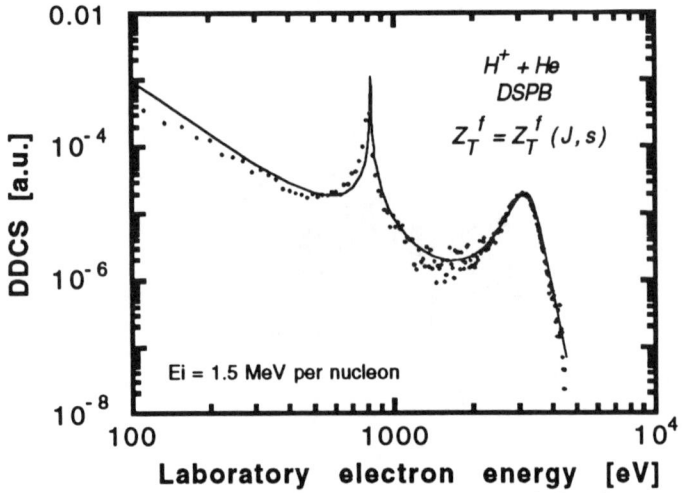

FIG. 7. The doubly differential cross section at $0°$ for the ionization of He by 1.5 keV proton impact. The solid curve is the DSPB theory and the dots are the measurements of Ref. (12).

IV. OTHER DEVELOPMENTS

While the story I have just related shows how Gene's work on the continuum capture mechanism was essential for the development of a comprehesive theory of ionization by charged particle impact, it by no means does justice to the wide impact this idea has had. Let me conclude by mentioning a few investigations that looked at different aspects of "continuum electron" capture. The unexpected connection between capture to high Rydberg states and ionization was investigated by Rødbro (15) for proton-Helium collisions. His work substantiated the basic premise of continuum capture. On the theoretical side Dettman (16) made the first attempt to include the Thomas double scattering mechansim in the theory of the cusp. This mechanism was much discussed in connection with capture to bound states, although firm experimental evidence for it was lacking. Shakeshaft and Spruch (17) recognized that the cusp could be used to test the double scattering theory and might provide the first experimental evidence for second Born terms in the theory of electron capture generally. They pointed out that the double scattering terms gave a cusp which had a step function discontinuity at $k = v$ for electrons ejected in the forward direction. Brenig et al. (18) found this discontinuity experimentally, thereby obtaining the first evidence for multiple scattering effects in electron capture. The cusp anisotropy then became a subject of much interest. Joachim Burgdorfer (19) developed expressions that connect the cusp anistropy with the density matrix of states of high principal quantum number produced by electron capture. Meckbach (20), amoung others, made extensive measurements of the cusp angular distribution to probe various theories (21) of ionization. Harrison and Lucas (9) and others noted that two electrons could be captured into autoionizing states. These states often decayed by emitting very slow electrons in the projectile frame giving rise to subsidary peaks on either side of the cusp. To investigate these, Stolterfoht (22) devised a tandem parallel plate analyser and was able to do high resolution spectroscopy of these very weakly bound autoionizing states.

Measurements specifically designed to exploit the Dopppler shift of electrons in the projectile frame are now quite common, and it is difficult for a mere theorist to do justice to the ingenuity of experimentalists (23) working in this field. The one sentence summaries of measurements and theory given above represent some of the developments that followed the seminal discovery of the CEC cusp. Undoubtedly I have omitted many important studies in my brief account, since there are now so many observations. That's one aspect of the cusp–its always there. If you make measurements of electrons ejected in the forward direction you cannot miss it. Indeed, the cusp feature has entered into the standard nomeclature of ionization by charged particle impact. For that reason this aspect of Gene's work will be recalled whenever the ionization process is involked. That is perhaps a fitting characterization all of Gene's work –the innovative interpretation of his carefully measured data stands the test of time.

ACKNOWLEDGMENTS

Support for this research by the National Science Foundation under grant number PHYS-8918713 is gratefully acknowledged.

REFERENCES

1. M. E. Rudd, *Phys. Rev. Lett.* **15**, 580 (1965).
2. U. Fano and W. Lichten, *Phys. Rev. Lett.* **14**, 627 (1965).
3. E. Everhart and Q. Kessel, *Phys. Rev. Lett.* **14**, 247 (1965).
4. M. E. Rudd, T. Jorgenson, and D. J. Volz, *Phys Rev. Lett.* **16**, 929 (1966).
5. M. E. Rudd, C. A. Sautter and C. L. Baily, *Phys. Rev.* **151**, 20 (1966).
6. W. J. B. Oldham, *Phys. Rev.* **140**, A1477 (1960).
7. J. Macek *Phys. Rev. A* **1**, 235 (1970).
8. J. Crooks and M. E. Rudd, *Phys. Rev. Lett.* **25**, 1599 (1970).
9. K. G. Harrison and M. W. Lucas *Physics Lett.* **33A**, 142 (1970).
10. M. E. Rudd and J. H. Macek, *Case Stud. At. Phys.* **3**, 47 (1972).
11. J. S. Briggs, *J. Phys. B* **10**, 3075 (1977).
12. D. H. Lee, P. Richard, T. J. Zouros, J. M. Sanders, J. L. Shimpaugh, and H. Hidmi, *Phys. Rev. A* **41**, 4816 (1991).

13. J. Macek and K. Taulbjerg, *J. Phys. B* **26**, 1353 (1992).
14. J. Madsen and K. Taulbjerg, *J. Phys B* **27**, L165 (1994).
15. M. Rødbro and L. Andersen, *J. Phys. B* **12** 1 (1974).
16. K. Dettman, K. G. Harrison and M. W. Lucas, *J. Phys. B* **7**, 269 (1974).
17. R. Shakeshaft and L. Spruch, *Phys. Rev. Lett.* **41**, 1037 (1978).
18. M. Brenig, S. Elston, I. Sellin, L. Liljeby, R. Thoe, C. R. Vane, H. Gould, R. Marrus and R. Laubert, *Phys. Rev. Lett.* **45**, 1689 (1980).
19. J. Burgdorfer, *Phys. Rev. A* **33**, 1598 (1986).
20. W. Meckbach, I. B. Nemirovsky and G. Garibotti, *Phys. Rev. A* **24**, 1793 (1981).
21. D. H. Jakubassa-Amundsen, *J. Phys. B* **16** 1767 (1983).
22. N. Stolterfoht, *Phys. Rept.* **146**, 315 (1987).
23. I. A. Sellin, Invited papers of the XII ICPEAC, Gatlinburg, July 1981, ed. by Sheldon Datz (North Holland, New York,1982) p195ff.

M. E. Rudd's Contributions To Autoionizing States In Ion-Atom Collisions

A. K. Edwards

Department of Physics and Astronomy
The University of Georgia
Athens, Georgia 30602-2451, U.S.A.

Abstract. The use of ion-atom collisions to study the autoionizing states of atoms was discovered by Rudd while a professor at Concordia College in Moorhead, MN. The technique has been used to study the autoionizing transitions in target atoms and projectile ions, and to study the collisions that form the excited states. The doubly excited states of helium are still being investigated in collision experiments within the framework of the few-body problem.

DISCOVERY OF He** IN ION-ATOM COLLISIONS

While a graduate student at the University of Nebraska, Gene Rudd measured the doubly-differential cross sections (DDCS) for the production of secondary electrons in 50-, 100-, and 150-keV H$^+$ on helium collisions (1). As part of the experimental procedure an electron energy analyzer was stepped over the low energy range in 10-eV steps; at 10, 20, 30, 40, etc. eV. In the process the interesting region of the electron spectrum around 35 eV was missed.

Upon graduating from the University of Nebraska Rudd returned to his alma mater Concordia College in Moorhead, Minnesota where he continued his work in ion-atom collisions. An undergraduate student at Concordia College, David Lang, began redoing the Nebraska measurements, but with a finer mesh and discovered that the point at 35 eV was always a high point. Ugo Fano had corresponded with Gene Rudd and suggested that he look closely at the electron spectrum in the neighborhood of 35 eV. Fano had recently published his classic paper (2) on the shape of resonances in inelastic electron scattering and was interested in the autoionizing states of helium. Smaller slits were inserted into the energy analyzer in order to improve its resolution and still smaller step-sizes in energy used. The results shown in Fig. 1 led to the first publication in Physical

© 1996 American Institute of Physics

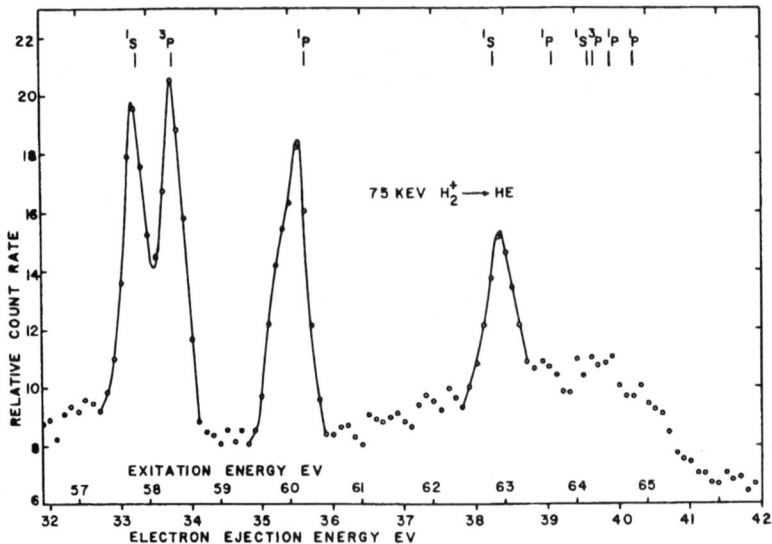

FIGURE 1. Autoionizing states of He excited in H_2^+-on-He collisions as observed by Rudd (3) in 1964.

Review Letters (3). The resolution was further refined and an X-Y plotter added to simplify data collection and more states were discovered (4).

Excitation of autoionizing states through ion-atom collisions brought new dimensions to the study of the states. Non-dipole transitions occurred which excited levels that were not formed in photoexcitation experiments, and as can be seen in Fig. 1, triplet states of He^{**} were readily excited through spin exchange. This was not the case in either inelastic electron scattering of electrons or photoexcitation.

In 1965 the University of Nebraska made the call to Gene Rudd and brought him back to build the highly successful program in the newly opened Behlen Laboratory of Physics. Much of the new work centered on Auger transitions, but several years later further measurements would be made on the autoionizing states of helium once again. One paper (5) dealt with the Fano shape of the peaks in the electron spectra and another (6) with the production of He⁻ and excitation of the helium autoionizing states by various projectiles and the angular distribution of the ejected electrons.

LINE SHAPES

It was 1961 when Fano published his famous paper (2) on the asymmetric line shapes seen in the inelastic scattering of electrons. The early work of Rudd

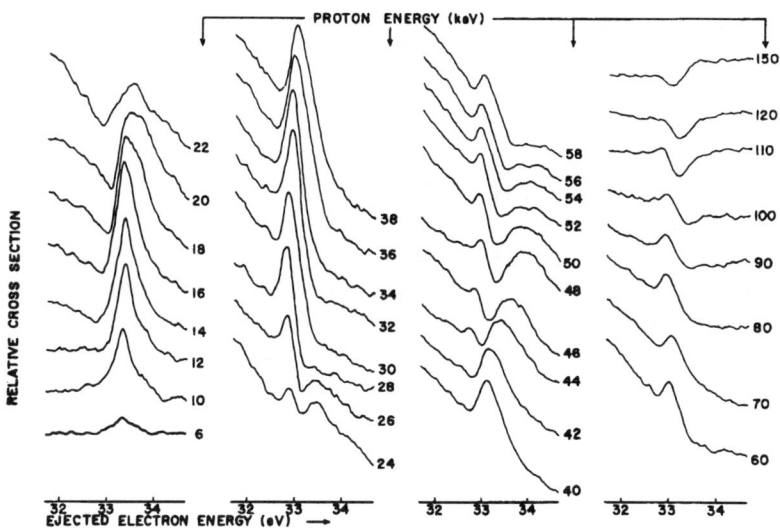

FIGURE 2. Electron spectra at an ejection angle of 10° for H$^+$-on-He collisions. The autoionizing state of helium is the 2s^2 (^{1}S). (Ref.5)

(3,4) displayed only symmetric line shapes for the autoionizing transitions of helium. These data had been collected at the backward angle of 160° from the beam direction. Subsequent measurements (7,8) made at other laboratories at smaller angles indicated that the autoionization transitions had asymmetric profiles. In 1972 Schowengerdt and Rudd (5) published a paper showing the autoionization profiles of electrons near the 2s^2(^{1}S) state ejected at 10° for collisions of 6- to 150-keV H$^+$ on He. A strong dependence of the line shape on the beam energy was discovered (Fig. 2). This effect found renewed interest (9,10) fifteen years later and is still being investigated (11,12).

CLASSIFICATION OF AUTOIONIZING TRANSITIONS

In an early edition of Case Studies in Atomic Physics, Rudd and Macek (13) teamed-up to write an article entitled *Electron production in ion-atom collisions*. States leading to autoionization were separated into five categories: multiple excitations, inner-shell excitation, excitation plus core rearrangement, inner-shell vacancy, and vibration-rotation plus electronic excitation. The He** states are examples of multiple excitations and Auger transitions occur from inner-shell vacancies. For the latter, studies in Rudd's laboratory centered mostly on argon. Measurements were made on the properties of the transitions; properties such as the energies and widths of the transitions and the angular distributions of the Auger electrons. And, in addition, measurements were made of the cross sections for the production of the inner-shell vacancies which lead to the Auger transitions.

AUGER TRANSITIONS

Don Volz, a Ph.D. student of Rudd's, took on the task of studying the autoionizing transitions found in H^+, H_2^+, and Ar^+ collisions with argon. Two regions of the electron spectra were found that were rich in structure. The low energy region (0-20 eV) consisted of states of double excitation and inner-shell excitation, while the higher energy region (150-210 eV) consisted of Auger transitions from vacancies in the L-shell. Using fine resolution and H^+ and H_2^+ projectiles Volz and Rudd (14) measured the energies, widths, and branching ratios of the Auger transitions and the angular distributions of the Auger electrons. Later, Rudd (15) extended the range of the H^+ beam energies and, using an analyzer of lower resolution, he measured the total cross sections for the production of the $L_{2,3}$ vacancy.

Studies of inner-shell ionization played a big roll in understanding the mechanisms for electron production in heavy-ion collisions. The Fano and Lichten (16) model of electron promotion was proposed and proved to be very successful in explaining the formation of inner-shell vacancies. This theory was tested in the work of Rudd, Jorgensen, and Volz (17) where they measured the DDCS for 100-keV Ar^+ on Ar.

An interesting discovery in this experiment (17,18) was the finding of a doppler-shifted electron spectrum. Electrons ejected by the projectile produce a spectrum superimposed upon the target spectrum. However, the electrons coming from the projectile are shifted in energy by an amount that depends on the beam velocity and the angle of observation. Figure 3 shows the shifted and unshifted spectra in the low-energy region for 100-keV Ar^+-on-Ar collisions (17).

In the 1960's a group at the University of Connecticut under the direction of Edgar Everhart was involved in energy-loss measurements in heavy-ion-atom collisions. A former student of Everhart's, Quentin Kessel, proposed a simple model to explain the Auger emission cross section. He had shown experimentally that for certain distances-of-closest-approach in ion-atom collisions, there were jumps in the amount of energy lost in the collision. This meant that at each step in energy loss a given shell of the projectile was penetrating the corresponding shell of the target and promotion of electrons from that shell could occur. A paper by Cacak, Kessel, and Rudd (19) demonstrated the success of the model and its general applicability.

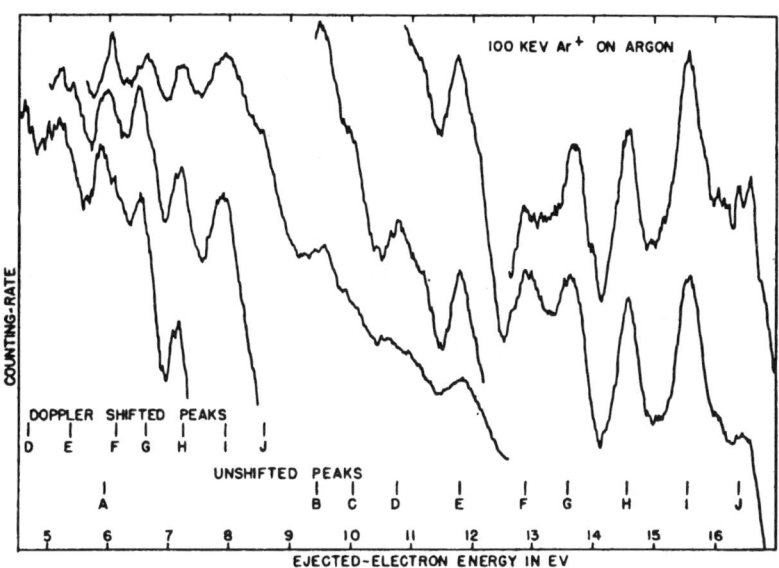

FIGURE 3. Electron spectrum at 160° for 100 keV Ar$^+$-on-Ar collisions. Both shifted and unshifted peaks produced by autoionizing states of argon are observed. (Ref. 17)

Rudd's work on inner-shell ionization lead to a collaboration with the group at Aarhus in the mid-70's. These investigations used the doppler shifting of the electrons as a tool and relied on the Fano-Lichten theory to explain the observations. A paper (20) contradicting the findings (21) of researchers in Leningrad was published on the double vacancy formed in heavy-ion-atom collisions. The Russian group had found peaks in the electron spectrum produced in Ar$^+$-Ar collisions that they attributed to a decay from a double vacancy with all of the energy given to a single electron. The Aarhus-Nebraska collaboration argued that the double vacancy state in Ar$^+$-Ar collisions is not very probable, but it is in P$^+$-Ar collisions with P having the double vacancy. And from this state the transition is a two-step process with two electrons carrying away the energy rather than one.

A few years later they published two comprehensive studies on the Auger spectroscopy of heavy-ion-atom collisions. The first paper (22) was a description of the apparatus used in the experiments, and detailed explanations of the effects that lead to line broadening were given. The second paper (23) reported on heavy-ion-Ar collisions where Z of the projectile was less than Z=18 of argon. The Auger-electron spectra arising from the projectiles were fairly well resolved and could be analyzed. The results gave a wealth of information on the formation of double-vacancy states and their subsequent transitions.

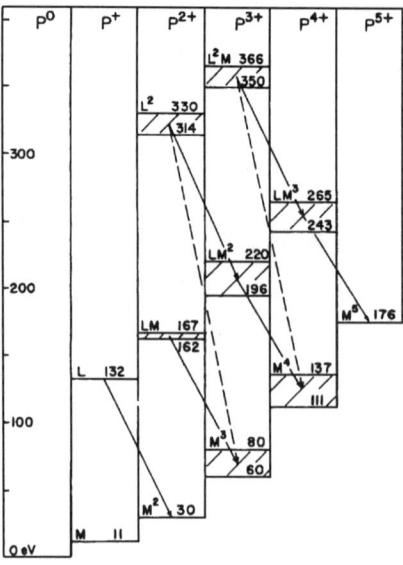

FIGURE 4. Rudd diagram showing Auger transitions. This example is for phosphorus and is taken from Ref. 20. Dashed lines are hypothetical one-step transitions for double vacancy states.

OTHER STUDIES

The autoionizing states of neon excited in ion-atom collisions were reported by Edwards and Rudd (24). The autoionizing states included inner-shell excitation and double excitation in the low-energy region of the electron spectra and the Auger transitions in the high-energy region. Examples of autoionizing states formed by excitation plus core rearrangement were found by Rudd and Smith (25). Molecular oxygen was bombarded by H^+ or He^+ ions and resulted in the molecular target dissociating into the autoionizing states of atomic O. A total of six series with either the $2p^3(^2P)$ or (^2D) excited core were identified.

In the course of discussing the many transitions that are possible from autoionizing states a new scheme of displaying the transitions was developed by Rudd (26) and is shown in Fig. 4. Different charge states are grouped into adjacent columns and transitions are represented by lines drawn from the initial state in one column to the final state in another column.

In all of Rudd's experimental work an electrostatic analyzer was used to measure the energy of the secondary electrons, hence he developed an expertise on the construction and operation of the analyzer. A very thorough discussion on the properties of electrostatic analyzers was written by Rudd and included in the book *Low Energy Electron Spectrometry* (27). The resolution, dispersion, transmission function and other characteristics of the various types of energy analyzers were covered in detail.

Spurred by the work from Rudd's laboratory during the late '60's and early '70's the study of autoionization and Auger transitions became a topic of great interest. So much so, that in 1971 Science Abstracts began listing "Auger effect" as a separate entity and the same for "Autoionization" in 1973.

RECENT INVESTIGATIONS OF He**

In the paper of Schowengerdt and Rudd (5) it was found that the shape of the autoionizing peak observed at small angles depended on the projectile velocity. It was suggested that this changing shape may be due to the interaction of the projectile with the continuum electrons. More recent work (9-12) has centered on the interaction of the autoionized electron with the fast beam particle. This post-collision interaction (PCI) has been studied as a function of the projectile velocity and charge and found to exist at quite high velocities. The work of Arcuni and Schneider (10) described this as a fast-ion post-collision interaction (FIPCI). This is unlike the more well-known PCI where the projectile travels slowly relative to the ejected electrons. An observable difference is that the FIPCI has a strong angular dependence while the usual PCI is more isotropic. The work of Bruch and coworkers (12) has extended the angular measurements down to zero degrees and with very fine energy resolution. Their conclusion is that more work needs to be done on the problem of the three-particle Coulomb interaction in the final state.

The cross section for double ionization of helium by equivelocity particles of positive or negative charge (viz, e^-, e^+, p^-, and p^+) is about a factor of two greater for the negatively charged particle compared to the positive one (28). The removal of both electrons from He can occur by either a single projectile-electron interaction plus an electron-electron interaction or through a double collision event. The first case is described by the first term of a Born expansion and has a Z_p^2 dependence while the second is described by the second term and has a Z_p^4 dependence. The observed difference between negative and positive charges must arise from a Z_p^3 term which indicates an interference between the first and second Born amplitudes. However, there have been arguments that this interference will not occur in an independent particle model (29,30).

The double excitation of helium by fast projectiles is another two-electron process similar to double ionization. Measurements (31,32) of the double-excitation of helium cross sections by negative and positive projectiles do not show the same factor of two that is seen in double ionization. There are experimental difficulties that arise in double excitation of helium that do not occur in double ionization. Namely, some states are not resolved, there is a strong angular dependence of the observed resonances, the unfolding of the asymmetric resonance from the background is troublesome, and the FIPCI. The factor-of-two due to projectile charge in the cross section has been observed in the double excitation of H_2 by Edwards, et al. (33). They detected the presence of the doubly-excited states by detecting the H^+ fragment ions following dissociation from the excited states rather than by observing the electron spectra as was done in the helium experiments. As pointed out by Briggs and Macek (29) more work needs to be done on the importance of the interference of the first and second born amplitudes in two-electron processes in ion-atom collisions. The doubly-excited states of helium formed in ion-atom collisions have played a major role in past investigations and will, no doubt, be an important contributor in future studies.

REFERENCES

1. Rudd, M. E., and Jorgensen, Jr.,T., Phys. Rev. **131**, 666-675 (1963).
2. Fano, U., Phys. Rev. **124**, 1866-1878 (1961).
3. Rudd, M. E., Phys. Rev. Lett. **13**, 503-505 (1964).
4. Rudd, M. E., Phys. Rev. Lett. **15**, 580-581 (1965).
5. Schowengerdt, F. D., and Rudd, M. E., Phys. Rev. Lett. **28**, 127-130 (1972).
6. Schowengerdt, F. D., Smart, S. R., and Rudd, M. E., Phys. Rev. A **7**, 560-566 (1973).
7. Bordenave-Montesquieu, A., and Benoit-Cattin, P., Phys. Lett. **36A**, 243-244 (1971).
8. Stolterfoht, N., Phys. Lett. **37A**, 117-118 (1971).
9. Arcuni, P. W., Phys. Rev. A **33**, 105-111 (1986).
10. Arcuni, P. W., and Schneider, D., Phys. Rev. A **36**, 3059-3070 (1987).
11. Godunov, A. L., Novikov, N. M., and Senashenko, V. S., J. Phys. B **23**, L359-L364 (1990).
12. Wang, H., Bruch, R., Hao, F., Fuelling, S., Xu, Z., Wang, Z., and Rauscher, E., Nucl. Instrum. and Methods Phys. Res. **B79**, 114-116 (1993).
13. Rudd, M. E., and Macek, J. H., Case Stud. At. Phys. **3**, 47-136(1972).
14. Volz, D. J., and Rudd, M. E., Phys. Rev. A **2**, 1395-1403 (1970).
15. Rudd, M. E., Phys. Rev. A **10**, 518-521 (1974).
16. Fano, U., and Lichten, W., Phys. Rev. Lett. **14**, 627-629 (1965).
17. Rudd, M. E., Jorgensen, Jr., T., and Volz, D. J., Phys. Rev. **151**, 28-31 (1966).
18. Rudd, M. E., Jorgensen, Jr., T., and Volz, D. J., Phys. Rev. Lett. **16**, 929-930 (1966).
19. Cacak, R. K., Kessel, Q. C., and Rudd, M. E., Phys. Rev. A **2**, 1327-1331 (1970).
20. Rudd, M. E., Fastrup, B., Dahl, P., and Schowengerdt, F. D., Phys. Rev. A **8**, 220-225 (1973).
21. Ogurtsov, G. N., Flaks, I.P., and Avakyan, S. V., Proceedings of the Sixth International Conference on the Physics of Electronic and Atomic Collisions (MIT Press, Cambridge, MA., (1969), pp. 274-276.
22. Dahl, P., R•dbro, M., Fastrup, B., and Rudd, M. E., J. Phys. B **9**, 1567-1579 (1976).
23. Dahl, P., R•dbro, M., Hermann, G., Fastrup, B., and Rudd, M. E., J. Phys. B **9**, 1581-1599 (1976).
24. Edwards, A. K., and Rudd, M. E., Phys. Rev. **170**, 140-144 (1968).
25. Rudd, M. E., and Smith, K., Phys. Rev. **169**, 79-84 (1968).
26. Rudd, M. E., Proceedings of the Second Oak Ridge Conference on the Use of Small Accelerators for Teaching and Research, Ed. Jerome Duggan, (Oak Ridge Associated Universities, 1970) p. 305.
27. Rudd, M. E., in *Low Energy Electron Spectrometry*, by Sevier, K. D., (Wiley-Interscience, New York, 1972) pp. 17-31.

28. McGuire, J. H., Adv. At. Mol. Opt. Phys **29**, 217-323 (1991).
29. Briggs, J. S., and Macek, J. H., Adv. At. Mol. Opt. Phys. **28**, 1-74 (1990).
30. Stolterfoht, N., Phys. Rev. A **48**, 2980-2985 (1993).
31. Pedersen, J. O. P., and Hvelplund, P., Phys. Rev. Lett. **62**, 2373-2376 (1989).
32. Giese, J. P., Schulz, M., Swenson, J. K., Schone, H., Benhenni, M., Varghese, S. L., Vane, C. R., Dittner, P. F., Shafroth, S. M., and Datz, S., Phys. Rev. A **42**, 1231-1244 (1990).
33. Edwards, A. K., Wood, R. M., Davis, J. L., and Ezell, R. L., Phys. Rev. A **42**, 1367-1375 (1990); *ibid.* **44**, 797 (1991).

Modeling Ionization Cross Sections: Two Decades of Dreams Come True

Yong-Ki Kim

National Institute of Standards and Technology, Gaithersburg, MD 20899

Abstract. Modeling of differential and total ionization cross sections by electron impact is reviewed. A new theoretical model that does not depend on any empirical or arbitrary parameters is described. The prototype of this new model was proposed by Rudd and was originally based on the binary-encounter theory. The model has been improved by replacing a part of the binary-encounter theory with the dipole contribution as prescribed by the Bethe theory. The current model, henceforth referred to as the binary-encounter-dipole (BED) model, reproduces known singly differential and total ionization cross sections for small atoms and molecules accurately. The possibility of extending the BED theory to doubly differential cross sections as well as to proton-impact ionization cross sections is discussed.

INTRODUCTION

In addition to being of fundamental interest, results from atomic physics have been used widely in basic and applied sciences. Today, atomic data and theory are used extensively in astrophysics, plasma physics, accelerator physics, and radiation and medical physics. In all of these fields, various atomic and molecular ionization processes play a prominent role. For example, they must be understood in order to accurately model plasmas in stars and tokamaks, or to model radiation damage in matter and biological tissues.

For most of these applications, radiation and plasma physics in particular, the primary demand has been for models of ionizing events meeting the following criteria:

(a) They should describe qualitatively correct gross features, even at the expense of fine details;

(b) They should be valid over a broad range of particle energies, rather than a narrow range at high resolution;

(c) They should be consistent with each other, *e.g.*, between differential and integrated cross sections;

(d) They should be expressed in terms of simple, analytic formulas, rather than extensive tables;

(e) They should include secondary (ejected) electron cross sections, regardless of the type of incident particles.

The last point arises from the fact that secondary electrons of all energies are generated regardless of the type of incident particles.

Although elastic scattering cross sections of electrons are of interest in some applications, *e.g.*, track structure modeling, the discussions in this article will focus on cross sections associated with the generation of secondary electrons.

SECONDARY ELECTRON CROSS SECTIONS

The most detailed secondary electron cross sections are known as the triply differential cross sections (TDCSs). For a given incident electron energy, the TDCS provides information on the angular distribution of the scattered electron and the angular and energy distributions of the ejected electron. Since the scattered and ejected electrons are indistinguishable, it is customary to call the faster one the *primary* electron and the slower one the *secondary* electron. When the incident particle is a bare ion, then there is no ambiguity about the identity of secondary electrons. However, a new ambiguity arises if the incident ion is dressed, *i.e.*, has its own bound electrons. In this case, electrons ejected from the incident ion can be distinguished by its "carrier" speed equal to the speed of the ion, but the distinction between the electrons ejected from the projectile and those from the target becomes blurred for slow incident ions.

To measure TDCS, one must measure the primary and secondary electrons in coincidence, varying the angles at which the electrons are detected as well as the energy of the secondary electron, while keeping the incident energy constant. For a multishell target, the energy of the primary electron must also be monitored to identify the subshell from which the secondary electron comes. For testing an ionization theory, TDCSs provide the greatest detail, but in reality, data can be taken only for limited combinations of parameters because of the low coincidence counting rates. For practical applications, these limited combinations of parameters make TDCSs difficult to use. When the incident particle is an ion rather than an electron, its deflections are very small and difficult to separate. One can take advantage of this forward peaking of the incident ions, however, and measure them all to integrate over all scattering angles of the ions.

When TDCSs are integrated over the angles of the primary or scattered particle, one gets doubly differential cross sections (DDCSs), which provides the energy and angular distributions of the secondary electron only. Singly differential cross sections (SDCSs) are then obtained by integrating DDCSs over the angles of the secondary electron. Finally, total ionization cross sections (TICSs) are obtained by integrating SDCSs over the energy of the secondary electron. The DDCS, SDCS and TICS are basic ingredients in modeling radiation effects and plasmas in fusion devices. The TICS is probably the most valuable form of atomic data used by the radiation and plasma physics community. We define the relationship between various ionization cross sections below:

TDCS: $\dfrac{d^3\sigma}{dW d\Omega_p d\Omega_s}$;

DDCS: $\dfrac{d^2\sigma}{dW d\Omega_s} = \int \dfrac{d^3\sigma}{dW d\Omega_p d\Omega_s} d\Omega_p$;

SDCS: $\dfrac{d\sigma}{dW} = \int \dfrac{d^2\sigma}{dW d\Omega_s} d\Omega_s$;

TICS: $\sigma_i = \int \dfrac{d\sigma}{dW} dW$,

where W is the secondary-electron kinetic energy, Ω_p is the solid angle for the primary electron, and Ω_s is the solid angle for the secondary electron.

Experimental Difficulties in Secondary Electrons

The difficulty with low counting rates in measuring TDCSs has already been mentioned. Even in measuring DDCSs, practical problems make it difficult to measure cross sections at extreme forward and backward angles. In addition, there is a general difficulty associated with detecting slow secondary electrons, say W < 10 eV. Unfortunately, slow secondary electrons constitute anywhere from one-half to three-quarters of electrons ejected in ionizing collisions, so it is very important to be able to measure DDCSs to as low W as possible.

For very fast secondary electrons, there is another kind of challenge to experimentalists. Fast secondaries are ejected in a particular direction, known as the binary peak, with a rather narrow angular width. Hence, experimental data measured with detectors in fixed angular positions may miss a binary peak completely. With increasing secondary-electron energy, the binary peak will shift to smaller angles, and eventually be detected by one of the detectors. The angular position of this peak is given by

$$\cos\theta_b = \sqrt{E/T}, \tag{1}$$

where θ_b is the binary peak angle, $E = W + B$ is the energy transfer with the binding energy B, and T is the incident electron energy. We shall discuss the changes necessary for a heavy ion, such as a proton, later.

Another difficulty is the lack of a reliable method to measure SDCSs directly. Integrating DDCSs over $d\Omega_s$ always leaves some uncertainty because of the difficulty in measuring cross sections in extreme forward and backward directions, though the uncertainty is considerably reduced due to the $\sin\theta$ factor in $d\Omega_s$. Integrating the ionization cross section over the angular distribution of the primary electron, without regard to secondary electrons, will in principle lead to the correct SDCS, but the primary electrons are sharply peaked in the forward direction, again resulting in considerable uncertainties.

The differential cross sections must be consistent with each other when appropriate integrations are carried out, and particularly with σ_i, which can often be independently measured with much better accuracy than differential ionization cross sections.

The role of theory is to provide qualitative understanding for prominent features seen in differential and integrated cross sections and then to offer quantitative formulae for the angles and energies of the secondary electrons that cannot be effectively covered by various experimental methods. A proper theory will also be able to predict cross sections for targets difficult to prepare or handle experimentally. Since the late 1960s, the search for this type of an "ideal" theory or model for ionization of atoms and molecules has consumed the efforts of many atomic theorists and experimentalists around the world, most notably Eugene Rudd and many of the participants of this symposium in his honor.

Prominent Features of Secondary Electron Distributions

We shall focus our discussions on electron-impact ionization for the moment, although many qualitative aspects will be common to both incident electrons and ions. Collisions between an incident electron and a target electron fall under two broad categories; *hard (=close) collisions* with small impact parameters and *soft (=distant) collisions* with large impact parameters.

In a hard collision, a large amount of momentum is transferred from the incident particle to the target particle, and it is most likely that one of the bound electrons will absorb the entire transferred momentum and be ejected. This is called a binary collision, and the energy and the angle of the ejected electron is determined by energy-momentum conservation laws. Equation (1) for the binary peak angle, θ_b, was derived by assuming the collision of an electron with another at rest using the conservation laws. Hard collisions are likely to produce fast secondary electrons with sharp binary peaks.

The Rutherford and Mott cross sections (1), though they are exact solutions only for the collision of two free particles, provide a solid basis for understanding the nature of hard collisions. When $W \gg B$, they even offer quantitatively reliable cross sections.

In contrast, a small amount of momentum is transferred in a soft collision, and the transferred momentum is absorbed by the atom as a whole. Hence, there is even the possibility that the nucleus moves forward and ejects an electron in the backward direction. The classical theory does not account for the ejection of an electron in the backward direction.

Soft collisions are mostly responsible for ejecting slow secondaries, and are dominated by the dipole interaction. The Bethe theory (2) predicts that the dipole interaction grows with the incident energy as lnT, and that it will dominate ionization and excitation cross sections at high incident energies. This is why the Bethe theory plays a crucial role in radiation protection from high-energy particles emitted by accelerators and other artificial radioactive sources. The Bethe theory also provides an explicit relationship between photon-impact and electron-impact cross sections. Since the interaction of photons with atoms are better understood (at least the cross sections are better known) both theoretically and experimentally, comparisons of electron-impact and photon-impact cross sections provide a powerful tool in assessing the reliability of electron-impact cross sections using this connection provided by the Bethe theory.

MODELING IONIZATION CROSS SECTIONS

Once the roles of hard and soft collisions are understood, it is natural to try to build a model for ionization cross sections by combining the Rutherford (or Mott) theory with the dipole interaction as predicted by the Bethe theory. Many researchers in the past few decades searched for a "magic" formula of the form

$$\sigma_{ion} = A\sigma_{soft} + B\sigma_{hard} , \qquad (2)$$

where σ_{ion} is either a differential or integrated ionization cross section, σ_{soft} represents soft collisions, and σ_{hard} accounts for hard collisions. The stumbling block in this seemingly straightforward task is that both A and B are complicated functions of the incident energy, T. Only a rigorous theory of ionization can provide the true T dependence of A and B.

Since our present capability in solving a many-electron Schrödinger equation is rather limited—*e.g.*, there is still no exact solution for the electron-hydrogen atom collision—many of us have had to resort to models that (a) included some empirical parameters; (b) worked only on specific targets; (c) worked only in a limited range of primary-electron energies; (d) worked only in a limited range of secondary-electron energies; (e) did not indicate how to extend the model to other cases; and/or (f) contained little physics. Up until now, popular models earned their reputation mostly for their simplicity rather than for their reliability.

For high incident electron energies T, the Bethe theory gives very reliable cross sections, provided that good wave functions are used. For intermediate T down to the ionization cross section peak, which occurs at $3 \sim 5$ times the binding energy of the ejected electron, some methods that go beyond the usual first-order perturbation theory provide reliable cross sections. However, from ionization threshold to the peak, no *ab initio* theory seems to work well. Those based on a perturbation theory are particularly unsuccessful, perhaps because the interaction is too strong near the peak to be treated perturbatively. Moreover, for low and intermediate T, the exchange interaction between the incident and target electrons is significant. Hence, a successful theory should account for the soft and hard collisions as well as the exchange interaction without using a perturbation scheme.

Binary-Encounter-Dipole Model for Singly Differential Cross Sections

After several attempts in modeling ionization cross sections, Rudd (3) has proposed a model based on the binary-encounter theory (4), which successfully reproduced SDCSs of H, He and H_2, and the corresponding TICSs. This model, however, had some difficulty in reproducing the shape of the SDCSs of targets with more complicated shell structures. This difficulty arose from the fact that the binary-encounter theory, which includes the Rutherford and Mott theories, lacks terms representing the dipole interaction.

This deficiency was corrected by replacing one of the terms in the binary encounter theory with the dipole interaction as prescribed by the Bethe theory. This new combination, the BED model (5), works very well on a variety of targets: H, He, He^+, Li^{++}, Ne, and H_2. The BED model predicts both SDCSs and corresponding TICSs. It provides a relatively simple formula for the SDCS of a specified subshell with binding energy B and average kinetic energy $U = \langle p^2/2m \rangle$:

$$\frac{d\sigma}{dW} = \frac{S}{B(t+u+1)} \left\{ -\frac{\Gamma}{t+1} \left[\frac{1}{w+1} + \frac{1}{t-w} \right] \right. \tag{3}$$
$$\left. + \Gamma \left[\frac{1}{(w+1)^2} + \frac{1}{(t-w)^2} \right] + \frac{\ln t}{N(w+1)} \frac{df(w)}{dw} \right\},$$

where $t = T/B$, $u = U/B$, $w = W/B$, $\Gamma = 2 - N_i/N$, $N_i = \int (df/dw)dw$, df/dw is differential dipole oscillator strength, and $S = 4\pi a_0^2 N(R/B)^2$. Here, N is the occupation number of the subshell, a_0 is the Bohr radius and R is the Rydberg energy.

The first term in the curly braces of Eq. (3) represents the interference between the direct and exchange interactions, the second term represents hard collisions both in the direct and exchange interactions (from the Mott cross section), and the last term represents the dipole interaction from the Bethe theory. The factor Γ was chosen so that the asymptotic forms of the TICS and its matching stopping cross section, $\sigma_{st} = \int (W+B)(d\sigma/dW)dW$, would agree with those predicted by the Bethe theory.

Also note that the denominator, $t+u+1$ which originated from the binary-encounter theory, is different from the usual one, t, that appears in most *ab initio* theories. Using this denominator has the net effect of reducing the cross section at low t, where most *ab initio* theories tend to overestimate cross sections. It differs negligibly from $1/t$ at high incident energies, where the *ab initio* theories are reliable.

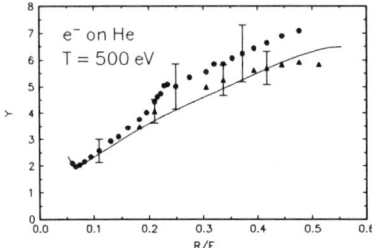

Figure 1. SDCS of He at T = 500 eV. Solid curve, the BED model; circles, experimental data by Opal et al. (6); triangles, those by Gorganthu and Bonham (7).

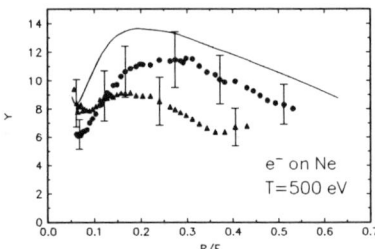

Figure 2. SDCS of Ne at T = 500 eV. Solid curve, the BED model; circles the experimental data by Opal et al. (6); triangles, those by DuBois and Rudd (8).

Experimental data on SDCSs are not as reliable as those for TICSs. Singly differential cross sections for He and Ne at an incident electron energy T = 500 eV are presented in Figs. 1 and 2, respectively (5). Although we have used experimental photoionization cross sections to deduce df/dw, we could have used theoretical df/dw to obtain the BED cross sections. In both figures, the solid curves are those predicted by the BED theory. The ordinates of the figures are the ratio Y of the SDCS to the Rutherford cross section for one bound target electron. Hence, the height represents the effective number of bound electrons participating in ionizing collisions, similar to the f-values representing the effective number of bound electrons absorbing incident photons. The abscissa of Figs. 1 and 2 is chosen such that the area under the curve is proportional to the matching TICS.

The differential oscillator strength, df/dE, where $E = W + B = h\nu$, can be deduced from photoionization cross sections:

$$\sigma_{h\nu} = 4\pi^2 a_0^2 \alpha R (df/dE), \tag{4}$$

where α is the fine-structure constant. Unfortunately, df/dE is not always easy to obtain, particularly for inner shells. For valence or near valence subshells, however, photoionization cross sections and branching ratios are available for some atoms and molecules (9). For atoms, one can probably use theoretical df/dE, since the level of accuracy expected for the BED model is of the order of 10 percent.

Total Ionization Cross Sections

The TICS for a subshell is obtained simply by integrating Eq. (3) over dW:

$$\sigma_i(t) = \frac{S}{t+u+1} \left[D(t) \ln t + \Gamma \left(\frac{t-1}{t} - \frac{\ln t}{t+1} \right) \right], \tag{5}$$

where

$$D(t) = \frac{1}{N} \int_0^{(t-1)/2} \frac{df/dw}{w+1} dw. \tag{6}$$

The quantity D(t) is related, in the limit $t \to \infty$, to a dipole quantity M_{ion}^2 familiar in the Bethe theory:

$$M_{ion}^2 = \frac{R}{B} \int_0^\infty \frac{df/dw}{w+1} dw. \tag{7}$$

Equation (5) has the threshold behavior $\sigma_i(t \to 1 + \Delta t) \propto \Delta t$.

The TICSs for He and Ne based on the BED model and matching the SDCSs in Figs. 1 and 2 are shown in Figs. 3 and 4, respectively (solid curves) (5). The remarkable agreement between the BED cross sections and the available experimental data gives confidence that the BED model will be useful in providing not just a consistent set of SDCS and TICS but rather a reliable set, covering a wide range of incident electron energies from threshold to a few keV.

For those cases for which information on df/dE is lacking, one can use a much simpler form of the BED theory, which we shall refer to as the Binary-Encounter-Bethe (BEB) model (5). In the BEB model, details of df/dE are discarded, and *one gets something for practically nothing!* For instance, the TICS in the BEB model is given by:

$$\sigma_i(t) = \frac{S}{t+u+1}\left\{\frac{\ln t}{2}\left[1-\frac{1}{t^2}\right] + \left[\left(1-\frac{1}{t}\right) - \frac{\ln t}{t+1}\right]\right\}. \tag{8}$$

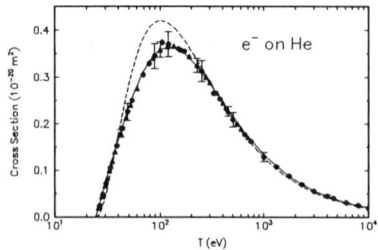

Figure 3. TICS of He. Solid curve, the BED model; dashed curve, Younger's distorted-wave Born cross section (10); circles, experimental data by Shah et al. (11); triangles, those by Montague et al. (12).

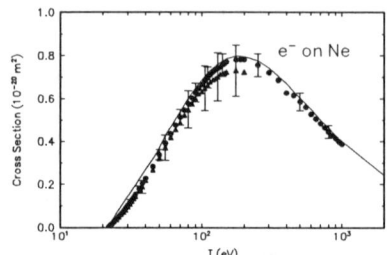

Figure 4. TICS of Ne. Solid curve, the BED model; circles, experimental data by Rapp and Englander-Golden (13); triangles, those by Wetzel et al. (14).

The BEB model, which uses only binding energies, average kinetic energies and occupation numbers of each subshell (or molecular orbital), is very simple to apply. It is particularly useful in the study of molecules since no reliable theoretical method exists for calculating molecular ionization cross sections. Indeed, the BEB model provides a TICS within ~10 percent of measured cross sections for H_2O (5), while various experimental TICSs for H_2O differ from each other by 10−20 percent.

A by-product of the BED model is following general analytic form for a TICS:

$$\sigma_i(t) = \frac{4\pi a_0^2}{t} \left[A\ln t + B\left(1 - \frac{1}{t}\right) + C\frac{\ln t}{t+1} + D\frac{\ln t}{(t+1)^2} + E\frac{\ln t}{(t+1)^3} + \cdots \right], \quad (9)$$

where A, B, C,..., are constants independent of t. The first logarithmic term in the square bracket of Eq. (9) represents the dipole interaction, the second term arises from the hard collision, and the second logarithmic term represents the interference between the direct and exchange interactions. The additional logarithmic terms provide an extra measure of flexibility, if needed. Most experimental and BED TICSs can be expressed by Eq. (9) using 3−5 terms. One can also use Eq. (9) for dipole forbidden cross sections, such as double ionization of He, by omitting the first logarithmic term.

FURTHER EXTENSION OF THE BED MODEL

Angular Distributions of Secondary Electrons

With the impressive success of the BED model in constructing SDCSs and TICSs, it is tempting to try a similar combination of soft and hard collisions to construct angular distributions of secondary electrons. The advantage of the BED theory is that it provides the absolute magnitudes of each component contributing to an SDCS.

For instance, Rudd has proposed (3) to model the binary peak by using a Lorentzian peak centered at the angular position given by Eq. (1) and with the width of the peak given by the ratio of the momentum of the ejected electron and its orbital momentum **p** before ionization. Rudd noted (3) that the Lorentzian peak model alone often had difficulty in reproducing the DDCS in the backward direction. These backward secondary electrons are most likely ejected by the dipole interaction, whose angular distribution is described by the photoelectron angular distribution (15):

$$\frac{d\sigma(E)}{d\Omega} = A(E) + B(E)\cos^2\theta, \quad (10)$$

where $A(E)$ and $B(E)$ are functions of the photon energy E only, and θ is measured from the incident photon beam direction.

Indeed, a preliminary test of this simple idea on the DDCS of He at T=200 eV produced angular distributions very close to the experimental shape (6). However, such a simple scheme to "synthesize" angular distributions is likely to succeed only when the dipole interaction is substantial. A better understanding of the angular behavior of the direct and exchange interactions will probably be needed for modeling DDCSs at low incident electron energies where the dipole interaction plays only a minor role.

Proton-Impact Ionization Cross Sections

The Bethe theory predicts that there are some similarities between electron-impact and proton-impact cross sections. These are listed below.

(a) When incident particle velocity is very high (~ 10 times or more) compared to the orbital velocity of the target electron, proton-impact cross sections are very close to the cross sections of incident electrons with the same speed;

(b) The leading term in both cases is the dipole interaction represented by the first logarithmic terms in Eqs. (5) and (8); and

(c) Proton-impact cross sections basically consist of contributions from soft and hard collisions as is the case for electron-impact cross sections.

On the other hand, there are also important differences, some trivial and some subtle:

(d) Although an incident electron can transfer all of its kinetic energy to a target electron, a proton or heavy ion cannot. The conservation of energy and momentum restricts the maximum energy transfer by a proton, $W_{max}(p)$, to a free electron at rest to $W_{max}(p) = 2mv_0^2 = 4T$, where m is the electron mass, v_0 is the speed of the incident particle, and T is the kinetic energy of an electron traveling with the speed v_0;

(e) In reality, however, a tightly bound electron can be ejected with a kinetic energy that exceeds 4T when the nucleus recoils and provides the additional momentum;

(f) Instead of Eq. (1), the binary peak for heavy-ion impact ionization is now given by

$$\cos\theta_b = \tfrac{1}{2}\sqrt{ME/mT_0}, \qquad (11)$$

where M is the mass of the incident ion, and $T_0 = Mv_0^2/2$ is the incident ion energy;

(g) Obviously, the Rutherford cross section is the appropriate starting point for hard collisions instead of the Mott cross section; and

(h) For a slow incident proton, the colliding system temporarily becomes a diatomic molecule, and two-center effects, which are the main topics of this symposium, dominate. These effects have no counterparts in electron-impact ionization.

Rudd *et al.* (16,17) have provided empirically fitted formulas to reproduce TICSs and SDCSs for proton-impact ionization of many atoms and molecules. These formulas, however, are not based on as firm an understanding of the underlying physics as those in the BED model for electron-impact ionization. The formulas in Refs. 16 and 17 are applicable to the targets for which they were intended, but they do not reveal any systematic behavior that can be used to extend them to other targets. A better approach should combine the two-center effects, which primarily affects the angular distribution in the forward direction, with the BED model.

CONCLUDING REMARKS

This article has outlined a new approach, the BED model and its simplified version, the BEB model, to predict electron-impact ionization cross sections. Judging from its impressive agreement with experimental SDCSs and TICSs, the BED model has succeeded in finding the correct "ratios" of various terms contributing to ionization cross sections. The BED model is very simple to apply since it depends on easily accessible atomic and molecular properties, yet it provides very reliable ionization cross sections for a variety of targets.

The BED model is exactly the type of theoretical model that has been sought by modelers of radiation and plasma effects for the last two decades. There is a good possibility that the BED model can be extended to generate reliable DDCSs. By incorporating the systematics of two-center effects discussed at this symposium, we hope that eventually a proton-version of the BED theory can be developed to predict proton-impact ionization cross sections, not only for fast protons but also for slow protons. Such a model will provide the atomic physics community as well as those who need ionization cross sections in their own applications with a powerful tool to integrate more than three decades of experimental and theoretical work pioneered by Eugene Rudd.

ACKNOWLEDGEMENTS

I have learned from Eugene Rudd experimental aspects of ionization cross sections and many subtle aspects of theory, particularly in regard to what works and what does not. He has been generous in sharing his wisdom with me and a patient partner who kept me (and other theorists) honest and on track. I am grateful to Eric Meyer for his careful reading of the manuscript and many helpful suggestions. This work was partly supported by the Office of Fusion Energy of the U.S. Department of Energy.

REFERENCES

1. For a brief discussion of the Rutherford and Mott cross sections, see Landau, L.D. and Lifschitz, E.M., Quantum Mechanics—Nonrelativistic Theory, 2nd ed. (Addison-Wesley, Reading, MA, 1965), p. 575.
2. For a review of the Bethe theory, see Inokuti, M., Rev. Mod. Phys. **43**, 297–347 (1971).
3. Rudd, M.E., Phys. Rev. A **44**, 1644–1652 (1991).
4. For a review of the binary-encounter theory, see Vriens, L., in *Case Studies in Atomic Physics*, Vol. 1, ed. McDaniel, E.W. and McDowell, M.R.C. (North Holland, Amsterdam, 1969), Chap. 6, pp. 335–398.
5. Kim, Y.-K. and Rudd, M.E., Phys. Rev. A **50**, 3954 (1994).
6. Opal, C.B., Beaty, E.E. and Peterson, W.K., At. Data **4**, 209–253 (1972).
7. Gorganthu, R.R. and Bonham, R.A., Phys. Rev. A **34**, 103–126 (1986).
8. DuBois, R.D. and Rudd, M.E., Phys. Rev. A **17**, 843–848 (1978).
9. See for instance, Gallagher, J.W., Brion, C.E., Samson, J.A.R. and Langhoff, F.W., J. Phys. Chem. Ref. Data **17**, 9–153 (1988).
10. Younger, S.M., J. Quant. Spectrosc. Radiat. Transfer **26**, 329–337 (1981).
11. Shah, M.B., Elliot, D.S., McCallion, P. and Gilbody, H.B., J. Phys. B **21**, 2751–2761 (1988).
12. Montague, R.G., Harrison, M.F.A. and Smith, A.C.H., J. Phys. B **17**, 3295–3310 (1984).
13. Rapp, D. and Englander-Golden, P., J. Chem. Phys. **43**, 1464–1479 (1965).
14. Wetzel, R.C., Baiocchi, F.A., Hayes, R.R. and Freund, R.S., Phys. Rev. A **35**, 559–577 (1987).
15. Starace, A.F., in *Handbuch der Physik*, Vol. XXXI, ed. W. Mehlhorn (Springer-Verlag, Berlin, 1982), pp. 1–121.
16. Rudd, M.E., Kim, Y.-K., Madison, D.H. and Gallagher, J.W., Rev. Mod. Phys. **57**, 965–994 (1985).
17. Rudd, M.E., Kim, Y.-K., Madison, D.H. and Gay, T.J., Rev. Mod. Phys. **64**, 441–490 (1992).

CONTRIBUTED PAPERS

Two-center effects in electron excitation

César A. Ramírez and Roberto D. Rivarola

Instituto de Física Rosario (CONICET and UNR), and Escuela de Ciencias Exactas, Ingeniería (UNR), Av. Pellegrini 250, (2000) Rosario, Argentina

Abstract. Reactions of target excitation by impact of bare ions are studied. Two-center distortion effects have been theoretically predicted at low intermediate collision energies for impact of heavy projectiles on light targets. Recent experiments support the existence of these effects.

The existence of binding and polarization effects in electron ionization of atomic targets by impact of lighter projectiles was discovered more than twenty years ago [1,2]. Binding has been determined to give subtractive Z_P^3-contributions to the total cross sections for these asymmetric $Z_P < Z_T$ systems, where Z_P and Z_T are the projectile and target nuclear charges respectively. It must be noted that the first order of the Born series presents a Z_P^2- dependence of impact parameter probabilities and total cross sections. The Z_P^3-behavior has been thus interpreted as coming from interferences between first and second orders of the Born approximation [3]. Basbas *et al.* [1,2] claim that the physical picture that supports the existence of this mechanism is that at low intermediate impact velocities ($v \lesssim v_e$, with v the collision velocity and v_e the electron orbital velocity in the entrance channel), where the effect appears, the ionization reaction is dominated by small impact parameter collisions, with larger initial electron binding energy and a consequently smaller corresponding total cross section. Binding effects have also been studied for asymmetric $Z_P > Z_T$ systems by using the two-center-Continuum Distorted Wave-Eikonal State approximation [4]. However, for these systems it has been shown that ionization is mostly produced at large internuclear distances [5,6,7]. The strong field of the projectile affects the dynamics of the electron even at large internuclear distances. The good agreement obtained with existing experimental data is due to the inclusion of an eikonal phase, associated with the projectile-electron interaction, which distorts the initial stationary bound wavefunction in the entrance channel. The representation of the electron in the simultaneous presence of the projectile and target fields thus plays a main role in the description of the reaction. Due to the large distance character of the process we prefer to invoke a *distortion* effect produced by the projectile charge instead of the binding one,

© 1996 American Institute of Physics

even if both effects could be related. In a recent work, Bugacov et al. [8] have predicted the possible existence of distortion effects for electron excitation to bound states of the target in $Z_P > Z_T$ systems. They have used the Symmetric Eikonal approximation (SE, [9,10]) to study excitation of H(1s) targets to final H(2l) (l=0,1) states. In the SE approximation the initial and final bound wavefunctions are distorted by multiplicative projectile-electron eikonal phases in both the entrance and exit channels. A detailed analysis on the applicability of the SE model to describe excitation of H(1s) atoms by impact of bare ions with $Z_P > 1$ can be found in references [8,9,10]. As in the ionization case, it has been proved that contributions to the total cross sections come mostly from impact parameters larger than the initial mean orbital radius of the electron. This impact parameter region moves to larger values as Z_P increases, at the collision energies for which the distortion effect is expected to appear.

Recent experimental data [11], shown in figures 1 and 2, support the existence of the effect. Measurements correspond to impact of protons and alpha particles on H atoms followed by formation of H(np) targets, with the principal quantum number n ranging from 1 to 6.

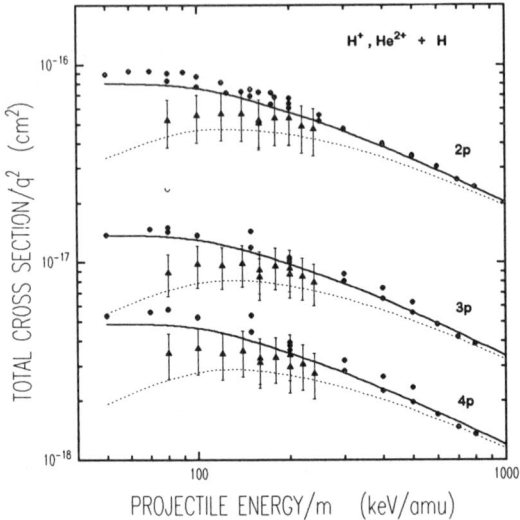

FIGURE 1. Total cross sections for single excitation of H to 2p, 3p and 4p states by impact of H^+ and He^{2+} ions at collisions energies from 50 to 800 keV amu^{-1}. SE total cross sections curves: full line, H^+; broken line, He^{2+}. Experimental data: circles, H^+; triangles, He^{2+}, from reference [11].

Total cross sections reduced by a factor q^2 (where q is the net charge of the projectile; in our case $q = Z_P$) are represented as a function of the laboratory collision energy per nucleon.

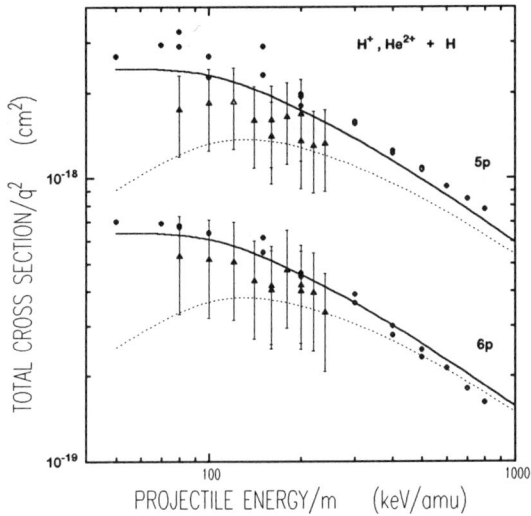

FIGURE 2. Total cross sections for single excitation of H to 5p and 6p states by impact of H^+ and He^{2+} ions at collisions energies from 50 to 800 keV amu^{-1}. SE total cross sections curves: full line, H^+; broken line, He^{2+}. Experimental data: circles, H^+; triangles, He^{2+}, from reference [11]. Cross sections for 6p are divided by a factor of 2.

At high impact energies theoretical SE results for H^+ and He^{2+}- projectiles converge one to each other for all the np cases analysed. So, at these high energies the Z_P^2-law predicted by the first Born approximation is valid. However, as the impact energy decreases the cross sections for He^{2+} ions separate from the corresponding ones for H^+ projectiles, giving lower cross sections than for the proton case. This behavior, which corresponds to the distortion effect, is clearly observed in experiments for n=2,3,4 and 5. The situation is less conclusive for n=6.

In conclusion, very recent experimental data [11] support the theoretical predictions from Bugacov et al. [8] on the existence of two- center distortion effects in electron excitation of the target by impact of more highly charged projectiles.

ACKNOWLEDGMENTS

We acknowledge to Prof. K. H. Schartner and Dr. D. Detleffsen for providing us with their experimental data previous to publication.

References

[1] Basbas G., Brandt, W. and Laubert R., *Phys. Rev. A* **7**, 983 (1973).

[2] Basbas G., Brandt W. and Laubert R., *Phys. Rev. A* **17**, 1655 (1978).

[3] Martin M. H., Ford A. L., Reading and Becker R. L., *J. Phys. B: At. Mol. Phys.* **15**, 1729 (1982).

[4] Fainstein P. D., Ponce V. H., and Rivarola R. D., *J. Phys. B: At. Mol. Phys.* **36**, 3639 (1987).

[5] Fainstein P. D., Ponce V. H., and Rivarola R. D., *Phys. Rev. A* **45**, 6417 (1992).

[6] Olson R. E. and Salop A., *Phys. Rev. A* **14**, 579 (1976).

[7] Grozdanov T. P. and Janev R. K., *Phys. Rev. A* **17**, 880 (1978).

[8] Bugacov A. Maidagan J. M., Rivarola R. D. and Shingal R., *Phys. Rev. A* **47**, 1052 (1993).

[9] Deco G. R., Fainstein and Rivarola R. D., *J. Phys. B: At. Mol. Phys.* **19**, 213 (1986).

[10] Reinhold C. O., Olson R. E. and Fritsch W., *Phys. Rev. A* **41**, 4837 (1990).

[11] Detleffsen D., Anton M., Werner A. and Schartner K-H., *J. Phys. B: At. Mol. Phys.* **27**, 4195 (1994).

EVIDENCE OF DOUBLE SCATTERING PROCESSES IN ION-ATOM IONIZATION

S. Suárez, W. Cravero, R. Barrachina and W. Meckbach.
Centro Atómico Bariloche, 8400 Bariloche, Argentina

R. Maier, M. Tobisch and K. O. Groeneveld
Institut für Kernphysik der J. W. Goethe Universität, D-60486 Frankfurt a.M., Germany.

Abstract

Evidences of double scattering mechanism in ion-atom ionization are presented. The observed structures are explained by a simple process in which an electron initially bound to the target suffers a binary encounter with the incoming projectile and then a second collision with the target core.

Double scattering mechanisms have been proposed in order to interpret a variety of processes occurring in ion-atom collisions. Double scattering processes, represented by second order terms in the Born expansion, are the dominant mechanisms of electron capture at very high collision velocities. For forward capture at high impact energies the mechanism of double scattering was first suggested by Thomas [1,2]. In a classical description, Thomas supposed that for capture to take place, an electron should be knocked by the projectile towards the target nucleus with a speed very close to v_p where it undergoes a second and elastic scattering. The electron emerges with velocity v_p in the direction of the projectile, being captured by Coulomb attraction. The experimental fingerprints of such a double scattering mechanism is a peak at $\theta=\sqrt{3}/2.(m/M_p)$ (m and M_p are the electron and projectile masses) in the angular differential cross section for the scattering of the charge-changing projectile. This "Thomas" peak has been confirmed experimentally by Pedersen et al [3] and Vogt et al [4]. Nevertheless, there exist the possibility of an additional electron-electron scattering giving rise to a mechanism of capture followed by ionization (TI). A quantum mechanical description of this double scattering process has been given by Briggs and Taulbjerg [5] and Ishihara and McGuire [6], whereas an experimental evidence was reported by Pálinkás et al [7]. In this paper, we investigate a double scattering process for target ionization which has some similarity to the Thomas mechanism, but with no restrictions on the velocity and angle of the emitted electron.

Let us consider the ionization of a hydrogenic target by the impact of a high energy ion of charge Z_p. The transition amplitude for this process in First Born approximation (FBA) reads [8]

$$T_{k,i}^{FBA} = \langle \varphi_{Kf} \Psi_k^- | V_p | \varphi_{Ki} \phi_i \rangle \qquad (1)$$

where V_p is the projectile Coulomb potential, φ is a plane wave describing the free relative internuclear motion, and ϕ_i is the initial bound state. The final scattering state Ψ^- satisfies the Lippmann-Schwinger equation:

$$|\Psi^-\rangle = (1 + G_T^- V_T) |\varphi_k\rangle \qquad (2)$$

where V_T is the target nucleus-electron interaction potential and G_T^- is the associated Green's operator. Inserting Eq.(2) in Eq.(1), we obtain

$$T_{k,i}^{FBA} = \langle \varphi_{Kf} \varphi_k | V_p + V_T G_T^+ V_p | \varphi_{Ki} \phi_i \rangle \qquad (3)$$

The first term in this equation describes a single scattering process between the projectile and a target electron, and gives rise to the well known binary peak [9]. The second term describes a double scattering mechanism: After the first projectile-electron collision, the electron undergoes another collision with the target nucleus and is finally ejected [10].

From Eq.3 we have calculated the TDCS, as a function of the electron velocity v_e and for a fixed deflection angle of the projectile. Figure 1a shows the TDCS calculated with the Plane Wave Born Approximation (PWBA) where the electron final state is represented by a plane wave, i.e., the transition amplitude is given by the first term of the right side of Eq.3. When the electron-target interaction is "switched-on" in the final state, the transition amplitude is given by the sum of the two terms in the right side of Eq.3 (Fig. 1b).

We find that the double scattering mechanism taken into account in the latter case, gives rise to observable structures which appear as two concentric circular rings centered at the position of the target (in velocity space). When the TDCS is integrated in the scattering direction of the projectile, these structures persist as a smooth shoulder of the double differential cross sections (DDCS) about $v_e \sim 2v_p$, for all angles of emission. In the forward direction, this structure merges into the binary peak and cannot be observed, while direct measurements of DDCSs for large angles of emission ($\theta > 90^0$) show no conclusive evidences. For this reason, we have investigated such an effect through a target dependence study of the electron emission induced by energetic ions.

A proton beam delivered by the 2.5 MV Van de Graaff accelerator of the J. W. Goethe University in Frankfurt am Main impinged with energy 0.7, 1.1 and 1.6 MeV on thin (600Å thick) **C** and **Au** -foils. A cylindrical mirror electron spectrometer described elsewhere [11] has been used to analyze the electron emission energy. Doubly differential electron distributions have been obtained for backward angles and for electron energies up to 7 keV. We paid attention to those electrons emitted with velocities v_e of the order of, and even higher, than that of the binary encounter electrons. Solid, instead of gaseous, targets have been used in order to obtain better statistics within the highest electron velocity range.

Figure 2 shows a doubly differential electron distribution for $\theta = 0^0$ as measured for 1.6 MeV $H^+ \rightarrow C$. Three distinctive features are observed: The **C** -Auger peak at about ~270eV, the convoy peak at the projectile velocity and the binary encounter peak at about ~3.2keV. In the same figure the ratio between the electron emission at $\theta = 165^0$ resulting from **Au** and **C** -foils is also shown. This ratio shows a prominent and broad peak at $v_e \sim 2.v_p$. Similar results were obtained for 0.7 and 1.1 Mev.

We tentatively interpret this finding as a preliminary experimental evidence of a double scattering mechanism in ion-atom ionization collisions. The first step of our proposed mechanism consists in the production of binary electrons by a direct knock-on collision between the projectile and a target electron, whereas the second step is an electron-target core interaction. This second scattering should give rise to a larger electron emission near $v_e \sim 2.v_p$ for all angles of emission. This latter electron-atom scattering cross section is larger for **Au** than for **C**[12], giving rise to the broad peak at $v_e \sim 2.v_p$ observed in the ratio of figure 2. The difference between the width of the initial momentum distribution

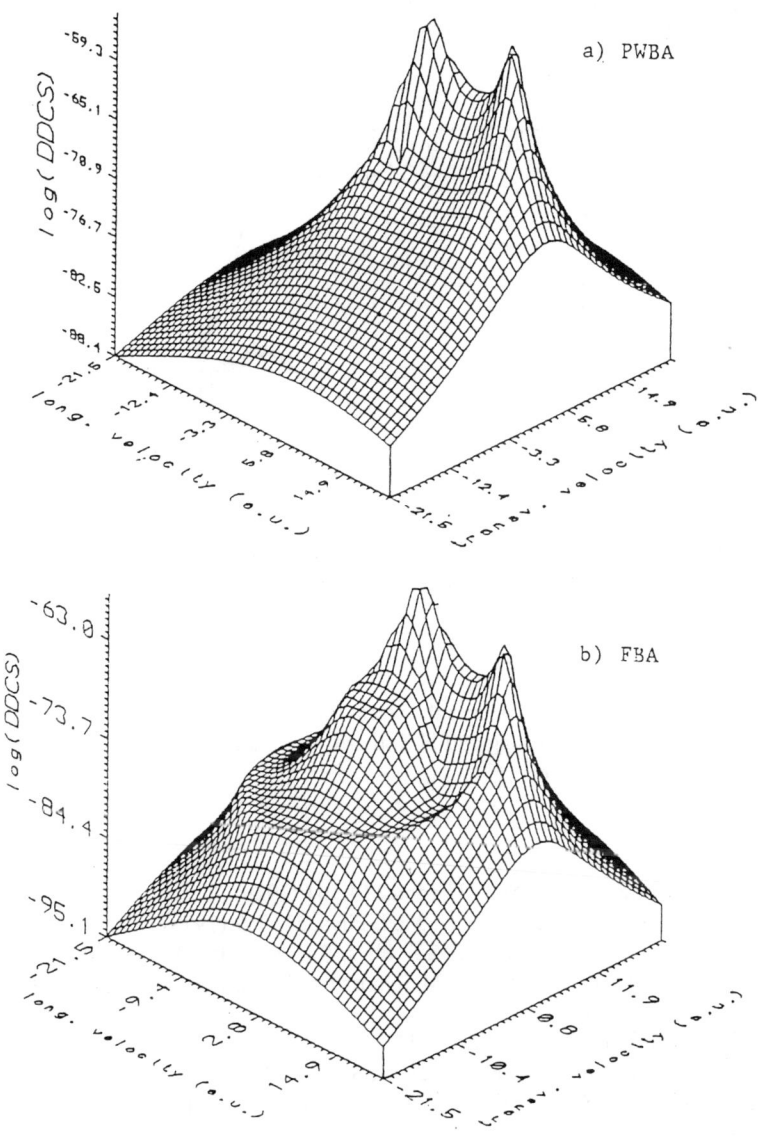

Figure 1: Triply Differential Cross Section (TDCS) for ionization of H by impact of 1.6MeV H$^+$, as a function of the electron transversal and longitudinal velocities. The deflection angle for the projectile is $\theta=0.0017^0$. a): Plane Wave Born Approximation, b): First Born Approximation.

of the emitted electrons for different targets could also contribute to highlight the effect. This target effect has also been found in calculated ratios and by using simple targets as **H** and **He** [13].

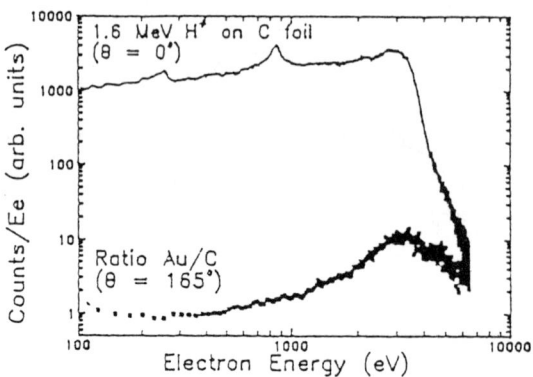

Figure 2: Upper curve: Forward spectrum obtained for 1.6MeV H⁻ on a C-foil. Bottom curve: Ratio between the electron emission from **Au** and C -foils for $\theta=165^0$.

Acknowledgments

One of us (S.S.) acknowledges the warm hospitality at the IKF at Frankfurt a.M. and helpful discussions with M.Jung and H.Rothard. W.C. thanks C. Garibotti for permitting the use of his FBA program.

References

[1] L.H.Thomas, Proc. Roy. Soc., London **114**,561 (1929).
[2] R.Shakeschaft and L.Spruch, Rev. Mod. Phys.,**51**, 369 (1979).
[3] E.H.Pedersen, C.Cocke and M.Stökli, Phys. Rev. Lett.**50,** 1910 (1983).
[4] H.Vogt, R.Schuch, E.Justiniano, M.Schultz and W.Schwab, Phys. Rev. Lett. **57**, 2256 (1986).
[5] J.Briggs and K.Taulbjerg, J.Phys.B **12**,2565 (1979).
[6] T.Ishihara and J.McGuire, Phys. Rev. A**38** 3311 (1988).
[7] J.Pálinkás, R.Schuch, N.Cederquist and O.Gustafsson, Phys. Rev. Lett. **63**, 2464 (1989).
[8] M.R.C.McDowell and J.P.Coleman, Introduction to the Theory of Ion-Atom Collisions, North Holland Publishing Co., Amsterdam 1970.
[9] J.Briggs and J.Macek, in Adv. At. Mol. Opt. Phys., Vol.**28**, (1990).
[10] R.Barrachina, unpublished (1992).
[11] H.Rothard et al., Nucl. Instrum. Meth. B**48**,616 (1990).
[12] M.Fink and J.Ingram, Atomic Data **4**, 129 (1972).
[13] S.Suárez et al., Nucl. Instrum. Meth. B**86**,197 (1994).

Electron Plate-impact Distortions in Electron Spectroscopy

V. D. Irby

Department of Physics
University of Louisville
Louisville, KY 40292

Abstract. There exist several inherent difficulties involved with measurements of low-energy electrons emitted in ion-atom collisions, such as stray magnetic fields and detector efficiencies. However, there exists one experimental problem that tends to be ignored in the literature: secondary-electron emission and reflection due to electrons impacting the back plate of an analyzer. A straightforward analysis reveals that even small amounts of electron plate-impact contamination can lead to significant spectral distortions in measurements of low energy electrons.

Let us assume that the function $C(E)$ represents an arbitrary incident electron energy distribution, that we wish to measure, using a typical electrostatic energy analyzer. Because the analyzer has a finite energy acceptance range, ΔE, the measured count rate versus electron energy, $I(E)$, or raw data, can be approximated by

$$I(E) \approx \varepsilon A_0 C(E) E \, , \qquad (1)$$

where ε is the analyzer detector efficiency (which we will assume is on the order of unity) and $A_0 = \Delta E / E$ is the analyzer resolution. In this case, $I(E)$ represents the actual signal one would obtain if plate-impact contaminations were not present.

Electrons entering the analyzer with energies E greater than

$$E > E_{min} = e V_p / \sin^2\theta \, , \qquad (2)$$

where e is the charge of an electron, V_p is the analyzer plate voltage, and θ is the analyzer entrance angle, will strike the back plate of the analyzer and may result in spurious signal due to secondary electron emission and electron reflection. (See Fig. 1).

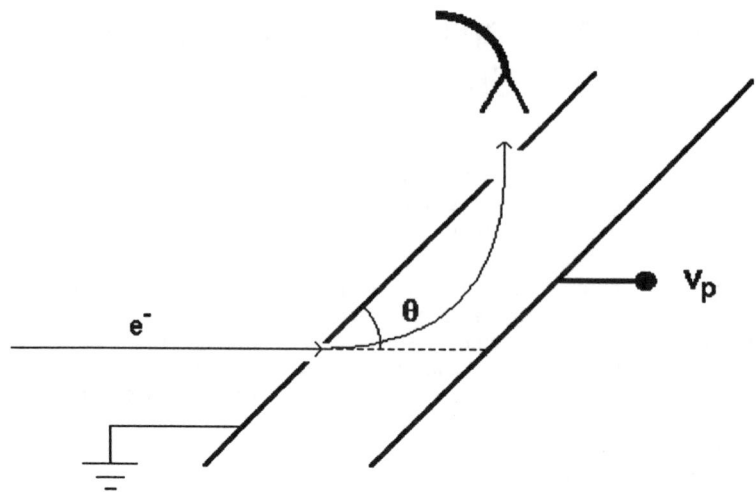

FIGURE 1. Schematic diagram for a single-stage parallel-plate, or cylindrical mirror, electrostatic analyzer. The analyzer entrance angle is denoted by θ and the curved line illustrates the central-ray path of the spectrometer.

Equation 2 can be rewritten as

$$E > E_{min} = C_o E_p / \sin^2\theta \,, \qquad (3)$$

where E_p is the spectrometer central-ray pass energy ($C_o = eV_p / E_p$).

Let us define the plate-impact contamination signal, $I_c(E_p, E)$, measured at an energy E_p, as

$$I_c(E_p, E) = a(E_p, E)\, C(E) \,, \qquad (4)$$

where $a(E_p, E)$ is the fractional contamination factor and is assumed to be much less than one. This contamination is caused by electrons of energy E impacting the analyzer back plate, with the requirement that $E > E_{min} = C_o E_p / \sin^2\theta$. To obtain the total plate-impact contamination at E_p, one must integrate $I_c(E_p, E)$ over all electron energies E. The total contamination signal I_T, at E_p, is then given by

$$I_T(E_p) = \int_{E_{min}=C_oE_p/\sin^2\theta}^{E_{max}} a(E_p, E')\, C(E')\, dE', \qquad (5)$$

where E_{max} is the maximum energy of the electrons under study. Since the uncontaminated signal at E_p can be obtained from Eq.(1), the ratio $R_c(E_p)$ of the contamination to the uncontaminated signal is therefore given by

$$R_c(E_p) = I_T(E_p) / I(E_p) \qquad (6)$$

$$R_c(E_p) = (A_oC(E_p)E_p)^{-1} \int_{E_{min}=C_oE_p/\sin^2\theta}^{E_{max}} a(E_p, E')\, C(E')\, dE'.$$

In order to understand how small amounts of electron plate-impact contamination can significantly affect measurements of electron energy spectra, let us examine the case when the incident electron energy distribution $C(E)$ is constant for energies ranging from 0 to 100 eV. If we make the additional approximation that $a(E_p, E)$ is essentially constant, that is $a(E_p, E) \approx a_o$, then Eq. (6) results in

$$R_c(E_p) = (a_o/A_oE_p)(E_{max} - (C_oE_p/\sin^2\theta)). \qquad (7)$$

Using Eq. (7), (and assuming typical values of $C_o = 0.65$, $A_o = 0.01$ and $\theta = 45°$) one can then estimate the required value of a_o for 50% plate-impact contamination at an electron energy $E_p = 1\text{eV}$, for $E_{max} = 100\text{eV}$. This results in $a_o = 5 \times 10^{-5}$, or 0.005 %. Although the assumption $a(E_p, E) \approx a_o$ is rather unrealistic, it does indicate that miniscule amounts of contamination at E_p, due to plate-impact of electrons at a particular energy E, can lead to significant contamination when the proper integration over electron energy is performed.

A more realistic estimate of total contamination can be made using experimental measurements of Bernardi and Meckbach.[1] In their work, scattered electron intensities were determined experimentally, from an energetic electron beam, and demonstrated that a(E_p, E) exhibited values ranging from 0.004 % to 0.08 %. Using their experimental estimates for a(E_p, E), and assuming a Rutherford distribution $C(E) = \text{const}/(E + B)^2$ (where B is the binding energy) in Eq. (6) above, results in 10 % total contamination at E_p = 1eV. However, one must note that $R_c(E_p)$ is extremely sensitive to the choice of C(E). As an example, if one uses the ejected-electron energy distributions measured by Irby *et al.*[2] (in which plate-impact contaminations were *not* present) Eq. (6) results in a total contamination of 40 to 50 % at E_p = 1eV.

In conclusion, the analysis presented here indicates that small amounts of electron plate-impact contamination can lead to significant low-energy spectral distortions in experimental measurements involving ejected electrons in ion-atom collisions. It is important to point out that this distortion results from the fact that the real signal depends upon the analyzer resolution, whereas the contamination signal does not. Since the real signal decreases with decreasing analyzer voltage, and the contamination signal does not, the ratio $R_c(E_p)$ begins to diverge as $E_p \rightarrow 0$. It is also important to point out that in order to obtain the actual contamination induced by electron plate-impact, one must have prior knowledge of the actual incident electron energy distribution C(E), which confounds the situation because C(E) is *precisely* what one is trying to measure. (One cannot subtract the plate-impact contamination from the measured signal because the contamination cannot be reliably estimated without knowledge of the actual real signal). Thus, it is imperative that subsequent research involving low-energy ejected electrons in ion-atom collisions properly addresses this problem. One method, used by some researchers to reduce plate-impact contamination, is to coat the analyzer back plate with carbon soot. On the other hand, other researchers further reduce this problem by cutting a slot, or milling a sawtooth profile, in the analyzer back plate. The method used in Reference 2 involves *preventing* electrons from hitting the analyzer back plate, subsequently eliminating the problem altogether, without modification of the electrostatic plates within the analyzer.

References
1.). G. Bernardi and W. Meckbach, Phys. Rev. A **51**, 1709 (1995).
2.) V. D. Irby, S. Datz, P. F. Dittner, N. L. Jones, H. F. Krause, and C. R. Vane, Phys. Rev. A **47**, 2957 (1993).

Charge State Distributions in Copper Following Multiple Ionization and One-Electron Capture in Collisions with H$^+$ Ions

M B Shah, C J Patton, J Geddes and H B Gilbody

Department of Pure and Applied Physics
The Queen's University of Belfast, Belfast, UK

Abstract. The charge state fractions F_q of slow Cu^{q+} ions formed by one-electron capture in single collisions between protons and Cu atoms for q=1 to 5 have been measured in the energy range 70-730 keV amu^{-1}. Measured fractions have been satisfactorily described in terms of an independent electron model of ionization of the 4s and 3d electrons.

Introduction

In earlier work in this laboratory (1),(2) we have used a crossed beam technique incorporating coincidence counting of the collision products to obtain total cross sections for ionization in H$^+$-H collisions over a wide energy range. These measurements provided a sensitive test of the range of validity of the many different theoretical predictions. More recently, the first studies of double differential scattering of electrons in H$^+$-H collisions by Rudd and his collaborators (3) have the potential for more stringent tests of theory including the role of two-center effects. In the case of multiple ionization of heavy atoms by protons, the theoretical picture involving multicenter effects is less well developed although the independent electron approximation (4) is useful in many cases. In the present work, we have adapted our previous crossed beam coincidence counting technique to allow measurements of the fractions F_q of slow Cu^{q+} ions formed by electron capture with simultaneous multiple ionization in single H$^+$-Cu collisions over the range 70-730 keV amu^{-1}. The results have been compared with our predictions based on an independent electron model.

Experimental approach

The basic apparatus and measuring procedure was similar to that used in our previous measurements (1),(5) and need only be summarised briefly here. The primary beam of H$^+$ ions was arranged to intersect (at right angles) a thermal energy beam of Cu atoms in a high vacuum region. Slow Cu^{q+} collision products were extracted from the crossed beam region by a transverse electric field (applied between two high transparency grids) strong enough to ensure complete collection and counted by a particle multiplier. Cu^{q+} ions in a particular charge state q were selectively identified by time-of-flight analysis. The primary beam was charge analyzed by electrostatic deflection beyond the crossed beam region and the fast H atoms formed by one-electron capture were counted in coincidence with the Cu^{q+} ions arising from the same collision events corresponding to processes

$$H^+ + Cu \rightarrow H + Cu^{q+} + ne \qquad (1)$$

with cross sections $_{10}\sigma_{0q}$ for specified values of q (where n = q - 1) ranging from 1 to 5. In these reactions n=0 corresponds to simple charge transfer while n>0 corresponds to transfer ionization.

Results and Discussion

Fig 1 shows the energy dependence of our measured Cu^{q+} fractions F_q for q = 1 to 5. While F_1 is dominant over the energy range considered, the increasing importance of transfer ionization at higher energies is evident.

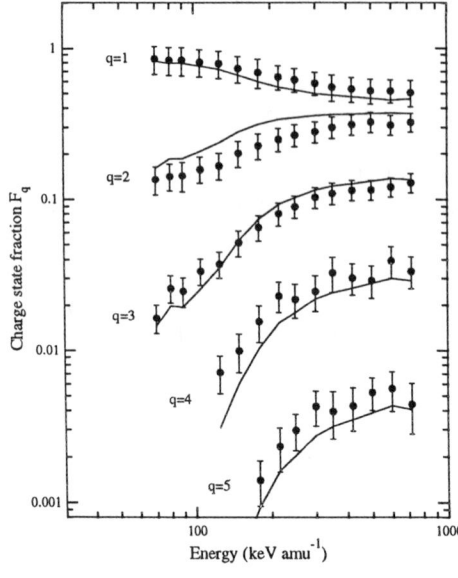

Fig 1 Measured charge state fractions F_q of Cu^{q+} ions (shown •) together with calculated ionization probabilities P_n (solid lines) where n = q - 1 based on binomial distribution.

In an attempt to describe these results in terms of the independent particle model, the probability of one-electron capture and simultaneous ionization in H^+ - Cu collisions is expressed as the product of a one-electron capture probability P_c and an ionization probability P_n for the removal of n electrons from the target where n≥0.

Then $_{10}\sigma_{0q} = 2\pi \int_0^\infty b P_c(b) P_n(b) \, db \qquad (2)$

where b is the impact parameter. If we assume that $P_n(b)$ remains constant over the relatively small range of impact parameters over which electron capture occurs then

$$P_n = {}_{10}\sigma_{0q}/\sigma_{10} \tag{3}$$

where the total electron capture cross section $\sigma_{10} = \sum_q {}_{10}\sigma_{0q}$. Measured fractions F_q can then be identified with the ionization probabilities P_n through equation 3.

In the present energy range, one-electron capture is expected to involve either the single 4s electron or one of the ten 3d electrons in Cu. If we also assume that ionization only involves these electrons, on the basis of the independent electron model, P_n can be expressed as

$$P_n = S_0 D_n + S_1 D_{(n-1)} \tag{4}$$

where S_i and D_j are the respective probabilities of removal of i electrons from the 4s level where $i = 0$ or 1 and j electrons from the 3d level where $j = 0$ to 9 if we assume that a 3d electron is most likely to be captured.

Binomial distributions apply if the single electron transition probabilities p_s and p_d are the same for an s and d shell respectively (4) so that $S_i = p_s^i (1-p_s)^{(1-i)}$ for $i = 0,1$ and $D_j = \binom{9}{j} p_d^j (1-p_d)^{9-j}$ for $j = 0,1,2 \ldots 9$.

Values of P_n predicted by equation 4 have been fitted to the experimentally measured values obtained from equation 3 using a weighted least squares fit. The agreement between experiment and the predictions of the binomial distribution can be seen (Fig 1) to be very satisfactory especially at higher energies where equation 3 should be increasingly valid as the electron capture cross section involving smaller impact parameters decreases.

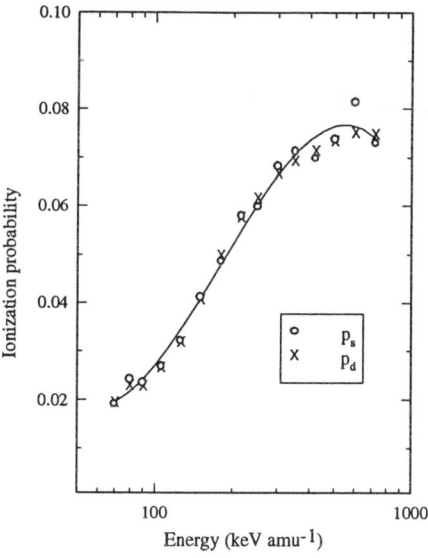

Fig 2 Ionization probabilities p_s and p_d derived from binomial fits to measured charge state distributions.

Our binomial distribution fits to the experimental data provide values of the ionization probabilities p_s and p_d at small impact parameters which are shown in Fig 2. The approximately equal values of p_s and p_d provide support for an ionization model based on a structureless electron cloud (6). Hence it is unnecessary to specify whether a 4s or a 3d electron has been captured.

Acknowledgements

This research was supported by the UK Engineering and Physical Sciences Research Council.

References

1. Shah M B and Gilbody H B, J Phys B, **14**, 2361-2377 (1981).

2. Shah M B, Elliott D S and Gilbody H B, J Phys B, **20**, 2481-2485 (1987).

3. Rudd M E, Gealy M W, Kerby G W and Ying-Yuan Hsu, Phys Rev Lett, **68**, 1504-1506 (1992).

4. McGuire J H, Advances in Atomic, Molecular and Optical Physics, Vol 29, Academic Press (1991) pp217-323.

5. Shah M B, McCallion P, Itoh Y and Gilbody H B, J Phys B, **25**, 3693-3708 (1992).

6. Meron M and Rosner B, Phys Rev **A30**, 132-135 (1984).

Radial Dose Distributions in the Delta-Ray Theory of Track Structure

Francis A. Cucinotta[+], Robert Katz[++], John W. Wilson[+] and Rajendra R. Dubey[*]

[+]*NASA Langley Research Center*
Hampton, VA 23681-0001

[++]*Behlen Laboratory of Physics*
University of Nebraska
Lincoln, NE 68588-0150

[*]*Old Dominion University*
Norfolk, VA 23708

Abstract. The radial dose distribution from delta rays, fundamental to the delta ray theory of track structure, is recalculated. We now include the model of Rudd for the secondary electron spectrum in proton collisions. We include the effects of electron transmission through matter and the angular dependence of secondary electron emission. Empirical formulas for electron range versus electron energy are intercompared in a wide variety of materials in order to extend the track structure theory to arbitrary media. Radial dose calculations for carbon, water, silicon, and gold are discussed. As in the past, effective charge is used to scale to heavier projectiles.

INTRODUCTION

The δ-ray theory of track structure attributes the radiation damage and detection in the passage of heavy ions through matter to the secondary electrons (δ-rays) ejected from the medium by the passing ion (1–5). The track-structure theory has a long history of providing the correct description of a wide variety of phenomenon associated with heavy ion irradiation. Track-structure theory provided the first description of the spatial distribution of energy deposition from ions through the formulation of the radial distribution of dose, as introduced by

© 1996 American Institute of Physics

Butts and Katz (1) and Kobetich and Katz (2), which led to many experimental measurements of this phenomena (6–10). The response of physical detectors to heavy ions, such as organic scintillators (5), TLD's (5), alanine (11), nuclear emulsion (12), and the Fricke dosimeter (5, 13), have been described using track theory. Many applications in describing biological effects have been made, including the prediction of thindown (5) nearly 20 years prior to the first experimental observation (14) in mammalian cells. More recently, researchers are utilizing track theory in developing improved lithography methods (15) for applications in microelectronics and microtechnology using ion beams.

The radial dose distribution and the geometry of a target site is used in track theory to map gamma-ray response to ion response. The radial dose for intermediate distances from the track is known to fall off as the inverse square of the radial distance to the ion's path, which has led to simplified formulas to be used in many applications (1, 5, 16, 17). It is more difficult to predict the radial dose both near to the ion's path and far from the path due to uncertainties in the electron range versus energy relation, the angular dependence of the secondary electron production cross section, and the effects of δ-ray transport in matter, especially in condensed phase. Many track-structure calculations have used simple, analytic forms for the radial dose from ions which ignore the electron transmission, the angular dependence of electron ejection, and also use simplified electron range-energy relations. In this paper, we consider these factors by following the method of Katz and Kobetich (2, 3, 17) and make new comparisons to recent experimental data for radial dose distributions. An improved model for the secondary electron spectrum in proton collisions with atoms and molecules due to Rudd (19) is used in calculations with the electron spectrum from heavy ions found using scaling by the effective charge of an ion.

RADIAL DOSE FORMALISM

We next review the calculation of the radial dose as a function of radial distance, t, around the path of an ion of atomic number Z, and velocity, β, as introduced by Kobetich and Katz (2–4, 18). In formulating the spatial distribution of energy deposition as charged particles pass through matter, it is assumed that the dominant mode of radial dose deposition is due to electron ejection from the atoms of the target material. The residual energy of an ejected electron (δ-ray) with energy W after penetrating a slab of thickness t is given by the energy to go the residual range $r - t$, as

$$W(r, t) = \omega(r - t) \tag{1}$$

where r is taken as the practical range (determined by extrapolating the linear portion of the absorption curve to the abscissa) of an electron liberated with energy ω. The residual energy is then evaluated by Eq. (1) once the range-energy relation in a given target material is known, as discussed below.

The energy dissipated, E, at a depth t by a beam containing one electron per cm² is represented in (2) as

$$E = \frac{d}{dt}\left(\eta W\right) \quad (2)$$

where η is the probability of transmission for the electrons. As noted by Kobetich and Katz (4) Eq. (2) neglects several effects. First, it may neglect backscattering, although it may be argued that the energy lost from a layer dt by backscattering is compensated by energy backscattered from later layers. Second, all electrons are represented by an under-scattered class, namely those which penetrate the characteristic distance. Third, the energy deposited by the least-scattered electrons, which penetrate to a thickness $t > r$, is neglected. Such shortcomings could be overcome by direct solution of the electron transport (20) or through the use of Monte-Carlo methods (21). However, the model of Kobetich and Katz has the advantage of simplicity while achieving reasonable accuracy.

The transmission function used will be based on the expressions of Depouy, et al. (22) as modified by Kobetich and Katz (4) and is given by

$$\eta(r,t) = \exp\left[-\left(qt/r\right)^p\right] \quad (3)$$

with

$$q = 0.0059\, Z_T^{0.98} + 1.1 \quad (4)$$

and

$$p = 1.8\left(\log_{10} Z_T\right)^{-1} + 0.31 \quad (5)$$

where Z_T is the atomic number of the target material, and r and t are in units of g/cm².

In order to estimate the number of free electrons ejected by an ion per unit length of ion path with energies between ω and $\omega + \delta\omega$, the formula given by Bradt and Peters (23) was used by Kobetich and Katz (2)

$$\frac{dn}{d\omega} = \frac{2\pi N Z^{*2} e^4}{mc^2 \beta^2} \frac{1}{\omega^2} \left[1 - \frac{\beta^2 \omega}{\omega_m} + \frac{\pi \beta Z^{*2}}{137} \sqrt{\frac{\omega}{\omega_m}} \left(1 - \frac{\omega}{\omega_m}\right) \right] \quad (6)$$

where e and m are the electron charge and mass, N is the number of free electrons per cm^2 in the target, and ω_m is the classical value for the maximum energy that an ion can transfer to a free electron given by

$$\omega_m = \frac{2mc^2 \beta^2}{1 - \beta^2} \quad (7)$$

In Eq. (6) Z^* is the effective charge number of the ion which is represented by Barkas (24) as

$$Z^* = Z \left[1 - \exp\left(\frac{-125 \beta}{Z^{2/3}}\right) \right] \quad (8)$$

The electron-binding effects are taken into account by Kobetich and Katz (2) following the experimental findings of Rudd et al. (25) who found that ω may be interpreted as the total energy imparted to the ejected electron whose kinetic energy is W, such that ω in Eq. (6) is replaced by

$$\omega = W + I \quad (9)$$

Eq. (6) must be summed for composite materials in which there are N_i electrons per cm^3 having mean excitation energy I_i with values of I_i from Berger and Seltzer (26) and Hutchinson and Pollard (27).

Recently, Rudd (19) has provided a parameterization of the electron spectrum following proton impact based on a binary encounter model modified to agree with Bethe theory at high energies and with the molecular promotion model at low energies. For water, the contributions from five shells are included (19). We also consider this model, scaling to heavy ions using effective charge. In figure 1, we compare the secondary electron spectrum from Eq. (6) to

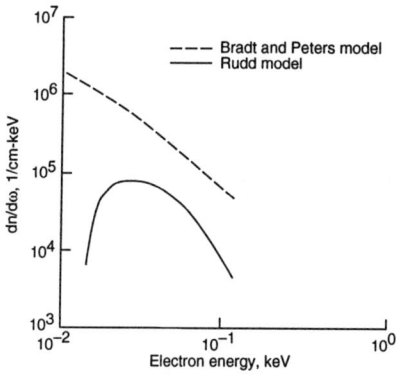
(a) Spectrum with 0.05 MeV proton in water.

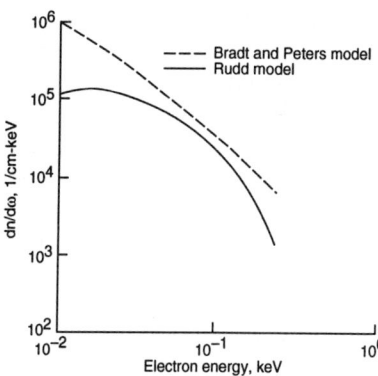
(b) Spectrum with 0.1 MeV proton in water.

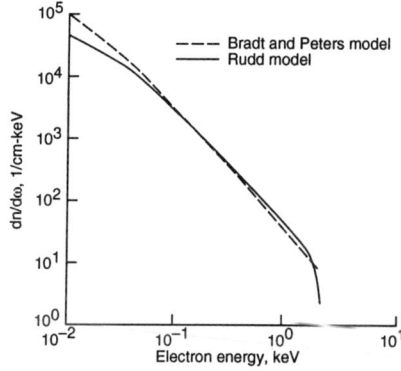
(c) Spectrum with 1 MeV proton in water.

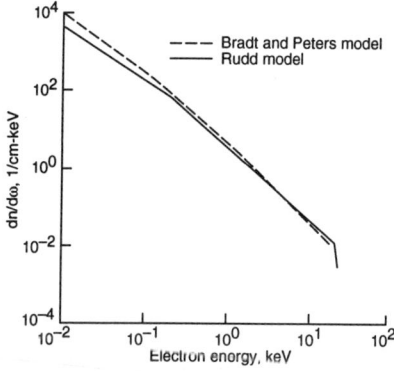
(d) Spectrum with 10 MeV protons in water.

FIGURE 1. Comparisons of models for secondary electron spectrum from incident protons in water. Shown are a dash-line model of Bradt and Peters (21) and solid-line model of Rudd (18).

the model of Rudd for several proton energies. Large differences between the models occur below proton energies of about 1 MeV and for small electron energies at all proton energies.

Using classical kinematics, electrons of energy ω are ejected at an angle θ to the path of a moving ion given by

$$\cos^2\theta = \frac{\omega}{\omega_m} \tag{10}$$

for the collision between a free electron and the ion. Eq. (10) indicates that close to the ion's path, where distances are substantially less than the range of δ-rays, i.e., $\omega \ll \omega_m$ and θ is near $\pi/2$, it is sufficient to consider that all δ–rays are normally ejected, and that their energy dissipation in cylindrical shells, whose axis is the ion's path, may be found from knowledge of the energy dissipation of normally incident electrons. Early calculations by Kobetich and Katz (2, 3), showed that assumptions on the angular distribution of δ-rays have little effect on radial dose calculations at intermediate distances where the radial dose falls off as $1/t^2$. If the δ-rays far from an ion's path have an important role on a particular response, then the angular dependence, as well as the dependence of electron range, on the ion's velocity becomes crucial.

If ε is the energy flux carried by δ-rays through a cylindrical surface of radius t whose axis is the ion's path, the energy density E deposited in a cylindrical shell of unit length and mean radius t is given by

$$E = \frac{-1}{2\pi t} \frac{d\varepsilon}{dt} \qquad (11)$$

The total energy flux is found by integrating the energy flux carried by a single electron, given by ηW, over the δ-ray distribution and summing over all atoms in the material

$$\varepsilon(t) = \sum_i \int_{\omega_t}^{\omega_m - I_i} d\omega \, W(t, \omega) \, \eta(t, \omega) \frac{dn_i}{d\omega} \qquad (12)$$

The integration limits in Eq. (12) are for the lower limit, ω_t, the energy for an electron to travel a distance t, and upper limit, $\omega_m - I_i$, the maximum kinetic energy that can be given to the electron by the passing ion. Using Eq. (11) and Eq. (12), the energy density distribution may be written as

$$E(t) = \frac{-1}{2\pi t} \sum_i \int_{\omega_t}^{\omega_m - I_i} d\omega \frac{\partial}{\partial t} \left[\eta(t, \omega) W(t, \omega) \right] \frac{dn_i}{d\omega} \qquad (13)$$

and $E(t)$ is identified as the radial distribution of dose.

In order to consider the angular dependence of the ejected electrons, the energy deposited at a depth t in a slab of material by normally incident electrons is modified by assuming the energy deposited in a cylindrical shell of radius t centered on the ion's path by a delta ray ejected at an angle θ is the same as the energy deposited by an electron normally incident on a slab at depth $t/\sin θ$ as shown in figure 2. Kobetich and Katz (3) assume that differences between the

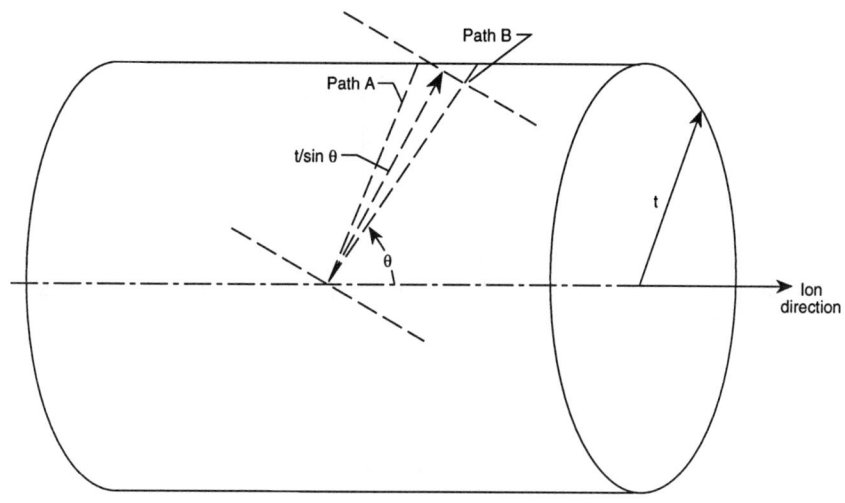

FIGURE 2. The transmission of electrons ejected at an angle to the ion's path through a cylindrical surface of radius t.

slab and cylindrical geometries do not greatly affect the energy density distribution, for the differences in the energy density at t caused by those electrons scattered as in path A of figure 2 are compensated by those scattered in path B of figure 2.

The energy density distribution, including an angular distribution of the ejected electrons, is assumed as

$$E(t) = \frac{-1}{2\pi t} \sum_i \int d\Omega \int_{\omega_t(\theta)}^{\omega_m - I_i} d\omega \frac{\partial}{\partial t} \left(\eta(t, \omega, \theta) W(t, \omega, \theta) \right) \frac{dn_i}{d\omega d\Omega} \quad (14)$$

The angular dependence in ω_t, η, and W is through Eq. (10). Experimental measurements for the double differential cross section for electron ejection are sparse and available for only a few ions and mostly modest ion energies (< 10 MeV/amu) (28–31).

A qualitative model for the angular distribution of the secondary electrons is to assume a distribution peaked about the classical ejection value of Eq. (10), such as

$$\frac{dn}{d\omega d\Omega} = \frac{dn}{d\omega} f(\theta) \quad (15)$$

with

$$f(\theta) = \frac{N}{\left[\theta - \theta_c(\omega)\right]^2 + \frac{K}{\omega}}$$ (16)

with $\theta_c(\omega)$ determined as the root of Eq. (10), N a normalization constant, and K a constant. The constant K may have some dependence on the incident ion's energy and target material, however, is estimated as 0.015 keV from the data of refs. (28–31). Illustrative results of Eqs. (15) and (16) are shown in figure 3, using the model of Rudd (19) for $\frac{d n}{d\omega}$.

RANGE-ENERGY EXPRESSIONS

The electron range-energy relationship is difficult to evaluate theoretically and, because of the complexity of the electron transport problem, empirical expressions based on experimental measurements have been developed by many authors (2, 4, 32–36). Over a limited energy range a power-law of the form, $r = k\omega^\alpha$ will be approximately correct and is used by Butts and Katz (1), Zhang, et al. (16), and Kieffer and Stratten (34). The power-law form is useful, since the residual range is easily found by inversion and leads to an analytic

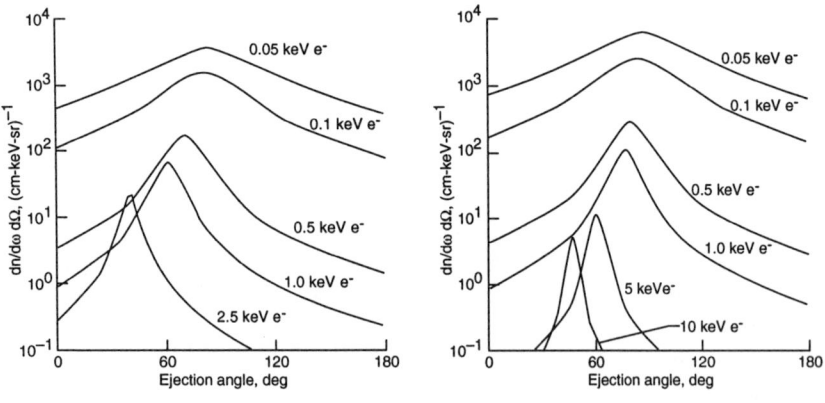

(a) Cross section for 2 MeV protons.   (b) Cross section for 10 MeV protons.

FIGURE 3. Model for double differential cross section for electron ejection in water for several proton energies.

form for the radial distribution of dose under the simplifying assumptions of normal ejection and unit electron transmission. A more accurate form as given by Weber (32) and modified by Kobetich and Katz (4) is the ABC formula

$$r = A\omega\left[1 - \frac{B}{1+C\omega}\right] \quad (17)$$

where

$$A = \left(0.81\, Z_T^{-0.38} + 0.18\right) \times 10^{-3}\, g/cm^2 \cdot keV \quad (18)$$

$$B = 0.21\, Z_T^{-0.555} + 0.78 \quad (19)$$

$$C = \left(1.1\, Z_T^{0.29} + 0.21\right) \times 10^{-3}\, keV^{-1} \quad (20)$$

as found by extensive comparison to experimental data for practical range in many materials. Equation (17) is inverted as a quadratic equation to provide $\omega = \omega(r)$.

As a final parameterization, we consider the range formula of Tabata, et al. (33)

$$r = a_1\left[\frac{1}{a_2}\log\left(1 + a_2 \tau\right) - a_3 \tau\left(1 + a_4 \tau^{a_5}\right)\right] \quad (21)$$

where $t = \omega/m$ and which reduces to Eq. (17) when $a_2\tau \ll 1$ and $a_5 = 1$.

The coefficients in Eq. (21) are

$$\begin{aligned}
a_1 &= b_1 A_T / Z_T^{b_2} \\
a_2 &= b_3 Z_T \\
a_3 &= b_4 - b_5 Z_T \\
a_4 &= b_6 - b_7 Z_T \\
a_5 &= b_8 / Z_T^{b_6}
\end{aligned} \quad (22)$$

with the b_i listed in Table 1. Tabata, et al. (33) provide a parameterization of the inversion of Eq. (21) as

$$\tau = c_1 \left(\exp\left\{ r\left[c_2 + c_3 / \left(1 + c_4 r^{c_5}\right) \right] / c_1 \right\} - 1 \right) \tag{23}$$

with

$$\begin{aligned} c_1 &= d_1 / Z_T \\ c_2 &= d_2 Z_T^{d_3} / A_T \\ c_3 &= d_4 - d_5 Z_T \\ c_4 &= d_6 / Z_T^{d_7} \\ c_5 &= d_8 / Z_T \end{aligned} \tag{24}$$

with the coefficients d_i listed also in Table 1.

A logarithm-polynomial relationship has been used by Iskev, et al. (35) and more recently by Zhang, et al. (36). This however is less useful for the radial dose model since the inversion formula for $\omega = \omega(r)$ is not found easily.

In figure 4, we compare the ABC formula to the expression of Tabata et al. (33) for water where we plot the maximum value of the electron

TABLE 1. Values of the constants b_i and d_i (R_{ex} in g/cm^2)

i	b_i	d_i
1	0.2335 ± 0.0091	(2.98 ± 0.30) x 10^3
2	1.290 ± 0.015	6.14 ± 0.29
3	(1.78 ± 0.36) x 10^{-4}	1.026 ± 0.020
4	0.9891 ± 0.0010	(2.57 ± 0.12) x 10^2
5	(3.01 ± 0.35) x 10^{-4}	0.34 ± 0.19
6	1.468 ± 0.090	(1.47 ± 0.19) x 10^3
7	(1.180 ± 0.097) x 10^{-2}	0.692 ± 0.039
8	1.232 ± 0.067	0.905 ± 0.031
9	0.109 ± 0.017	0.1874 ± 0.0086

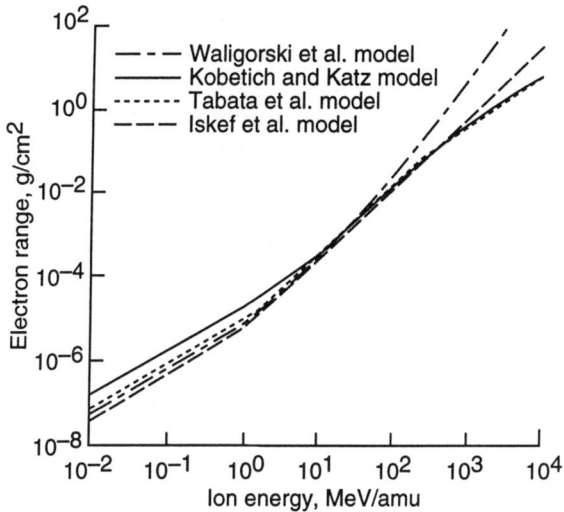

FIGURE 4. Comparison of maximum electron range versus ion energy in water with model of reference (35) shown by dash-dot line, the model of Kobetich and Katz (4) shown by a solid line, model of Tabata et al. (29) shown a by dash line, and model of reference (16) by dotted line.

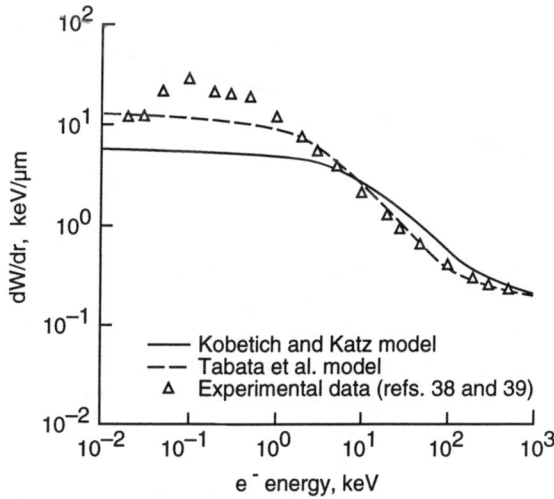

FIGURE 5. Comparison of dW/dr in water for model of Kobetich and Katz (4) shown by solid line and of Tabata, et al. (29) shown by dash line to data for electron LET (35, 36).

range versus ion energy using Eq. (7) to relate to the electron energy. The model shown in fig. 4 is for water vapour. We plot versus the ion energy rather than the electron energy to display the maximum width of the ion track for ions of different energies. In the low to intermediate energy range, the formula agrees closely; however, large differences occur below 1 MeV/amu, and above 1000 MeV/amu. These ion energies correspond to electron energies of about 500 keV and 5000 keV, respectively. In figure 4, the formula of Iskev, et al. (35) for water is also shown. In figure 5, we compare dW/dr found from Eq. (16) or Eq. (21) to experimental data (37, 38) for e^- stopping power in water. The model of Tabata et al. (33) agrees well with experiment down to about 1.0 keV, and we will use this model for radial-dose calculations. The differences between the model and experimental data at low electron energy should lead to some uncertainty in radial dose calculations at small impact parameters. Some of the differences that occur between liguid or vapor water have been discussed in ref. (39) and references cited therein and are well known to become important only at low electron energy.

CALCULATIONS OF RADIAL DOSE

In figure 6, we illustrate the effects of electron transmission on calculations of radial dose in water. Calculations are for proton projectiles; however, we note that the radial dose scales approximately as Z^{*2}/β^2 from which results for other ions can be found. The comparison in figure 6 illustrates that the transmission-factor affects the radial dose calculation only very close to and very far from the ion's path with the normalization and expected fall-off as $1/t^2$ unchanged by including the transmission factor.

In figures 7–14, we compare the radial dose calculations to experimental data from refs. (6–9) for several projectiles for ion energies from 0.25 to 377 MeV/amu. In most cases the radial dose measurements are made in tissue equivalent gas. The comparisons to experiments in figs. 7–14 illustrate the fall-off in radial dose of $1/t^2$ in the intermediate distance range. Close to the ion track ($t < 10$ nm) a contribution to the radial dose from molecular excitations contributes as discussed in ref. (39) is expected. It is important to keep the contributions from excitation and ionizations distinct, since it is the secondary electron dose which is assumed to be responsible for most physical effects by heavy ions. At large distances, the inclusion of angular dependence offers a substantial improvement in calculations. The use of the model of Eq. (15) provides an improvement over the classical ejection angle model at the lowest energies (< 2 MeV/amu). At higher energies the model of Eq. (15) appears to underestimate the radial dose at the largest distances. Clearly, more information on the double differential cross section for electron ejection is required.

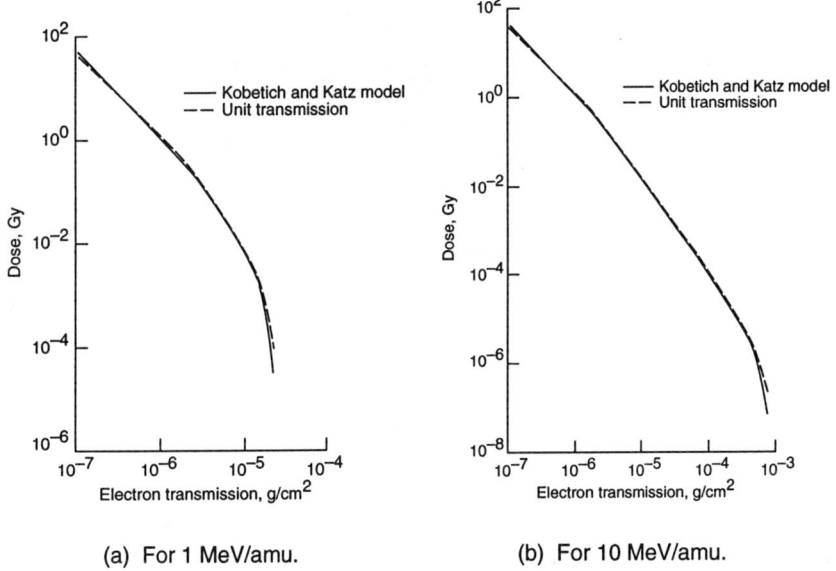

(a) For 1 MeV/amu.

(b) For 10 MeV/amu.

FIGURE 6. Comparison of effects of electron transmission function on calculation of radial dose in water for several proton energies. The solid line is with transmission function of Kobetich and Katz (4), and the dash line assumes unit transmission.

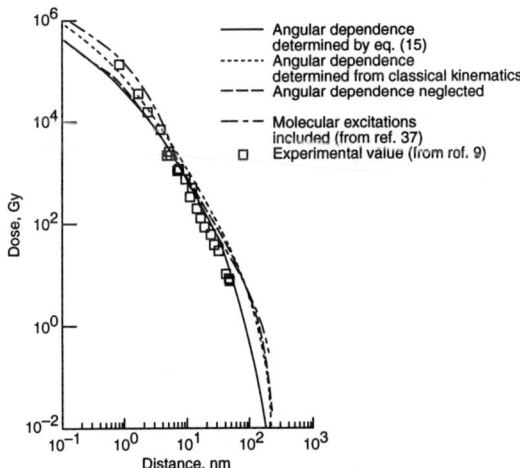

FIGURE 7. Comparison of calculations of radial dose in water for 1 MeV protons to experiment (9). The solid line is with angular dependence of Eq. (15), the dotted line, assuming classical kinematics, the dashed line neglects angular dependence, and the dash-dot line is the result of reference (37), which includes contributions from molecular excitations to radial dose.

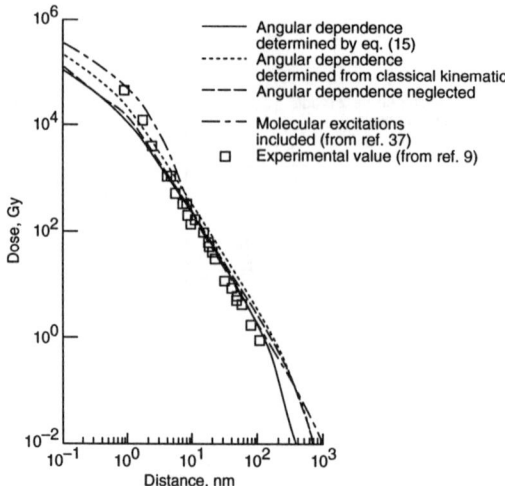

FIGURE 8. Comparison of calculations of radial dose in water for 3 MeV protons to experiment (9). The solid line is with angular dependence of Eq. (15), the dotted line, assuming classical kinematics, the dashed line neglects angular dependence, and the dash-dot line is the result of reference (37), which includes contributions from molecular excitations to radial dose.

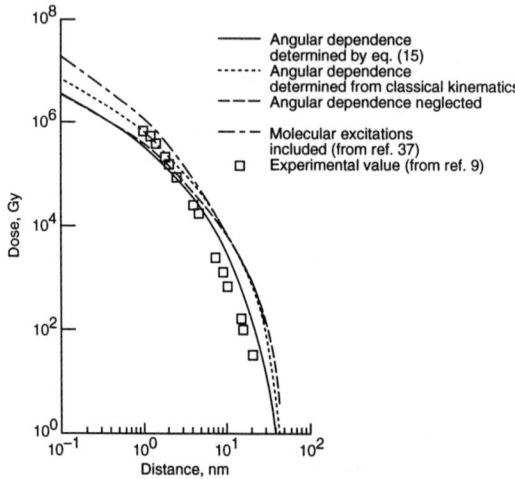

FIGURE 9. Comparison of calculations of radial dose in water for 0.25 MeV/amu ^{4}He ions to experiment (9). The solid line is with angular dependence of Eq. (15), the dotted line assuming classical kinematics, the dashed line neglects angular dependence, and the dash-dot line is the result of reference (37), which includes contributions from molecular excitations to radial dose.

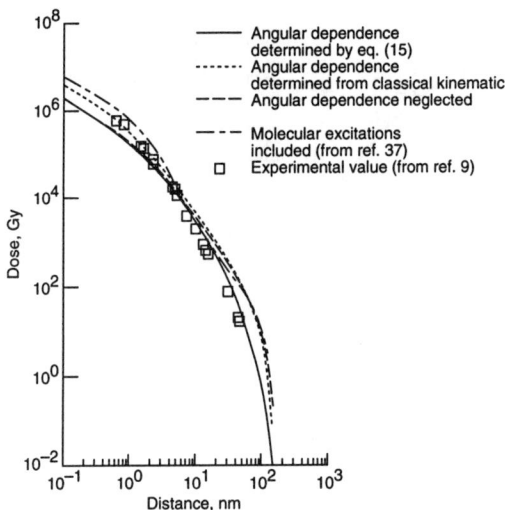

FIGURE 10. Comparison of calculations of radial dose in water for 0.75 MeV/amu ^{4}He ions to experiment (9). The solid line is with angular dependence of Eq. (15), the dotted line assuming classical kinematics, the dashed line neglects angular dependence, and the dash-dot line is the result of reference (33), which includes contributions from molecular excitations to radial dose.

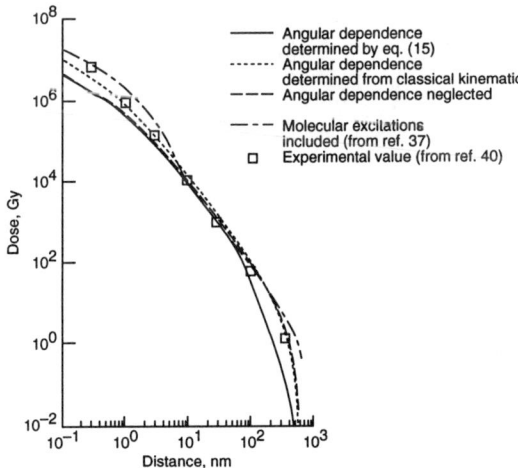

FIGURE 11. Comparison of calculations of radial dose in water for 2.0 MeV/amu ^{12}C to experiment (40). The solid line is with angular dependence of Eq. (15), the dotted line, assuming classical kinematics, the dashed line neglects angular dependence, and the dash-dot line is the result of reference (37), which includes contributions from molecular excitations to radial dose.

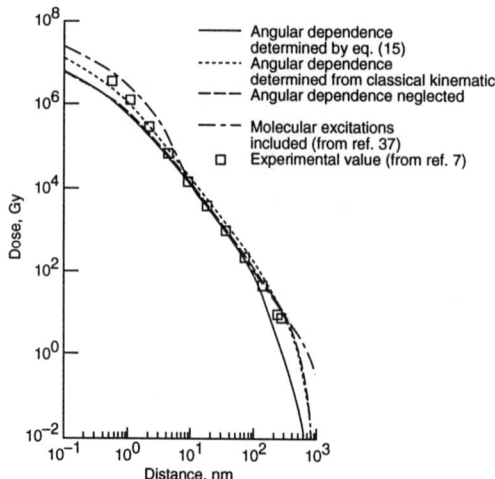

FIGURE 12. Comparison of calculations of radial dose in water for 2.57 MeV/amu ^{16}O to experiment (7). The solid line is with angular dependence of Eq. (15), the dotted line assuming classical kinematics, the dashed line neglects angular dependence, and the dash-dot line is the result of reference (37), which includes contributions from molecular excitations to radial dose.

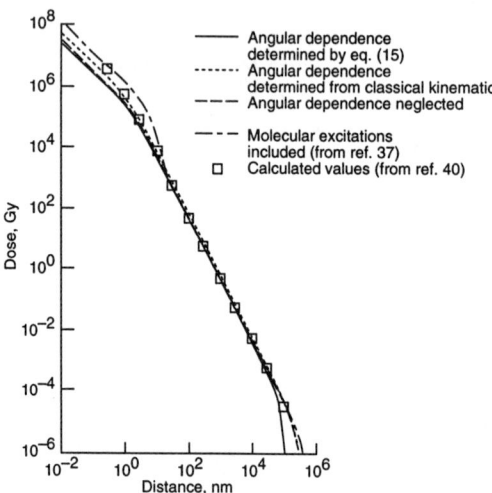

FIGURE 13. Comparison of calculations of radial dose in water for 90 MeV/amu ^{56}Fe with calculations from reference (40). The solid line is with angular dependence of Eq. (15), the dotted line assuming classical kinematics, the dashed line neglects angular dependence, and the dash-dot line is the result of reference (37), which includes contributions from molecular excitations to radial dose.

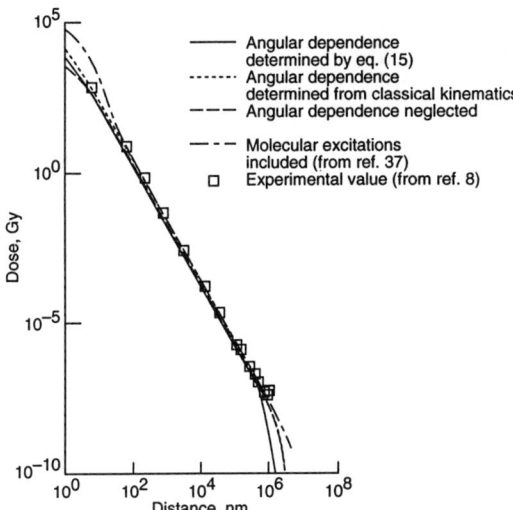

FIGURE 14. Comparison of calculations of radial dose in water for 377 MeV/amu ^{20}Ne to experiment (8). The solid line is with angular dependence of Eq. (15), the dotted line assuming classical kinematics, the dashed line neglects angular dependence, and the dash-dot line is the result of reference (37), which includes contributions from molecular excitations to radial dose.

In figure 15, we illustrate the effects of the radial dose calculations for several velocities in carbon, silicon, and gold. The calculations in figure 15 were made with the secondary electron spectrum of Eq. (6) and assuming classical angular ejection. The model presented in this paper is capable of providing the radial dose for an arbitrary ion in a wide variety of materials, as illustrated by figure 15.

LINEAR ENERGY TRANSFER FROM DELTA RAYS

The contribution to the linear energy transfer (LET) of an ion from delta-rays is evaluated from the radial dose distribution as

$$LET = \int 2\pi\, t\, E(t)\, dt \qquad (25)$$

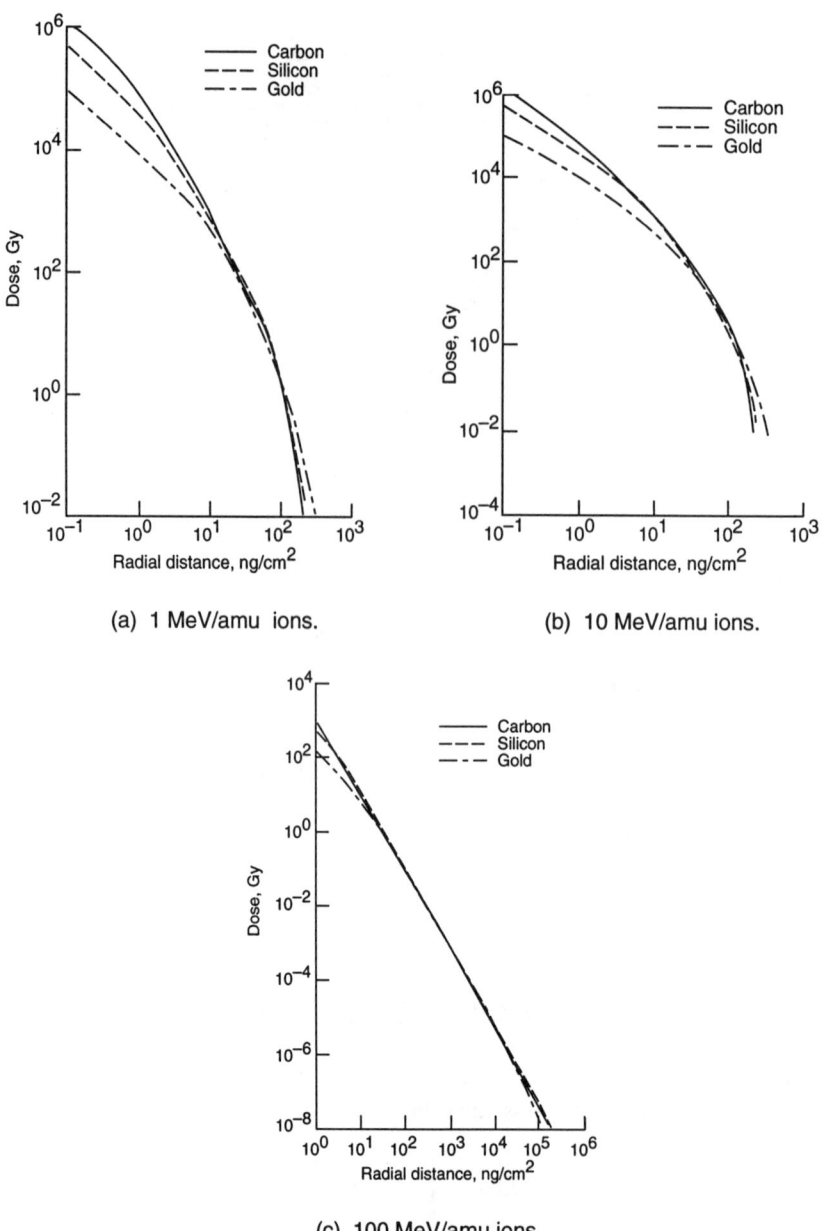

FIGURE 15. Comparisons of radial dose divided by Z^{*2}/β^2 in carbon, silicon, and gold for several ion energies. Solid line is for carbon, dash line is for silicon, and dash-dot line is for gold.

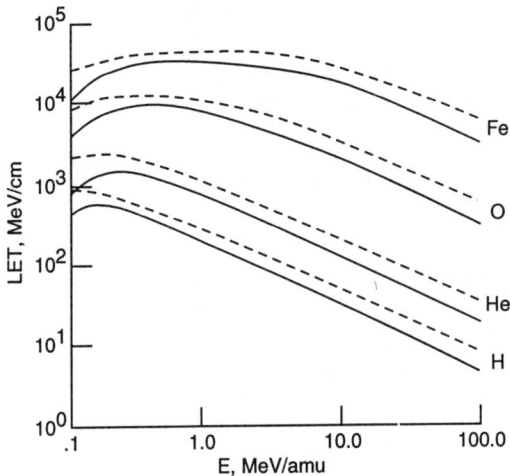

FIGURE 16. Comparisons of the contribution of delta-rays to linear energy transfer (LET) for several ions versus ion energy. Calculated contribution to LET from delta-rays is solid line and fits to measurements of Ziegler (41) is dashed lines.

Calculations of the LET from delta rays are compared to fits to the measurements of Ziegler (41) for several ions in fig. 16. The calculations performed with the present model find about 55–70 percent of the linear energy transfer to be due to the secondary electrons with only a small variation with ion velocity, except below 1 MeV/amu.

CONCLUDING REMARKS

A model for the radial distribution of energy deposited about the path of a heavy ion developed in 1968 by Kobetich and Katz, prior to most experimental measurements of this distribution, is updated with improved physical inputs and compared to experimental data for a variety of ions. Improved models of the electron-range energy and stopping power and the electron-ejection spectra and angular distribution are used in calculations. Excellent agreement with experiment is found. Calculations of the radial dose from heavy ions in water and several materials of interest for spacecraft design and microelectronics were discussed.

REFERENCES

1. Butts, J. J., and Katz, R., *Radiat. Res.* **30**, 855 (1967).
2. Kobetich, E. J., and Katz, R., *Phys. Rev.* **170**, 391 (1968).
3. Kobetich, E. J., and Katz, R., *Phys. Rev.* **170**, 405 (1968).
4. Kobetich, E. J., and Katz, R., *Nucl. Instrum. and Meth.* **71**, 226 (1969).
5. Katz, R., Sharma, S. C., and Homayoonfar, M., in *Topics in Radition Dosimetry. Supplement 1*, New York: Academic Press, 1972.
6. Baum, J. W., Stone, S. L., and Kuehner, A. V., in *Proc. Symp. Microdosim.*, 1968, Ispra, Italy, p. 269.
7. Varma, M. N., Baum, J. W., and Kuehner, A. V., *Radiat Res.* **62**, 1 (1975).
8. Varma, M. N., and Baum, J. W., *Radiat. Res.* **81**, 355 (1980).
9. Wingate, C. L., and Baum, J. W., *Radiat. Res.* **65**, 1 (1976).
10. Metting, N. F., Rossi, H. H., Braby, L. A., Kliauga, P. J., Howard, J., Zaider, M., Schimmerling, W., Wong, M., and Rapkin, M., *Radiat. Res.* **116**, 183 (1988).
11. Hansen, J. W., and Olsen, K. J., *Radiat. Res.* **104**, 15 (1989).
12. Katz, R., and Kobetich, E. J., *Phys. Rev.* **186**, 344 (1969).
13. Katz, R., Sinclair, G, L., and Waligorski, M. P. R., *Nucl. Tracks Radiat. Meas.* **11**, 301 (1986).
14. Kiefer, J., *Int. J. Radiat. Biol.* **48**, 873 (1985).
15. Spohr, R., *Ion Tracks and Microtechnology, Principles and Applications*, Friedr. Vieweg and Son, 1990.
16. Zhang, C., Dunn, D. E., and Katz, R., *Radiat. Protect. Dos.* **13**, 215 (1985).
17. Kobetich, E. J., Ph.D., Thesis, University of Nebraska (1968).
18. Chatterjee, A., and Schaeffer, H. J., *Radiat. Environ. Biophys.* **13**, 215 (1976).
19. Rudd, M. E., *Nucl. Tracks Radiat. Meas.* **16**, 213 (1989).
20. Spencer, L. V., and Fano, U., *Phys. Rev.* **93**, 1172 (1954).
21. Paretzke, H. G., *Proc. Fourth Symposium on Microdosimetry*, 1974, Verbania Pallanza, Italy, p. 141.
22. Depouy, G., Perrier, F., and Arnal, F., *Compt. Rend. Acad. Sci.* (Paris) **258**, 3655 (1964).
23. Bradt, H. L., and Peters, B., *Phys. Rev.* **74**, 1828 (1948).
24. Barkas, W. H., *Nuclear Research Emulsions*. Academic Press Inc., NY, 1963, Vol. 1, 371.
25. Rudd, M. E., Sauter, C. A., and Bailey, C. L., *Phys. Rev.* **151**, 20 (1966).
26. Berger, M. J., and Seltzer S. M., *Natl. Acad. Sci. —Natl. Res. Council, Publ.*, 1964, 1133, 205.
27. Hutchinson, F., and Pollard, E., in *Mechanisms in Radiobiology*, New York: Academic Press, 1961, vol. 1, p. 1.
28. Toburen, L. H., and Wilson, W. E., *J. Chem. Phys.*, **66**, 5202 (1977).
29. Schmidt, S., Kelbch, C., Schmidt-Böcking, H., and Kraft, G., in *Terresterial Space Radiation and its Biological Effects*, New York: Plenum Press, 1988.
30. Rudd, M. E., Toburen, L. H., and Stolterfoht N., *Atomic and Nuclear Data Tables* **18**, 413 (1976).
31. Toburen, L. H., *Phys. Rev. A* **9**, 2505 (1974).
32. Weber, K. H., *Nucl. Inst. Meth.* **25**, 261 (1964).
33. Tabata, T., Ito, R., and Okabe, S., *Nucl. Instrum. Meth.* **103**, 85 (1992).
34. Kieffer, J., and Stratten, H., *Phys. Med. Biol.* **31**, 1201(1986).
35. Iskef, H., Cunningham, J. W., and Watt, D. E., *Phys. Med. Biol.* **28**, 535 (1983).
36. Zhang, C. X., Liu, X. W., Li, M. F., and Luo, D. L., *Radiat. Prot. Dosim.* (1993).
37. Rao, A. R. P., and Fano, U., *Phys. Rev.* **162**, 68 (1967).
38. Cole, A., *Radiat. Res.* **38**, 7 (1969).
39. Waligorski, M. P. R., Hamm, R. N., and Katz, R., *Nucl. Tracks Radiat. Meas.* **11**, 309 (1986).

40. Fain, J., Monnin, M., and Montret, M., *Radiat. Res.* **57**, 379 (1974).
41. Ziegler, J. F., *Nucl. Instrum. and Meth.* **168**, 17 (1980).

Elastic Scattering Model of the Binary Encounter Electrons in Ion-Atom Collisions

C.P. Bhalla and S.R. Grabbe

J.R. Macdonald Laboratory, Department of Physics
Kansas State University, Manhattan, Kansas 66506 USA

Abstract. The elastic scattering model in the calculations of the double differential cross sections of the binary encounter electrons is described for fast ion collisions with light targets. There is an excellent agreement between theory and experiment when the electron exchange contribution is included in the calculations.

ELASTIC SCATTERING MODEL

Introduction

The essence of the elastic scattering model (ESM) in the description of the binary encounter electron (BEe) produced in the fast ion-atom collision is that the target electron can be considered as quasi-free with a momentum distribution around the target nucleus. The ionization of the target is described as quasi-free elastic electron scattering from the projectile ion. The quasi-free electrons have a momentum distribution that is usually taken as the Compton profile of the target electrons. Several authors, Burch et al. (1), Duncan and Menendez (2), and Brandt (3) have used ESM. More recently, theoretical work, for example Reinhold et al. (4), Shingal et al. (5), Schultz and Olson (6), Taulbjerg (7), and Bhalla and Shingal (8) has appeared in the literature initially to explain the inverted q dependence of the double differential BEe cross section for F^{q+}-molecular hydrogen collisions at zero degree laboratory angle, Richard et al. (9). We note that a similar model has been used in the high energy neutron-deuteron scattering by Chew (10).

Miraglia and Macek (11), and Macek and Taulbjerg (12) using the distorted-wave strong potential Born approximation show that ESM is an excellent model for the description of BEe. Furthermore, the electron exchange contribution must be included in ESM, as it was also shown earlier by Taulbjerg (7). An excellent review of theoretical developments is given by Macek (13).

ESM Calculations

The BEe production cross section can be related to the differential cross section for the electron-ion elastic scattering by eq. (3) below.

The spherical symmetric potential of the ion consists of the static potential, $V_s(r)$, due to the interaction of the incident electrons with the nucleus and the electrons of the ion, A^{q+}, and asymptotically is given by $-q/r$. In addition to the static potential, the non-local exchange contribution, $\chi_\ell(r)$, is due to the interaction of the continuum and bound orbitals. The radial wavefunction $u_\ell(r)$ of the scattered state of the incident electron (energy $E = k^2/2$ and orbital angular momentum ℓ) is a solution of the second-order coupled integro-differential equation of the form:

$$\left(\frac{d^2}{dr^2} - \frac{\ell(\ell+1)}{r^2} - 2V_s(r) + k^2\right) u_\ell(r) = \chi_\ell(r)$$

The scattering amplitude for such a potential can be expressed as

$$f(\theta) = f_c(\theta) + f_s(\theta)$$

where $f_c(\theta)$ is the static Coulomb scattering amplitude for the asymptotic charge q and $f_s(\theta)$ is the scattering amplitude that includes both V_s and the exchange contributions.

$$f_s(\theta) = (2ik)^{-1} \sum_{\ell=0}^{\infty} (2\ell+1) \exp(2i\sigma_\ell)(\exp(2i\delta_\ell) - 1) P_\ell(\cos\theta) \quad (1)$$

where σ_ℓ is the Coulomb phaseshift for the asymptotic charge q and δ_ℓ is the additional phaseshift due to screening and exchange effects.

The static potential, V_s, the non-local exchange contribution, χ_ℓ, and the radial wavefunction, $u_\ell(r)$, can be calculated in a self-consistent Hartree-Fock atomic model, Bhalla and Shingal (8). The phaseshifts should be calculated up to a maximum ℓ value beyond which there is negligible contributions to the scattering amplitude. Where appropriate, the phaseshifts are calculated for both singlet and triplet cases, and the differential elastic scattering cross sections, DCS, are obtained as follows

$$\frac{d\sigma(\theta)}{d\Omega} = \frac{1}{4}\left(\frac{d\sigma_S(\theta)}{d\Omega} + 3\frac{d\sigma_T(\theta)}{d\Omega}\right)$$

The double differential electron production cross section, DDCS, in the ion-atom collisions can be related to DCS in the electron-ion collision. In the projectile rest frame,

$$\frac{d^2\sigma(E,\theta)}{dEd\Omega} = \left(\frac{d\sigma(\theta)}{d\Omega}\right)\left(\frac{J(|Q|)}{V_p + Q}\right) \quad (2)$$

where $J(|Q|)$ is the Compton profile of the target atom with $Q = \sqrt{2}(\sqrt{(E+E_b)} - \sqrt{t})$. Here, E_b is the binding energy of the electron, and $t = \frac{1}{2}V_p^2$ is the cusp energy, where V_p is the projectile velocity.

The DDCS of the BEe in the laboratory frame is obtained by the standard transformation

$$\frac{d^2\sigma_{Lab}(E_L,\theta_L)}{dE_L d\Omega_L} = (\frac{E_L}{E})^{1/2} \frac{d^2\sigma(E,\theta)}{dEd\Omega} \qquad (3)$$

where E_L and θ_L are respectively the electron energy in the laboratory frame and the electron emission angle with respect to the incident projectile direction

$$E_L = (t^{1/2}\cos\theta_L + (E - t\sin^2\theta_L)^{1/2})^2$$

$$\cos\theta = \frac{V_p - (2E_L)^{1/2}\cos\theta_L}{(2E_L + V_p^2 - 2V_p(2E_L)^{1/2}\cos\theta_L)^{1/2}}$$

It should be noted that the factor multiplying DCS appearing in eq. (2) needs modification when the target has many electrons in different orbitals. This should be replaced with

$$\sum_i (\frac{J_i(Q_i)}{V_p + Q_i})$$

where the summation is over different orbitals that satisfy the condition that the projectile velocity is much larger than the typical electron orbital velocity.

Fig. 1 contains a comparison of theoretical DDCS of BEe for 30 MeV O^{8+} - Ar collisions at zero degree laboratory angle when (a) all shells (K, L, and M), (b) only L and M shells, and (c) only the M shell are included in the Compton profile. Hagmann et al. (14) give a similar plot for Au^{11+} projectiles.

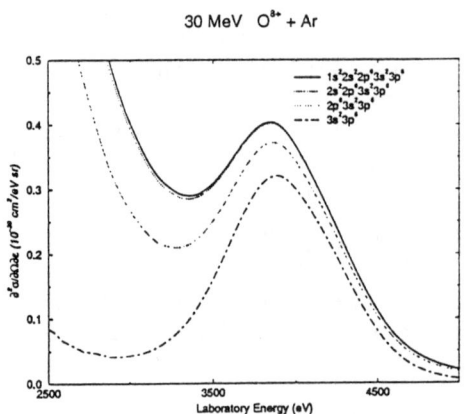

FIGURE 1. DDCS for O^{8+} - Ar collisions at 30 MeV at zero degree laboratory angle versus electron energy showing the effects of the inclusion of different orbitals of argon in ESM.

Fig. 2 shows the dependence of the ratio, defined as DCS divided by the corresponding Rutherford cross section with Z equal to the atomic number, as a function of the scattering angle, Bhalla and Shingal (8). These calculations were performed with the static Hartree-Fock potential, and with the inclusion of the electron exchange contributions. We note that this ratio increases starting from $(q/Z)^2$ at zero scattering angle to larger than unity for large scattering angles, in particular at 180° that corresponds to zero degree laboratory angle in ion-atom collisions.

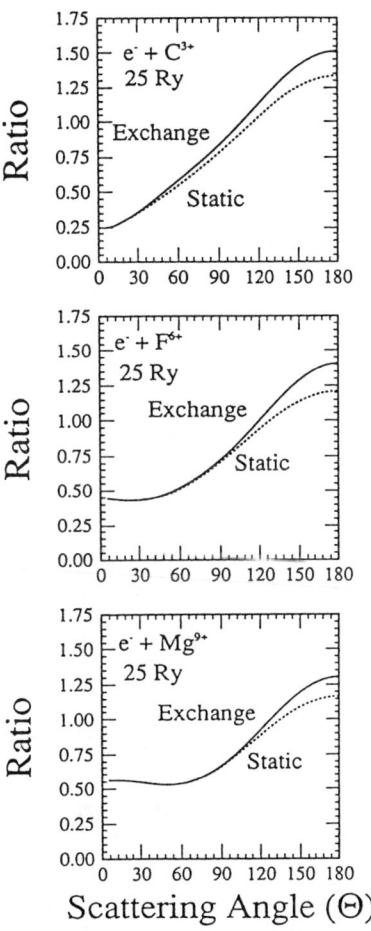

FIGURE 2. Ratio, DCS(q)/DCS(Z), versus scattering angle at 25 Ry electron energy for various three-electron ions. From Bhalla and Shingal (8).

The effects of the inclusion of electron exchange on the ratio is shown in Fig. 3 for a scattering angle of 180° (zero degree laboratory frame for BEe) as a function of the incident electron energy for carbon and magnesium ions.

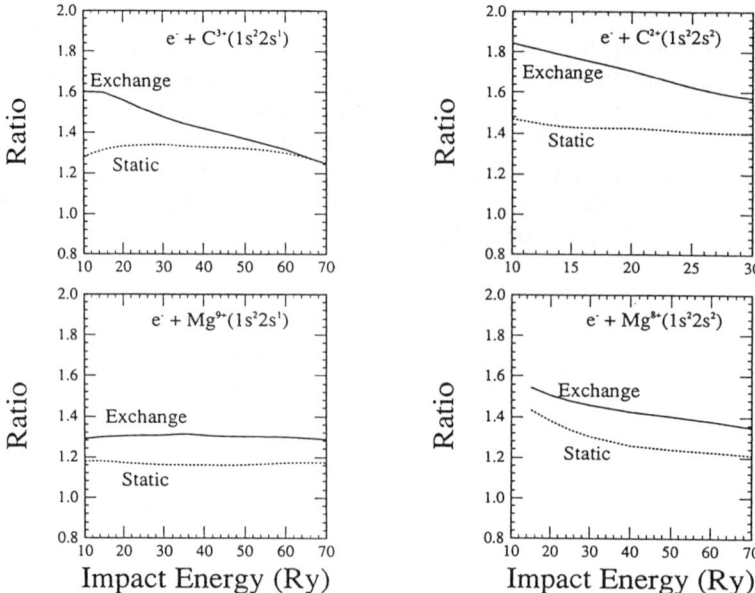

FIGURE 3. Calculated rations of differential elastic cross section for carbon and magnesium ions and Rutherford cross section for bare ions at a scattering angle equal to 180° plotted versus electron impact energy. The full and broken curves represent respectively the ratio with the inclusion of exchange contribution and the ratio with only static potential. From Bhalla and Shingal (8).

The exchange contribution of the continuum with the bound orbitals of the ion is more pronounced at large electron scattering angles and it is insignificant at smaller θ values. Such contributions increase in all cases with decreasing charge state of the ion.

COMPARISON OF ESM WITH EXPERIMENTS

Since several papers in this volume contain such comparisons for example, by Macek (13), Richard (15), and Stolterfoht (16), we present here a limited number of cases to illustrate ESM. Figure 4 contains the DDCS for C^{q+} - H_2 collisions at about 0.75 MeV/u at zero degree laboratory angle as reported by Hidmi et al. (17). The predictions of ESM that includes the electron exchange contribution in the Hartree-Fock atomic model are in excellent agreement with the data. Similarly good agreement between ESM and experiment are reported for 1.0 MeV/u F^{q+} colliding with H_2 for different laboratory angles (18).

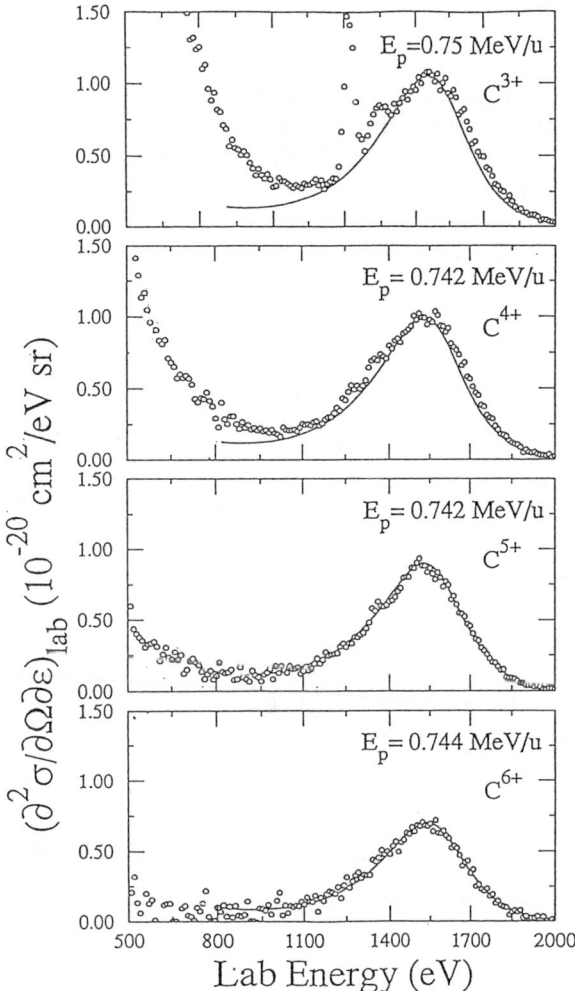

FIGURE 4. BEe DDCS for $C^{q+} + H_2$ collisions. The energy of the projectile is shown. The solid lines are the results of ESM with the inclusion of electron exchange contribution. From Hidmi et al. (17).

It is important to include the Compton profile only for those orbitals that satisfies the basic assumption of ESM when many-electron targets are used, for example molecular oxygen. Such an example is collision of 30 MeV O^{q+} with molecular oxygen. The K-shell does not satisfy ESM criterion. The Compton profile was taken as

$$J = 1.9[2J_{2s} + 4J_{2p}]$$

where J_{2s} and J_{2p} are respectively the Compton profiles of the 2s and 2p shells of atomic oxygen calculated with a Hartree-Fock model. The factor of 1.9 was obtained so that J is in agreement with experimental data for molecular oxygen. Fig. 5 shows good agreement of ESM with experiment of Zouros et al. (18).

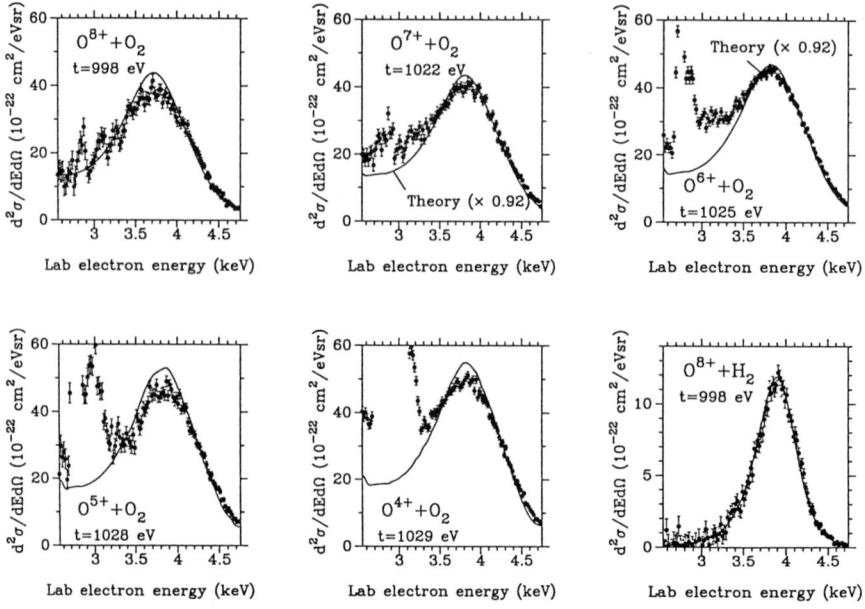

FIGURE 5. 0° laboratory BEe DDCS for 30 MeV O^{q+} + O_2 collisions after background subtraction. The solid lines are the ESM results from contributions of 2s and 2p electrons only. Experimental results have been normalized to the ESM results of 30 MeV O^{8+} + H_2 (lower right corner). The cusp energy, t, which was used to obtain the projectile velocity is indicated in each spectrum. From Zouros et al. (19).

It is worth while to point out that the zero degree laboratory measurements of BEe that correspond to an angle of 180° in the projectile frame provides a severe test of the theoretical electron-ion elastic cross sections. The reason is that $f_s(\theta=180°)$, defined in eq. (1), has significant cancellation in the summation over ℓ because $P_\ell(\theta=180°) = (-1)^\ell$.

ACKNOWLEDGEMENTS

This work was supported by the Division of Chemical Sciences, Office of Basic Energy Science, Office of Energy Research, U.S. Department of Energy.

REFERENCES

1. Burch, D., Wieman, H., and Ingalls, W.B., Phys. Rev. Lett. **30**, 823 (1973).
2. Duncan, M.M., and Menendez, M.G., Phys. Rev. A **16**, 1799 (1977).
3. Brandt, D., Phys. Rev. A **27**, 1314 (1983).
4. Reinhold, C.O., Shultz, D.R., and Olson, R.E., J. Phys. B **23**, L591 (1990).
5. Shingal, R., Chen, Z., Karim, K.R., Lin, C.D., and Bhalla, C.P., J. Phys. B **23** L637 (1990).
6. Shultz, D.R., and Olson, R.E., J. Phys. B **24**, 3409 (1991).
7. Taulbjerg, K., J. Phys. B **24**, L617 (1991).
8. Bhalla, C.P., and Shingal, R., J. Phys. B **24**, 3187 (1991).
9. Richard, P., Lee, D.H., Zouros, T.J.M., Sander, J.M., and Shinpaugh, J.L., J. Phys. B **23**, L213 (1990).
10. Chew, G.F., Phys. Rev. **80**, 196 (1950).
11. Miraglia, J.E., and Macek, J., Phys. Rev. A **43**, 5919 (1991).
12. Macek, Joseph., and Taulbjerg, Knud., J. Phys. B **26**, 1353 (1993).
13. Macek, J., see this issue.
14. Hagmann, Siegbert., Liao, Chun-lei., Bhalla, Chander., Shingal, Rajiv., Shinpaugh, Jeff., Wolff, Wanja., Wolf, Hans., Mann, Rido., Olson, Ron., and Schmidt-Böcking, Horst., Radiation Effects in Solids **126**, 35 (1993).
15. Richard, P., see this issue.
16. Stolterfoht, N., see this issue.
17. Hidmi, H.I., Bhalla, C.P., Grabbe, S.R., Sanders, J.M., Richard, P., and Shingal, R., Phys. Rev. A **47**, 2398 (1993).
18. Liao, C., Richard, P., Grabbe, S.R., Bhalla, C.P., Zouros, T.J.M., and Hagmann, S., (1994), Phys. Rev. A (to be published).
19. Zouros, T.J.M., Richard, P., Wong, K.L., Hidmi, H.I., Sanders, J.M., Liao, C., Grabbe, S., and Bhalla, C.P., Phys. Rev. A **49** 3155 (1994).

Auger Electron Spectroscopy of Free Argon Clusters

A. Knop,* D.N. McIlroy,[†] P.A. Dowben[†] and E. Rühl*

*Institut für Physikalische und Theoretische Chemie, Freie Universität Berlin,
Takustr. 3, D-14195 Berlin, Germany*

[†]*Department of Physics and Astronomy, Behlen Laboratory of Physics,
University of Nebraska, Lincoln, Nebraska 68588-0111*

Abstract. Auger electron spectra and Auger yields of free argon clusters in the Ar(2p) excitation regime are reported. The Auger yield spectra show characteristic changes as a function of cluster size. The results indicate that the Auger yield signal originates primarily from the surface of the clusters. The results are compared to bulk-sensitive experimental techniques, such as total electron yields (TEY), zero kinetic energy electron (ZEKE) spectra for variable size clusters, as well as Auger yield spectra of condensed argon multilayers.

Introduction

Free clusters have been investigated in the past as microscopic model systems for investigating size-dependent properties (1, 2). Rare gas clusters are ideal test cases for the investigation of surface and bulk properties of variable size species. Resonant excitation of surface and bulk excitons in the VUV - and x-ray-regimes can be used to distinguish between surface and bulk properties (3, 5). Structural information from free clusters can also be obtained in the soft x-ray regime from extended x-ray absorption fine structure (EXAFS) spectroscopy (5, 6). Nearest neighbor distances and coordination numbers have been determined through the application of the EXAFS analysis. Information about structural and dynamical disorder in free clusters, due to the increased Debye-Waller factor in comparison with the condensed phase, can also be obtained (5). The structure of clusters are of considerable importance because this affects electronic structure and configurational energies. Apart from calculations and a very few number of electron diffraction experiments (7), there is little known about the structures of clusters. Virtually nothing is known from experiments about the size dependent variations in cluster structure, besides first results from EXAFS spectroscopy (5, 6).

Most elements are not as likely as the noble gases to show differences between the surface and the bulk. This is because the other elements are dominated by band structure effects which smear the local signature of the electronic structure, unlike rare gases which form Einsteinian condensates. Within the perspective of bands, increasing the screening charge surrounding a core hole shortens the lifetime of a core-exciton or permits a core exciton to unbind (8). The consequence of this

© 1996 American Institute of Physics

picture is that with increased electron delocalization, resonant photoemission intensities decrease intra-atomic excitations leading to the formation of a core-exciton followed by Auger decay. Such behavior has been observed with core levels across the nonmetal to metal transition for thin films and 2-dimensional overlayers (8, 9) as well as Hg clusters (10). With rare gas-metal mixtures this effect is particularly pronounced. Wannier excitons of xenon in Hg/Xe mixtures are broadened and decrease in intensity with the onset of metalicity and short range screening (11). Thus the characterization of the size dependence of cluster excitons of rare gas clusters is important to understand in detail the phenomena in the solid state.

In this paper we report the results of Auger electron spectroscopy to variable size argon clusters in the regime of the Ar(2p)-edge (240-260 eV). The goal of this work is to investigate characteristic changes of Auger electron yields as a probe of electronic relaxation in the regime of core-excitons as a function of cluster size in comparison with zero kinetic electron energy (ZEKE) spectroscopy, total electron yields, and results from absorption spectroscopy of condensed argon.

Experimental

The experimental setup consisted of a continuous supersonic jet expansion, synchrotron radiation as a tuneable soft x-ray source, a cylindrical mirror electron analyzer (CMA), and a time-of-flight mass spectrometer. Most of the details of this setup have been described earlier (4, 5).

The electron energy analyzer was a 3 cm diameter (inner diameter of the outer cylinder) single pass cylindrical mirror analyzer schematically shown in Figure 1. The analyzer design was based upon the scheme outlined by Aksela and others (12). Due to the very compact size of this analyzer, it was mounted on a linear

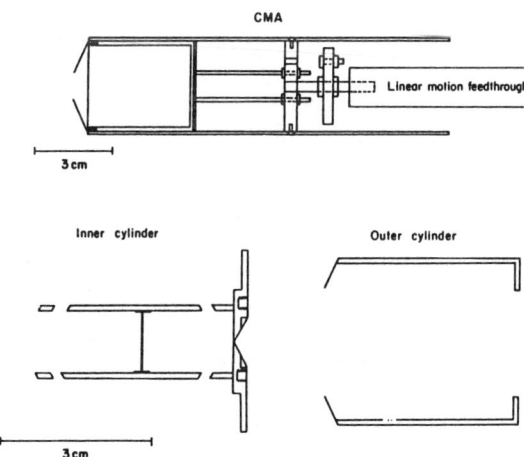

Figure 1. Schematic outlines of the inner and outer cylinders of the CMA. The drawing of the composite assembled of the CMA is also shown.

feedthrough for ease of optimizing the sample-plane to analyzer distance. Under conditions close to ideal (small sample spot size i.e. nearly point-like source for generated electrons and nearly flat sample plane of narrow sample width of a few Ångstroms), the analyzer was found to have the instrumental line width ($\Delta E/E$) of 1.5% based upon electron scattering from the surface of a single crystal of Ni(100). Using less focused incident excitation sources and sample volumes of finite width (on the order of 1 mm), the effective experimental resolution was found to be substantially worse.

Free clusters of an average cluster size between 1 and 700 atoms were excited with monochromatized synchrotron radiation from the HE-TGM-2 beam line at the storage ring BESSY (Berlin, Germany). Average cluster sizes $\overline{N}$ were estimated by correlating the reduced scaling parameter Γ^* (13) with experimental average cluster sizes (14, 15). Further details which characterize the cluster size distribution in our experiment can be found in ref. (4). A time-of-flight mass spectrometer as well as a total electron yield detector was used for aligning the jet expansion with respect to the soft x-ray beam. The total electron yield detector was then replaced by a small cylindrical mirror analyzer (CMA) for measurements of Auger electron spectra and Auger yields as a function of excitation energy. The advantage of this device is its compact size and high transmission despite limited energy resolution.

Results and Discussion

Figure 2 shows typical Auger spectra from the CMA for expansion conditions where no clusters are present in the jet. The broad feature centered at approximately 210 eV kinetic electron energy is found independent of the photon energy and is

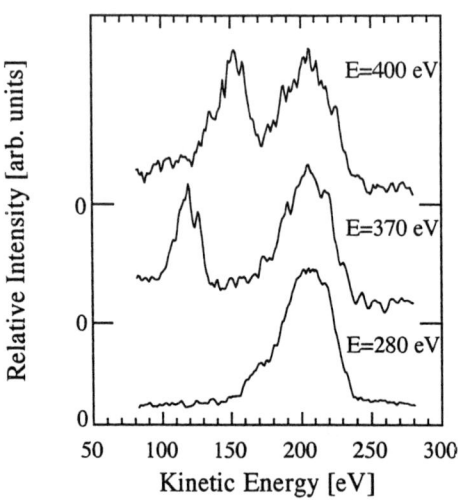

Figure 2. Auger electron spectra of atomic argon acquired at different excitation energies.

therefore assigned to the well-known LMM-Auger process (16). The broad structure indicates that the energy resolution of the CMA is too low to resolve individual Auger lines. Another signal occurs at lower kinetic energy which is dependent on the excitation energy. This is assigned to the Ar(2p) photoelectron signal, which occurs for excitations at 370 eV and 400 eV and is within the range of kinetic energies shown in Figure 2. The spectral shape is found to be unchanged, within the given resolution of the CMA, as the average cluster size is increased. This is in agreement with the results of photoelectron and Auger spectroscopy of condensed and clustered argon, where small shifts relative to the atomic value are found for the Ar(2p) ionization energies (17). For solid argon the Auger lines broaden, but do not shift (18).

Figure 3 shows Auger electron yield spectra recorded at different cluster sizes $\overline{N}$. The atomic spectrum ($\overline{N}=1$) is similar in shape to the total electron yield (4). Discrete resonances are found in the near-edge regime at 244.4 eV and 246.9 eV, corresponding to resonant excitations into Rydberg states (Ar(2p)$_{3/2}$→4s and Ar(2p)$_{3/2}$→3d, respectively), as well as high intensity in the Ar(2p) continuum.

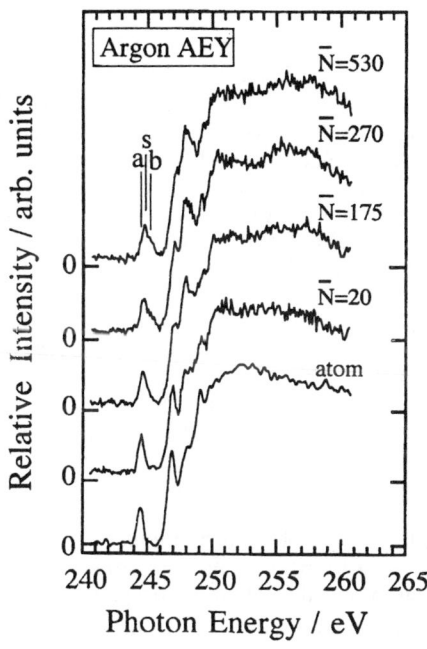

Figure 3. Auger electron yields (AEY) recorded in the regime of the Ar (2p) excitation at different average cluster sizes $\overline{N}$. The labels 'a,' 's,' and 'b' denote the energy positions of the atomic, surface, and bulk Ar(2p)$_{3/2}$→4s transitions.

Spectral changes are observed as the average cluster size is increased. These concern (i) the energy position of the Ar(2p)$_{3/2}$→4s transition, (ii) the appearance of a broad exciton line at 248 eV ($Ar(2p)_{3/2} \rightarrow 3d$) connected with the disappearance of the atomic $Ar(2p)_{3/2} \rightarrow 3d$–Rydberg transition, and (iii) weak oscillations in the Ar(2p) continuum. These changes are discussed in the following:

Total electron and partial cation yield spectra of variable size argon clusters have indicated that surface and bulk excitons can be identified by their energy positions (4, 19). This is especially true for the lowest $Ar(2p)_{3/2} \rightarrow 4s$ transition, since it is an isolated Frenkel-type exciton (20). Small blue-shifts of ≈0.35 eV are observed for surface excitons, whereas bulk core excitons are blue-shifted by ≈1 eV, as observed for the solid (20). For large clusters ($\overline{N}$=530) the maximum of the lowest exciton state at 244.8 eV shows a shift which is typical for surface excitons, contrary to total electron yield and ZEKE yields recorded under identical expansion conditions where the contribution of bulk excitons dominates (4, 17). We point out that total electron yield spectra, as well as ZEKE yield spectra, do not reflect the actual surface-to-bulk ratio of the neutral cluster size distribution since both techniques are particularly sensitive to those electrons which have low kinetic energy. We have shown earlier that low kinetic energy photoelectrons stem from inelastic scattering processes within the clusters (17), i.e. from the bulk. This is evidently not the case for Auger electrons. The maximum of the $Ar(2p)_{3/2} \rightarrow 4s$ transition at 244.8 eV in the Auger yield spectra shows therefore clear evidence for the surface sensitivity of the Auger yields (cf. Figure 3).

A blue-shift of the atomic $Ar(2p)_{3/2} \rightarrow 3d$–Rydberg transition occurring at ≈247 eV is also observed as a function of cluster size. However, this blue-shift is less clear than for the atomic $Ar(2p)_{3/2} \rightarrow 4s$–transition because of the underlying continuous intensity of Auger electrons in this energy regime. We observe, as in the case of TEY and ZEKE spectra for large cluster sizes a distinct resonance at 248 eV, corresponding to a blue-shift of 1 eV, which is typical for bulk excitons. Interestingly, the intensity of this feature is weaker for Auger yields, as compared to TEY and ZEKE spectra (4, 17), which is another indication for the surface sensitivity of Auger yields.

Another characteristic difference to TEY and ZEKE spectra concerns weak EXAFS oscillations in the Ar(2p) continuum, where the first pronounced maximum is observed at 256 eV (4, 17). This finding points also to the surface sensitivity of the Auger electron yields, since surface bound atoms are expected to have less neighbors than argon atoms in the bulk of the cluster. As a consequence, a single scattering process, such as EXAFS, is expected to result in weaker intensity oscillations for surface bound atoms compared to atoms which are located in the bulk of the clusters.

We have also considered possible contributions of Auger electrons which come from the outer part of the cluster beam containing mostly atoms. In order to estimate this possible contribution of the atomic component, we have subtracted a contribution of 40% of the atomic spectrum from that corresponding to $\overline{N}$=530. The resulting spectrum is compared in Figure 4 to the multilayer spectrum of condensed argon recorded by Rocker et al. (21). Both spectra are now more similar in shape, however differences are still evident: (i) in the regime of the lowest exciton (E≈245 eV) which is still more intense and less blue-shifted for clusters, compared to the multilayer spectrum, and (ii) in the region around 250 eV. Differences in energy scales between the solid and cluster spectra are discounted since both spectra were brought to a common energy scale, according to our energy calibration using atomic argon as an energy reference. Pavlychev et al. have pointed out, according to theoretical work, that the regime around 250 eV is sensitive to the local environment of excited atoms (21). Similar findings are reported by Björneholm *et al.* from experiments on free argon clusters (19).

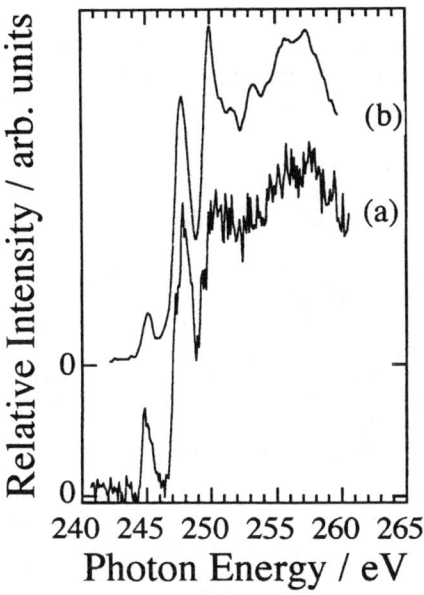

Figure 4. (a) Auger electron yield spectrum of argon clusters ($\overline{N}$=530) where a contribution of 40% of the atomic component has been subtracted; (b) Auger electron yield for thick multilayers of argon (adapted from ref. (21)).

These differences in shape are likely related to the dominant contribution of the surface of the clusters to the Auger electron yields as compared to the condensed phase. Argon microclusters are known to have polyicosahedral structures, in contrast to the fcc lattice of the solid (14). According to estimates by Hoare, one expects for six closed icosahedral shells, corresponding to Ar_{561}, that 45% of the atoms are located on the surface of the cluster (23). The smaller blue-shift of the 4s-exciton, compared to the multilayer spectrum, is therefore an indication that primarily the surface atoms contribute to the Auger yield of microclusters. Further experiments with improved spectral resolution using undulators in conjunction with high resolution x-ray monochromators are proposed in order to confirm this result.

Conclusions

The results indicate that Auger electron yields of variable size argon clusters are highly sensitive to the environment of the excited atom within a cluster. The intensities of the near-edge features indicate that Auger electrons originate primarily from the surface of clusters rather than from the bulk. This shows that Auger yields of free clusters are a valuable complement to investigations of the direct photoionization process studied by ZEKE photoelectron spectroscopy, as well as total electron yield spectroscopy where intense inelastic scattering is used as a probe of bulk properties of clusters.

Acknowledgment. Financial support by the Bundesministerium für Bildung, Wissenschaft, Forschung und Technologie (BMBF) (grant no.: 05 5KEFXB5-TP3) and National Science Foundation (INT-9300238 and DMR-92-21655) is gratefully acknowledged.

REFERENCES

1. "Structure and Dynamics of Weakly Bound Molecular Complexes," NATO-ASI Ser. C-212, A. Weber, ed., Reidel, Dordrecht (1987)
2. "Reactions in and with Clusters," Ber. Bunsenges. Phys. Chem. **96**, 1091 (1992)
3. T. Möller, HASYAB internal report 91-1 (1991)
4. E. Rühl, C. Heinzel, A.P. Hitchcock, H. Baumgärtel, J. Chem. Phys. **98**, 2653 (1993)
5. E. Rühl, C. Heinzel, A.P. Hitchcock, H. Schmelz, C. Reynaud, H. Baumgärtel, W. Drube, R. Frahm, J. Chem. Phys. **98**, 6820 (1993)
6. E. Rühl, C. Heinzel, H. Baumgärtel, W. Drube, A.P. Hitchcock, Jap. J. Appl. Phys. **32** (Suppl. 2), 791 (1993)
7. J. Farges, M.F. deFeraudy, B. Raoult, G. Torchet, J. Chem. Phys. **84**, 3491 (1986)
8. D. Li, J. Zhang, S. Lee, P.A. Dowben, Phys. Rev. B **45**, 11876 (1992)
9. P.A. Dowben, D. LaGraffe, Phys. Lett. A **144**, 193 (1990); P.A. Dowben, D. LaGraffe, D. Li, G. Vidali, L. Zhang, L. Dottl, M. Onellion, Phys. Rev. B **43** 10677 (1991); J. Zhang, D. Li, P.A. Dowben, Phys. Lett. A **173**, 183 (1993); J. Zhang, D. Li, P.A. Dowben, J. Phys. Cond. Matter **6**, 33 (1994)
10. C. Bréchignac, M. Broyer, Ph. Cahuzac, G. Delacreteaz, P. Labastie, J.P. Wolf, L. Wöste, Chem. Phys. Lett. **120**, 559 (1985); Phys. Rev. Lett. **60**, 275 (1988)
11. B. Raz, A. Gedanken, U. Even, J. Jortner, Phys. Rev. Lett. **28**, 1643 (1972)
12. S. Aksela, Res. Sci. Instrum. **42**, 810 (1971); B. Wannberg, Nucl. Instrum. Meth. **107**, 549 (1973); H.Z. Sar-El, Rev. Sci. Instrum. **38**, 1210 (1967); H.Z. Sar-El, Rev. Sci. Instrum. **43**, 259 (1972); S. Aksela, M. Karras, M. Pesa, E. Suoninen, Rev. Sci. Instrum. **41**, 351 (1970); S. Aksela, Rev. Sci. Instrum. **43**, 1350 (1972)
13. O.F. Hagena, Z. Physik D **4**, 291 (1987)
14. J. Farges, M.F. deFeraudy, B. Raoult, G. Tochet, J. Chem. Phys. **84**, 3491 (1986)
15. J. Wörmer, V. Guzielsky, J. Stapelfeldt, G. Zimmerer, T. Möller, Phys. Scr. **41**, 490 (1990)
16. K. Siegbahn *et al.*, "ESCA Applied to Free Molecules," North Holland, Amsterdam (1971); J. Väyrynen, S. Aksela, J. El. Spectrosc. Relat. Phenom. **16**, 423 (1979)
17. A. Knop, H.W. Jochims, A.L.D. Kilcoyne, A.P. Hitchcock, E. Rühl, Chem. Phys. Lett. **223**, 553 (1994)
18. J.D. Nuttall, T.E. Gallon, Phys. Stat. Sol. B **71**, 259 (1975)
19. O. Björneholm, F. Federmann, F. Fössing and T. Möller, Phys. Rev. Lett. **74**, 3017 (1995)
20. R. Haensel, G. Keitel, N. Kosuch, U. Nielsen, P. Schreiber, J. Phys. (Paris) C **4**, 236 (1971)
21. G. Rocker, P. Feulner, R. Scheuerer, L. Zhu, D. Menzel, Phys. Scr. **41**, 1014 (1990)
22. A.A. Pavlychev, A. Barry, Phys. Scr. **41**, 157 (1990)
23. M.R. Hoare, Adv. Chem. Phys. **40**, 49 (1979)

Charge Transfer and Ionization in Ion Collisions with Circular Rydberg Atoms

D. M. Homan, M. J. Cavagnero, and D. A. Harmin

Department of Physics and Astronomy
University of Kentucky
Lexington, Kentucky 40506-0055

Abstract. Explorations of the classical phase space for ion collisions with Rydberg atom targets provide insight into mechanisms for ionization and charge transfer. Thomas's analysis of charge transfer at high impact energies [Proc. Roy. Soc. London **114**, 561 (1927)] is extended here to intermediate- and low-energy capture and ionization processes through numerical integration of the classical equations of motion. Parameter space maps correlating initial conditions with final outcomes have well-defined zones corresponding to Thomas capture, direct capture, binary encounter ionization, saddle-point ionization, and ionization by S superpromotion. The relative contributions to charge transfer (or ionization) cross sections from these distinct classical collision channels are qualitatively interpreted by analysis of the parameter space maps.

We have studied classical collisions between ions and one-electron atoms, assuming that the initial circular orbit of the electron and the rectilinear orbit of the projectile lie in the same plane. The electron trajectory is then determined by three parameters: a phase angle, ϕ, of the electron in its circular orbit, the impact parameter, b, and the reduced velocity, $v = v_{\text{ion}}/v_n$, where v_n is the Bohr velocity of the target electron. (The phase angle, ϕ, is chosen as the angular position of the electron at closest approach of the projectile, assuming no interaction with the projectile.) The classical collision channels for ionization, excitation, and charge transfer form well defined zones in this 3-dimensional parameter space that evolve smoothly from simple forms at high projectile velocity to more complex structures at low velocity. Figure 1 shows the evolution of excitation (two darkest shades of gray), capture (next two lighter shades), and ionization (three lightest shades) channels as the reduced velocity is varied from 10 to 0.5. The shade of each pixel in the figures labels the outcome of a specific trajectory with impact parameter, b, and orbital phase, ϕ, of the target electron. While these maps become very complicated at low ion velocities, we find that for reduced velocities greater than unity the zones of parameter space associated with different reaction processes are *few in number* and *easily distinguishable*.[1]

© 1996 American Institute of Physics

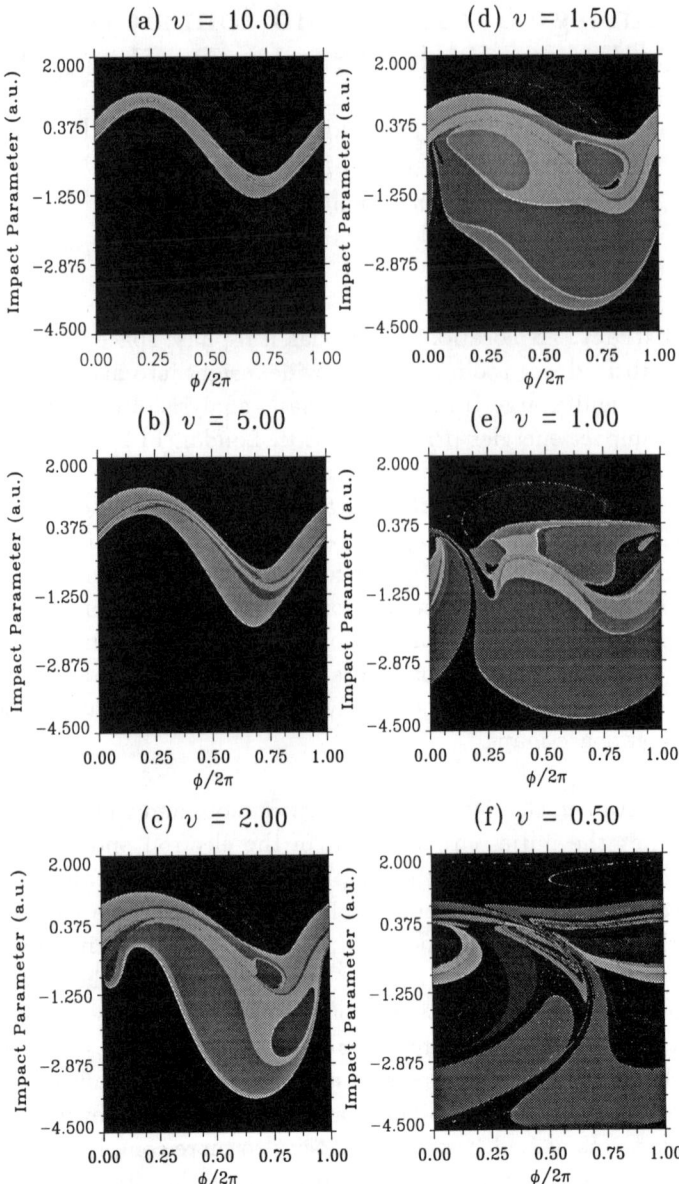

Fig. 1

As described below, a number of the classical collision channels have quantum analogs that have been identified in previous investigations of ion-atom collisions. We have isolated regions of parameter space that correspond to each of these processes and identified the dominant classical collision channels contributing to charge transfer and ionization in ion collisions with circular-state atoms.

After each trajectory is integrated from $-t_o$ to t_f, where t_o and t_f are forty times the orbital period, exit tests are performed to determine the outcome of the collision event. The first test determines whether the electron is closer to the target or to the projectile after the collision. If the electron is closer to the target, then the energy ϵ_t (kinetic plus potential) of the electron relative to the target rest frame is compared with the saddle point potential energy $\epsilon_s \equiv -4/R$ (a.u.), with R the internuclear separation. If $\epsilon_t < \epsilon_s$, then we conclude that neither capture nor ionization occurs for this trajectory and we label the trajectory as an inelastic scattering event. If the electron is closer to the projectile, its energy ϵ_p relative to the projectile rest frame is compared with ϵ_s. If $\epsilon_p < \epsilon_s$, then the collision event results in capture. If the energy relative to the closest center is positive, then the event results in ionization of the electron.

For a small number of trajectories, the energy relative to the closest center will be negative but greater than ϵ_s. In this case, the equations are integrated to larger internuclear separations to determine whether the electron is subsequently captured by either nucleus. The exit tests are then repeated at times equal to integral multiples of the final time, t_f, until a definite outcome can be assigned to the trajectory. This procedure is repeated, up to time $t = 6t_f$, and if convergence is not achieved at that point the outcome is labelled unknown and is represented in the parameter space maps by a *white* pixel.

Capture regions of the parameter space maps are differentiated according to the number of times (or swaps) that the electron crosses the mid-plane between the nuclei during the collision. Figure 2 shows an enhanced image of the $v = 2$ parameter space map; two types of capture zone are readily apparent. The lighter shade of capture events are 3-swap trajectories in which the electron scatters from both nuclei prior to capture. At high velocity, these events become localized about phase angles $\phi = 120°$ and $\phi = 240°$, as expected from Thomas's original analysis.[2] The darker 1-swap zones correspond to "direct capture." Direct capture occurs at negative impact parameters where the initial electron orbit co-rotates with the internuclear line. (Note that use of positive and negative impact parameters permits us to focus on initial orbits whose angular momentum vectors lie in the same direction; i.e., which circulate counter-clockwise as viewed in the parameter space maps.)

The ionization region in Fig. 2 is subdivided into three regions defined by the coordinate center—target, projectile, or saddle point—with respect to which the electron moves most slowly following the collision. The darkest

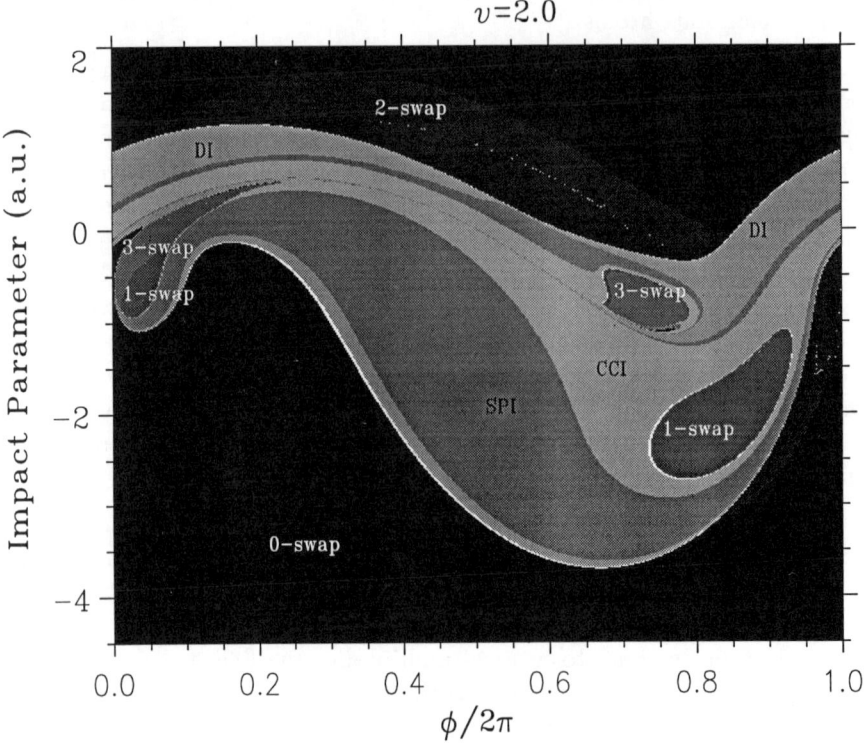

Fig. 2

shade of gray represents those trajectories for which the final electron speed is slowest with respect to the saddle point; we refer to these events as "saddle-point ionizations" and label them SPI. The next lightest shade corresponds to those trajectories for which final electron speeds are slowest with respect to the target; these are referred to as "direct-ionizations" (DI). The lightest shade represents those trajectories for which the final electron speed is slowest with respect to the projectile ion; this is generally referred to as "capture to the continuum" (CCI). Note that 1-swap captures are embedded in the CCI zone, while 3-swap captures occur near the approximately sinusoidal DI zone.

Extensive studies of the parameter space maps for fully three dimensional (non co-planar) collisions reveal the same regularities in the classical collision channels as described above. We have calculated the contributions of each of these channels to total transfer and ionization cross sections for ion collisions with circular Rydberg states.[3] Our elementary atomic model consisting of an electron in a single circular orbit results in classical cross sections remarkably similar to the experimental values reported in Refs. 4 and 5. By resolving the classical capture cross section into contributions from direct capture processes and Thomas-like captures (1-swap and 3-swap, respectively), our calculations (at $v = 1.65$) indicate that the 3-swap mechanism is essentially independent of the orbit's orientation, while 1-swap captures strongly favor the coplanar geometry.

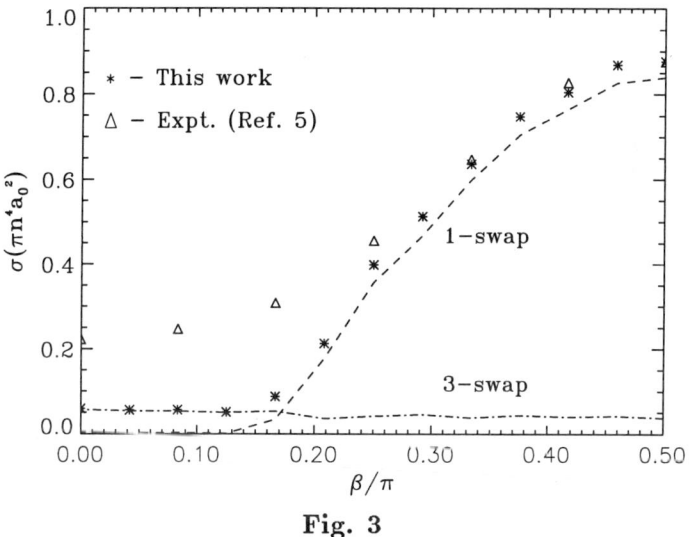

Fig. 3

Figure 3 displays our calculations relative to the experimental data for a reduced velocity $v = 1.65$. Cross sections are plotted versus the angle β between the target electron's angular-momentum vector $\vec{L}$ and the beam axis $\hat{k}$: $\beta = \cos^{-1}(\hat{L} \cdot \hat{k})$. The experimental data points are relative, so we have normalized them to our own cross section at its maximum, which corresponds to an initial orbital angular momentum perpendicular to the beam direction ($\beta = 90°$). The 1-swap and 3-swap contributions to the total cross section are shown as dashed and dot-dashed lines in Fig. 3.

We gratefully acknowledge helpful conversations with K. B. MacAdam and J. C. Day. This research is supported by the Division of Chemical Sciences, Offices of Basic Energy Sciences, Office of Energy Research, U. S. Department of Energy.

[1] D. M. Homan, M. J. Cavagnero, and D. A. Harmin, Phys. Rev. A **51**, 2075 (1995).
[2] L. H. Thomas, Proc. Roy. Soc. London **114**, 561 (1927).
[3] D. M. Homan, M. J. Cavagnero, and D. A. Harmin, Phys. Rev. A **50**, R1965 (1994).
[4] T. Ehrenreich, J. C. Day, S. B. Hansen, E. Horsdal-Pedersen, K. B. MacAdam, and K. S. Mogensen, J. Phys. B **27**, L383 (1994).
[5] S. B. Hansen *et al.*, Phys. Rev. Lett. **71**, 1522 (1993).

Author Index

B
Barrachina, R., 233
Bhalla, C. P., 266
Burgdörfer, J., 115

C
Cavagnero, M. J., 281
Cravero, W., 233
Cucinotta, F. A., 245

D
Dowben, P. A., 274
Dubey, R. R., 245

E
Edwards, A. K., 205

F
Fainstein, P. D., 147

G
Gay, T. J., 19
Geddes, J., 241
Gilbody, H. B., 241
Grabbe, S. R., 266
Groeneveld, K. O., 233

H
Harmin, D. A., 281
Homan, D. M., 281

I
Irby, V. D., 237

K
Katz, R., 245
Kim, Y-K., 214
Knop, A., 274

L
Lin, C. D., 135

M
Macek, J. H., 5, 193
Maier, R., 233
Manson, S. T., 59
McIlroy, D. N., 274
Meckbach, W., 233
Meyerhof, W. E., 103
Montenegro, E. C., 103

N
Niehaus, A., 41

O
Olson, R. E., 84
Ovchinnikov, S. Yu., 5, 41

P
Passovets, S. V., 5
Patton, C. J., 241
Pieksma, M., 41
Ponce, V. H., 147

R

Ramírez, C. A., 229
Reinhold, C. O., 84, 115
Richard, P., 69
Rivarola, R. D., 147, 229
Rühl, E., 274

S

Schultz, D. R., 84
Shah, M. B., 241
Stolterfoht, N., 163
Suárez, S. G., 29, 233

T

Tobisch, M., 233
Toburen, L. H., 181

V

van Eck, J., 41

W

Westerveld, W. B., 41
Wilson, J. W., 245